ROGER HEPPLESTON

MEMES, SOCIETIES AND HUMAN EVOLUTION

HOW HUMANS CAME TO DOMINATE THE EARTH AND THEN THREATEN ITS ECOLOGY

First printed in this edition 2015

Printed and Published by CreateSpace

ISBN-13: 978-1507802007

ISBN-10: 1507802005

Cover design Richard Saunders at Bacroom Design

CONTENTS

Acknowledgements

This book would not have been possible without the common sense, questioning and patience of my wife Judith and the helpful advice and constructive criticism of my editor Carole Pearce. They both have contributed greatly to the style, readability and development of the theories that underline this book.

INTRODUCTION

Many of the religions of the world have tried to justify the fact that their all-powerful God has allowed so much human suffering. Frequently they explain that suffering is due to human disobedience at the time of creation. One version of this idea is found in the creation story of the Kiganda religion of the Buganda tribe in Uganda. In this religion the original man, Kintu, marries God's daughter. Having disobeyed God, the couple are cursed by the presence on Earth of Walumbe, the spirit of disease and death. At the end of the story Kinto proclaims 'I will keep having children, the more Walumbe kills, the more I will have. Walumbe will not be able to remove them all from the face of the Earth.'

In the Garden of Eden story, the punishment of the first man, Adam, for his disobedience is recorded in the *Old Testament* as follows:

> cursed is the ground for thy sake; in sorrow shalt thou eat of it all the days of thy life; Thorns also and thistles shall it bring forth to thee; and thou shalt eat of the herb of the field; In the sweat of thy face shalt thou eat bread, till thou return unto the ground; for out of it was thou taken: for dust thou art and unto dust shalt thou return.

At the time the Bible was written, most people were peasant farmers and you can feel the incessant toil of a peasant's life in these words. As in the Uganda story, this Jewish vision of human existence is of an unrelenting struggle for survival in the face of adversity.

The world has moved on from the times these myths were composed. We now know an astonishing amount about how humans came to exist. We know the universe evolved from a big bang some 13.8 billion years ago. The Earth evolved from particles revolving around the sun 4.5 billion years ago. The first primitive, single cell life-forms emerged between 3.5 and 4 billion years ago. Multicellular bodies formed around 1 billion years ago in the sea; eventually they colonised land and developed into plants, insects, amphibians, birds, reptiles and mammals. The first apes appeared 15–20 million years ago and as little as 200,000 years ago man, *Homo sapiens*, first evolved in Africa.

The answer to the question as to why there is suffering is now different. Suffering is seen as part of a natural struggle for existence. Charles Darwin's theory of evolution envisaged life as a struggle for survival between all life-forms from viruses to blue whales, each trying to find its own niche in the environment. Out of competition and the resultant suffering, the best adapt and thrive.

Following the discovery of DNA in 1952, we also know the biological mechanism by which life-forms are copied from one generation to the next. Specific lengths of DNA, known as genes, are the code for the formation of proteins which are the building blocks of life. Very much in line with the Kiganda view that mankind's purpose in life is to create sufficient children to ensure the survival of the species, it is clear that mankind's biological purpose in life is to perpetuate the species by passing their genes onto future generations.

This book is about how, in this struggle for survival, one animal, *Homo sapiens*, came to dominate all other animals. The book's inspiration came from Richard Dawkin's assertion in *The Selfish Gene* that it is the transmission of ideas from generation to generation that distinguishes man from other animals. The dissemination of these ideas, which he called memes, has resulted in that myriad of forms of behaviour and skills which distinguish the human animal. Memes are a parallel concept to genes; they are subject to the same laws of survival of the fittest. Successful memes will thrive; unsuccessful ones will die off.

Almost all of mankind's astonishing development in the last 200,000 years is due to memes. The transmission of memes not only resulted in technical developments such as Stone Age tools or steam engines, memes also determined how we communicated, cooperated and organised ourselves. Memes have enabled us to write, to create nation-states and to develop religions.

The Selfish Gene was published in 1976. Since then I have expected historians and evolutionists to come forward and rewrite the history of mankind by tracing the memes critical to human development. None has appeared. I am still puzzled as to why. The only reason I can surmise that this has not been attempted before is because the subject crosses so many academic disciplines. The early part of the story comes from paleoanthropologists, anthropologists, geneticists and zoologists. There is a large body of composed work covering their own field of interest but no scientific texts sufficiently inspirational to allow historians to re-evaluate historical developments.

I, therefore, set myself the task of telling the story of the advance of mankind through the history of meme development. I envisaged the human story as a succession of breakthrough memes that changed the course of human evolution and set about compiling a list of these memes. Some memes, like the development of the alphabet by the Phoenicians around 1200 B.C., would be in everyone's list. Other more mundane developments, like the invention of Portland cement by the builder, Joseph Aspdin, would not be everyone's choice. I based my choice on their practical effect. Philosophy is scarcely mentioned. The advance of science concentrates on the important breakthroughs rather than the history of the development of scientific ideas. However, because of the fundamental importance of religion in human development, I have had to include some discussion of abstract concepts.

However as I delved deeper into the subject it became clear to me that the underlying story of human evolution is of how human society developed. It is the story of many types of communities evolving and interacting. We have moved from the simple structure of hunter–gatherer bands to our complicated world of states, companies, religions and political systems. On the journey we have formed clans, tribes, kingdoms and an aristocracy. To make sense of human evolution I had to show how the development of memes resulted in the development of different types of human interaction. Turning for guidance to books on memetics I quickly discovered that memetics as a science is still in its infancy. There has been some progress since Dawkins proposed his idea but the role of memes in evolution is far from achieving any academic consensus. I have therefore had to speculate more than I expected about the evolutionary mechanism of memes. *In Memes, Societies and Human Evolution* I have set out my own theories of how human society evolved to reach where we are today.

Chapters 1 and 2 introduce the concept of memes and tell how humans evolved from ape forebears to become hunter–gatherers. These chapters set out the evolutionary theory of meme development and the relationship between the evolution of memes and human communities on which the rest of the book is based.

Chapters 3 and 4 tell the story of human development from the Neolithic revolution to writing and the formation of the first states. I am dependent on two great works to piece together this part of man's story. Firstly *War in Human Civilisation* by Azar Ghat shows the important role war played in the evolution of human society. Secondly, *Guns, Germs and Steel*, by Jared Diamond, traces the early development of human technologies and shows why Western states have been so successful.

Chapters 5 to 12 are a summary of world history from the early empires to the twenty-first century. It revisits well-known events in human history and places them in a new context. Only a small part mentions the traditional story about wars, kings and political machinations. Instead, I have tried to describe the critical memes that advanced human society and give due credit to those who, by being the first creator of the breakthrough memes, were really responsible for human progress. I have also tracked the changes in human organisation that resulted.

In order to measure this human progress I have recorded population levels and life expectancy at each stage of human history. Population levels are a measure of success of any animal species. Uniquely, in the case of humans, their quality of life has also improved. I have taken the increase in life expectancy as an indication of this advance.

Part of the motivation of the study of human evolution is that it not only gives an insight into the reason for human successes and failures but also allows a prediction of human destiny. Chapter 13 entitled, 'Where are we going', looks at how the very success of humans is threatening their survival as a species. It reviews the ecological, financial, organisational and population issues that will affect humans in the future.

Chapter 14 was not planned at the start. I did not want to finish the book on a negative note with another Jeremiah view of the threat to man's survival as a species. For this reason it is entitled, 'Is there any hope?' This is entirely my own opinion and makes a plea for a new philosophy of life, called Eco-humanity, aimed at lengthening mankind's tenure on the planet.

An article in *The Guardian* newspaper on 19 August 2014 mentioned the famous British television presenter Jeremy Paxman's thoughts about religion:

> Paxman said he was an atheist, but 'I wish I wasn't '. He said that being involved with homeless charities he had observed how many of them had religious foundations. 'We atheists, humanists and others have signally failed, it seems to me, to harness that belief in humanity and the innate appropriateness of treating each other properly. It seems to be much more easily done if you have some kind of religious imperative telling you what you should do.

I have argued that structured religions and philosophies of life are the only organisations in history that have allowed humans to set aside our natural materialistic instincts for the benefit of a cross-section of humanity. Current religions with their ancient canons are struggling to keep up with modern developments, let alone to look towards the future. To preserve the planet for the benefit of succeeding generations we need new moral imperatives. They would have to be based on a new, ecologically friendly philosophy of life, a positive atheism which builds on our scientific knowledge and is dedicated to preserving our genes and memes for generations to come.

This concept, however, is merely a coda at the end of the book. The main thrust is to demonstrate how memes have changed the structure of human society over the ages. I am not an academic. My sources of information are books, newspapers, television and radio broadcasts, the internet and my own experience. This is history from a great height. My sources are not detailed studies of original documents and academic papers but broad compendiums of knowledge. As a result, my referencing is not specific to detailed comments, as would appear in academic texts. Instead, in the notes at the end I have listed the broad sources of information I have used.

For reason of space and continuity of narrative I have restricted myself to reviewing those historical civilisations that have direct relevance to modern developments. Hence I have made only passing reference to pre-Columbian civilisations in America, and similarly ignored Japanese or other non-Chinese far eastern civilisations before modern times. I also have to admit that, although this is a study of the advance of humans across the world, my knowledge of Chinese and Indian history is weak and the importance of developments in these areas may be understated. In contrast, because I know more of UK history, the influence of developments in the UK is probably given more global importance than it merits. This is a natural result of my educational grounding. However, I believe this bias is not sufficient to invalidate the overall truth of the story.

I hope the reader finds that this broad approach to history from an evolutionary perspective gives a fresh understanding of human behaviour. In this I am supported by Edward O. Wilson in his book *The Meaning of Human Existence*, p. 173:

> What does the story of the species tell us? ... I believe the evidence is massive enough and clear enough to tell us much: We were created not by a supernatural intelligence but by chance and necessity as one species out of millions in Earth's biosphere. Hope and wish for otherwise as we will, there is no evidence of ... [any] demonstrable destiny or purpose assigned to us.... We are it seems completely alone ... we are completely free. As a result we can more easily diagnose the etiology [cause] of the irrational beliefs that so unjustifiably divide us. Laid before us are new options scarcely dreamed of in earlier ages. They empower us to address with more confidence the greatest goal of all time, the unity of the human race.

As a summary of the purpose for studying human evolution, I would substitute only the word: 'future' for 'unity' in the last sentence. My fervent hope is that this book is a start of a new phase in the analysis of the human story. It will only be successful if others from the relevant academic disciplines can validate the underlying theory, expand on the ideas and further explore their implications.

CHAPTER 1:
MAN THE ANIMAL (TO 70,000 B.C.)

In 1860 there was a famous debate at the Oxford Union on the truth of Charles Darwin's theory of evolution between the Bishop of Oxford, 'Soapy' Sam Wilberforce and Thomas Huxley. Soapy Sam supposedly enquired whether it was through his grandmother's or his grandfather's side that Huxley descended from a monkey. Thomas Huxley is said to have replied that he would not be ashamed to have a monkey for his ancestor, but he would be ashamed to be connected with a man who used his great gifts to obscure the truth. Even, a century and a half later, many people are still in denial about the truth of man's animal origins.

If you have ever established eye contact with a chimpanzee, or observed how chimpanzee mothers care for their young, or noticed how they use their hands to eat, you will be instantly aware that humans and apes have much in common. However, at the time Darwin's theory that humans had an ape ancestry was deeply shocking. Religion taught that humans were created separately from animals and that their destiny was to rule over the Earth. They had refined feelings, artistic skills and were sophisticated and cultured. How could they be related to something as crude and beastly as an animal? Our history shows however, that for all the great successes of mankind, we have been more vicious and murderous to our own species than any other animal. Violence has been intrinsic in the human evolutionary story from the start.

When Darwin introduced his theory of evolution, no-one knew how the characteristics of a life-form were passed from one generation to the next. Only in the 1950s was it discovered that the 'code of life' is held in a molecule called DNA, which is divided into genes. In *The Selfish Gene*, Richard Dawkins sets out the argument that all life-forms are survival machines for their genes. Life-forms have finite lives but genes can survive indefinitely. Evolution is a battle between survival machines. Those genes that provide the code for the best survival machines in a particular environment will have the best chance of success.

The human life-form has evolved from those of the same forebear as today's great apes. 98.4% of our DNA is the same as our nearest relative, the chimpanzee. We are the 'naked ape' that Desmond Morris so graphically described. However, human development has not been defined by genes alone. More significant has been the evolution of their memes. Memes are an important concept described by Dawkins. He imagined them as ideas that could be copied from one brain to another. To quote Dawkins:

Examples of memes are tunes, catch-phrases, clothes fashions, ways of making pots or building arches. Just as genes propagate themselves in the gene pool by leaping from body to body via sperms or eggs, so memes propagate in the meme pool by leaping from brain to brain via a process which, in the broad sense, can be called imitation.

These transmitted memes result in animal behaviour being replicated from generation to generation. In other parlance, memes are responsible for behaviours that are induced by nurture, rather than nature.

Insects do not have the ability to pass on memes and their behaviour is always determined genetically. For example, the monarch butterfly is an amazing insect which, over three generations, flies away from its breeding site in Mexico to places as far away as Canada. Then the next generation returns to exactly the same nesting site. Since the young monarch butterflies make different journeys from their parents, imitation is impossible and their behaviour must be genetically determined.

The behaviour of animals, on the other hand, can be learned. Dawkins gives the simple example of the saddleback bird which lives on islands close to New Zealand. Each saddleback has its own repertoire of birdcalls which is shared with its neighbours. Occasionally a new form of song is created, seemingly as a random event. This new form of song is then taken up by other birds in the area, creating a new distinct repertoire. This is an example of a meme development that has no evolutionary benefit.

The memes discussed in this book are crucial to evolution and are carried forward by groups of animals. Evolution is usually envisaged as individual development, but in the battle for 'survival of the fittest,' group behaviour is important as well. An example of how group behaviour confers competitive advantage is the cooperative hunting technique of lions. A group of lions will set up an ambush whereby one lion will drive a herd of prey towards the hidden location of the others. Whether this particular example is due to genetic instincts or learned is difficult to determine. Whereas physical evolution in all animals is just due to genetic change, in higher animals behavioural evolution could be due to changes in memes. The fact the behaviour is complicated does not rule out a genetic cause. We know from insect behaviour in bee colonies that the most complicated organisational behaviour can be instinctive.

If the copied behaviour contains an element of choice then it is more likely to be due to a meme. The ability to choose distinguishes a higher animal from an insect. If a bee is locked in a room it will continue to fly towards the light, bumping into a window until it is exhausted. However, a cat locked in a room, once it has determined that there is no exit, will choose to save its energy and wait until the door is opened. Choice gives animals the ability to imitate successful behaviour and ignore those that fail. Hunting lions face many choices: when to hunt, where to set the ambush, who will act as the decoy, how to approach upwind and so on. To me, therefore, cooperative hunting by lions is more likely to be memetically than genetically determined.

Meme development is more flexible and rapid than genetic development. Whereas higher animals can develop new behaviour by experiment and teach it to succeeding generations, Monarch butterflies will have no chance to choose new, more appropriate nesting sites, until a random mutation in DNA happens to induce new behaviour.

Human memes must have developed from those of our ape forebears. The chimpanzee is our closest relative and the best model we have for the behaviour of our ape ancestors. Chimpanzees live in the forests in bands of typically 10 to 80 animals. One of the most important characteristics the band displays is group bonding. Chimpanzees will cooperate with members of their own band and exhibit violent behaviour towards competing bands.

There is an evolutionary context to this. If an individual has a rare gene there is at least a 50% chance that brothers, sisters or children will have the same gene. There is also a 25% chance that any cousins, nephews or nieces will have the same rare gene. It, therefore, makes evolutionary sense for the individual to cooperate with the extended family and enhance the survival prospects of his or her genes. There is no DNA test in the animal world. As there is a fair chance a chimpanzee's genes are also carried by other band members, chimpanzees simply perceive their entire band as kin irrespective of whether they are directly related. Humans have inherited the same ability to cooperate with those they feel a bond towards and exhibit violence to those they regard as foreigners.

A successful chimpanzee band will have organisational discipline, foraging skills and the ability to exclusively control entry to its territory. Organisational discipline is achieved by the establishment of a social hierarchy. Typically there is a dominant male, the alpha male, who controls the band, maintains order and leads it to new nesting sites. In chimpanzee society the alpha male does not always have to be the largest or strongest but must be manipulative and politically aware so that he can influence the group. Male chimpanzees typically attain dominance through cultivating allies, particularly among the females. As females care for the young, it is in their vital interest that the alpha male leads the band well. In some cases, a group of dominant females will oust an unsuccessful leader.

In order to avoid undue in-fighting there is a clear hierarchy in the band.

Lower ranking male chimpanzees will show respect by making submissive gestures. Female chimpanzees will show deference to the alpha male by presenting their hindquarters. Female chimpanzees also have a hierarchy and higher status females have better access to food.

Within the band individual chimpanzees also need allies, these are cultivated by grooming. Mutual grooming appears to build up an emotional bonding between animals. The larger the band the more time is spent on grooming as the social tensions increase. In large bands of around 80 individuals as much as 10% of time can be spent grooming, whereas in small bands of 20 this may only be around 3%.

A female will leave her band for another more successful band. Since alpha males have the mating rights, success attracts more females and the opportunity to sire more children and thus enhance their survival prospects. Acquiring 'foreign' females is important for the bands future, as it prevents inbreeding.

Up to 60% of a chimpanzee's day in the wild is spent foraging for food. Chimpanzees are omnivores; their diet includes fruits, leaves, blossoms, seeds, stems, bark, resin, honey, insects, eggs and meat. Typically, about 10% of the diet is meat, say from young monkeys or small antelope. This broad diet increases their survival prospects, but collecting it also requires great skill. During the daytime, bands divide into foraging groups and collect food from different sources. Diet varies according to the season; knowledge of the right locations to eat at the right times of the year is crucial to the band's success and must be widely shared.

Successful behaviour is learned during child development. Chimpanzees give birth every 4 to 5 years. Following gestation, which takes 7 to 8 months, chimpanzees have a long dependency period. The young are nursed for 5 years and stay with their mother for several more years. During this time they learn to care for younger siblings, communicate appropriately with others in the band and acquire different types of food. They are one of the few species known to use tools. The most well-known example is the use of a small stick to extract termites from their nests in order to prevent termites from biting their hands.

Chimpanzees maintain exclusive feeding rights to their own territory by ambushing individuals from an opposing band and maiming or killing them. Ambushing continues between bands until one band is too weak to compete. Females join the winning band and males are driven away. When this behaviour was first discovered in the 1960s by Jane Goodall it caused a sensation in academic circles. She observed two bands of chimpanzees operating in adjoining territories. Over 3 years raiding parties from one group killed six males and two females, raped three more females and eventually took over their territory.

Again, it is very difficult to tell whether all this chimpanzee behaviour is genetically or memetically determined. My guess is that behaviour involving tool use, battle tactics, and food-acquiring skills are memes, as these vary according to the band and the environment. Submissive and dominant interactions between apes involve behavioural choices and therefore could also be due to memes, possibly exploiting some genetic predisposition. For example, it could be that the procedure of grooming is learned but the pleasure gained from grooming is genetically determined.

It is important to emphasise the evolutionary process here. On the lowest level many genes are competing with rival genes and cooperating with synergistic genes to produce a successful chimpanzee. Success in evolutionary terms is measured by the number of replications. On the next level, individual chimpanzees are both competing and cooperating with members of their band. Their aim is to improve their overall status and as a result enhance their mating prospects. On the highest level, bands compete against other bands for the best feeding grounds. In other words, there are three levels of evolutionary competition: the genetic, the individual and the band. The success of the band depends both of the genetic health of its members and the robustness of its memes. Just as chimpanzees are survival machines for their genes, bands are survival organisations for their memes.

This concept of multilevel evolutionary competition, with animal organisations as the survival mechanisms for their memes, should be true for all animals that are territorial and live in large groups. However, it is a theory that I have not seen explicitly stated elsewhere. The fact is that still, almost 40 years after the concept was introduced, memetics has remained an interesting idea rather than a scientifically attested principle of evolution. The similarity to the word gene has been a distraction. People have tried to define the unit of a meme; whether it is a syllable, a word, a sentence, a description or a whole set of dependent ideas. They have tried to consider its physical form by attempting to deduce where and how it is stored in the brain. Also unhelpfully, memes have been defined as the evolutionary mechanism for cultural evolution. This implies that memes are only concerned with artistic developments like songs, stories and dress, or conjectures and ideas like philosophy and religion. Memes are concerned with all ideas. These include technologies, like how to make Portland cement, laws that define what we can and can't do and how we wage war. These aspects of human behaviour are not normally considered as cultural developments.

The evolutionary mechanism of memetic evolution described in this book rests on three key premises:

1. There are two replicators at work in higher animals, genes and memes.

2. The same meme can be shared between the brains of a group of animals.

3. The behaviour of groups of animals that are territorial is subject to the evolutionary principles of the survival of the fittest.

In evolutionary terms a replicator is anything that can be copied from one life-form to another that satisfies three criteria. It must be capable of being copied reasonably faithfully; it must also be susceptible to change over a period of time and it must confer an evolutionary advantage. Genes fulfil these criteria. Their evolutionary advantage is obtained through the success or failure of the life-form they are contained in. Dawkins uses the word vehicle for the gene-containing life-form.

Memes also satisfy the definition of a replicator. Most studies in memetics have assumed that the vehicle of a meme is also an individual life-form. In this view memes are seen as alternative replicators to genes. Memetics has stalled because it is very difficult to prove that individual behaviour is a result of genetic or memetic replication. It seems to me that concentrating on individuals misses the point. The critical aspect of memes is that they are shared across a group of individuals and result in common forms of behaviour. The insight expounded in this book is to consider a group of territorial animals as the vehicle of propagation of memes. Successful groups will thrive and propagate their memes.

Note this is not the same as what evolutionists call group selection and which has been much derided. Steven Pinker describes group selection as

> a version of natural selection which acts on groups in the same way that it acts on individual organisms, namely, to maximize their inclusive fitness ... to increase the number of copies that appear in the next generation.

Thus, for example, in group selection a pack of wolves would be considered as a replicator. However, just as humans cannot copy themselves and therefore cannot be a replicator, a pack of wolves cannot be a replicator for the same reason. In memetic evolution, however, memes are the replicators and groups of animals are merely the means to their survival.

Memes function as replicators in a completely different way to genes:

- Whereas genes can only be transmitted by sexual reproduction, memes can be transmitted in many ways, particularly in humans e.g., by demonstration or by speech.

- Genes can only be transmitted from parents to children. Memes can be transmitted to other non-related individuals.

- Genes can be physically identified and their chemical expression can be analysed. Memes on the other hand cannot be measured and they are only physically present as a configuration of neurons in the brain. We only know of their presence by the behaviours and skills they engender.

Memes and genes have co-evolved. Gene development enabled meme development; the genetic evolution of the brain was necessary for humans to converse using language. Similarly, some memes could not be fully exploited without genetic change; so that, as an example, tolerance to lactose in milk developed after humans had discovered how to domesticate cattle.

Gene and meme development is also independent. The custom of Roman Catholic priests to remain celibate clearly inhibits any genetic development of catholic priests but the custom has survived for centuries. Similarly, autism as a gentically determined disease has survived despite the resultant limitation on communicative skills. Memes, however, have the advantage of being able to tap into genetically based instincts to arouse individual and communal feelings. Feelings of beauty can be gained by looking at a picture, feelings of joy can be felt by singing and dancing and, most importantly for humans, feelings of kinship can be aroused by indentifying an individual as one of us.

Meme develpoment can also cause individuals to act differently in a group than when they are on their own. An example is participation in communal fighting against rival bands. In evenly matched individual combat, animals rarely kill their own species. The risk of a serious injury for the winner is far too great. Animals prefer to aggressively posture and skirmish in a way that does not overly compromise their future. However, successful bands of chimpanzees attack and kill individuals from other bands. Dawkins has conclusively shown in The Selfish Gene that there is no gene that causes sacrificial behaviour in a life-form. It appears, however, that behavioural memes can, by tapping into feelings of kinship, affect changes in emotional responses which result in more aggressive animal behaviour. Natural caution can be overridden by instincts which are related to family protection.

In addition to the power of speech and the use of tools, a unique characteristic of human animals is their ability to create and live in complicated organisational structures. I shall call a human group sharing a number of memes in common a community. A community can be as simple as a family or as complicated as a religion or a state. Each different community will have different sets of shared memes which define, among other things, its rituals and its skill sets. Whereas other animals live in at most one community, humans have the ability to be a member of several communities at the same time; they can be concurrent members of their families, the companies they work for, the clubs they join and the states they live in. All these communities have different sets of shared memes and humans are able to hold all these meme sets at the same time.

The human story told in this book is of how developing technologies have affected human communities and how these communities have competed, developed and interacted to form the complicated structure of society that we have today.

Both genetic changes and new memes were important in human development from its ape ancestor. The first uncontested fossils of an ape that walked on two legs are members of the Australopithecine family from about 6 million years ago. The reason why the species first walked on two legs is still a matter of conjecture. The old hypothesis that it was because it allowed humans to see prey across the savannah is now discredited. Australopithecine fossils appear to have been particularly associated with lake and river side environments. I like the aquatic ape hypothesis of human origin put forward by Elaine Morgan in her book of the same name. In this hypothesis our forebears evolved in a lakeside environment and, while walking in water, they developed the ability to stand upright.

Several human attributes indicate an aquatic origin to our species. Being almost hairless, the human body can move faster through water and is less susceptible to parasites. This absence of hair is also found in other aquatic and semi-aquatic mammals like the walrus, sea lion and hippopotamus. Body temperature is maintained by increased levels of subcutaneous fat and a greasy skin with an abundance of sebaceous glands secreting lubricating oil into the hair follicles. There are vestigial indications of webbing between the fingers, allowing a stronger swimming stroke. The hooded shape of the nose (not flat like apes) matches the recessed part of the upper lip to block water ingress into the nose. The most important factor, however, is breath control. Humans have a descended larynx (a hollow muscular organ forming an air passage to the lungs and containing the vocal cords). This allows humans to breathe in large amounts of air through the mouth, rather than being restricted to breathing through the nose like all other apes. As a result of this humans have a greater range of vocalisations.

Whatever the earlier origins of bipedal apes, after a further 4 million years of evolution, the second major transition occurred. *Homo ergaster* fossils from 1.9 million years ago have been found in South and East Africa and the closely related *Homo erectus* was the first of the genus *Homo* to emerge from Africa; skeletal remains have been found in Georgia, Java, China and India. These two species had larger brains and longer legs than the Australopithecines, and used stone tools. They are the first uncontested species of the same genus as humans.

Their brain size was in the range 600–1200 cm³, which compares to 300–600 cm³ for australopithecines. A large brain size is an indication of increasing social ability. It makes it possible to interact more cooperatively with larger numbers of the same species and as a consequence, to become more effective and inventive in acquiring sources of food.

Homo erectus is the longest lived hominid species, having survived on Earth for about 1.5 million years. During that time hands adapted to allow the use of tools; the thumb became stronger and the position of the thumb in relation to the little finger changed allow better manipulation of small objects. Stone tools dating to this period of existence include crude hand axes, cleavers, pounders and scrapers. We can assume these hominids also were able to use wooden clubs and throw rocks. This gave them a much greater ability to kill large animals. They were also able to pound food to make it more digestible and would have been able to extract bone marrow. Long legs would certainly have allowed *Homo erectus* to stride out. Some say these hominids could jog or even run in pursuit of a hunted animal.

Homo erectus was similar in size and shape to humans. It is interesting to speculate whether, if we met a *Homo erectus* hominids today, we would consider them human. It would naturally depend on how they looked and were clothed but crucially on how they communicated. There is evidence that in the third transition, which took place around 500,000 years ago, the physical changes necessary to communicate in language were beginning to take place. A new species, *Homo heidelbergensis*, emerged in Africa which had a modification to the ear canal that improved the ability to detect human vocalisations.

The difference between being able to emit and hear sounds and full language depends on crucial developments in the function of the brain. How the brain functions is still a mystery. We have discovered the secret of DNA, have delved in to the basic formulation of matter, can send a message with the speed of light and have a view on the creation of the universe. Yet this small lump of grey matter in our skull still remains a mystery. We know the brain monitors and regulates the body's actions by continuously reacting to sensory information. The brainstem controls breathing, heart rate and other processes that are independent of conscious control. The neocortex is the centre of higher order thinking, learning and memory. The cerebellum is responsible for the body's balance, posture and the coordination of movement. There are about 85 billion neurons in the brain and an equal number of non-neuron (glial) cells. These cells pass information to each other via trillions of synaptic connections. We know quite a lot about its physical structure and which type of activity takes place in which part of the brain, but we know very little about how this incredible organ actually functions. Fundamentally, we do not know how the brain stores and retrieves information.

My specific interest here is to demonstrate that the facility of speech requires genetic evolution of the brain. We know humans have a limited capacity for short-term memory; there is a process in the brain for transferring information to long-term memory and sleep is crucial in this process. Brain patterns stored in our long-term memory matched with signals from the senses must allow us to identify an object. By way of an example, first consider the identification of an animal that does not speak, say a cat identifying a mouse. In the cat's long-term memory there must be coded information that determines the concept of a mouse. On seeing a mouse, a cat must have an animal equivalent of a computer browser which scans the brain for a match with its visual senses to identify the mouse as prey it wants to eat. We know, from experience with computers, that developing visual recognition software is incredibly complicated; yet cats seem to have a brain browser that works quickly and accurately.

Now consider two humans, one of whom is telling another that a mouse has been seen. Firstly, there has to be the same visual identification process as for the cat. The brain browser has to find the right word for this image, and then it has to find the right instruction to say the word mouse. The second person on hearing the word has to again browse his or her brain to match the sound against the concept of a mouse. In the human brain there must be links between the mouse image, the sound of the word mouse and the instructions to say mouse and this must be true of all words in our vocabulary.

That the function of the brain is altered by speech was illustrated in a Horizon programme in 2011 on BBC2 about colour. It showed the Himba people from northern Namibia, who have only five basic words for colour, as opposed to a reported eight for Western humans. The astonishing thing was that Westerners were unable to distinguish two shades of blue-green which appeared quite different to the Himba people. It seemed that because Westerners did not have a word for the colour they were unable to identify it. The programme also demonstrated that, once a specific colour is identified by a word, this changes the place in the brain where colour is examined. Thus, the human brain not only stores words for colour separately from colour images but also learns which signals from the eye correspond to which word for the colour. In this way someone who is red-green colour blind will still call some colours red and others green, based on the shade of the colour.

We also know from personal experience that the brain has a word-based browser keyed on the first letter of the word. How many times do we say: 'I can't remember the name but I think it begins with ...'. Word-based browsers are vital in order to communicate information by speech and must be unique to humans.

In addition, the brain has to learn the whole syntax of language to impart meanings to strings of sounds. When we learn a language later in life, we have to learn how verbs conjugate and nouns decline, how tenses are formed and so on. Language syntax is incredibly complicated, taking years of study to master a new language, yet the human brain of a 2-year-old is able to do this naturally. It is highly likely that the languages we first learn are hardwired into the brain during early development.

My point here is that just articulating a sound that can be understood by another brain is a long way from having a brain that is fully enabled by speech. The current consensus is that the full power of language was not available to hominids at this stage. A limited vocabulary linked to specific actions would have been perfectly possible, however. There is also speculation that humans could also have engaged in musical activities, like drumming and humming in unison. Such improved communication ability would have given methods for bonding between hominids in addition to the grooming rituals developed by apes. Larger social networks could now be maintained is less time-consuming ways. Larger social groups are associated with larger brains. The cranial capacity of modern humans is half as big again as that of *Homo erectus,* in the range 1200–1850 cm^3, indicating that *Homo sapiens* was capable of interrelating with larger groups beyond the band itself.

During the time of this third transition the hunting ability of humans improved dramatically. There is evidence that from 400,000 B.C. early hominids had learned to create fire and cook food. Raw meat is fairly indigestible to humans. Cooking the meat allows humans to extract up to 50% more nutrients. We also have the first concrete evidence of the development of the spear. Now humans could attack animals from a distance. They found themselves a new niche in evolution, as the predators of large mammals, and were able to thrive. They relied on endurance to outrun larger animals. I happened to see this process on a television documentary a few years ago. The BBC showed a Bushman hunter chasing a large antelope in the Kalahari Desert. Eventually the animal stopped, exhausted, and was dispatched with a spear. Except for the better quality of spear, this technique of hunting had remained unchanged for hundreds of thousands of years.

The difference in size between males and females reduced progressively as hominids developed from earlier bipedal apes. *Homo erectus* males were about 25% larger than females, for modern humans the difference is less than 10%. A more equal size between males and females is an indication that the social structure was changing. Bands were no longer dominated by an alpha male; pair-bonding was becoming more common. My view is that this was necessary to improve the training of children. As mankind developed technically and culturally, the number of memes that children had to learn multiplied greatly. In particular, those associated with male activities like hunting and tool creation became much more complicated. Similarly, cooking and gathering skills became more demanding, requiring the mother to spend more time with the daughter. Bands in which the alpha male continued to take no part in training his sons would have lost out in the evolutionary struggle to those with pair-bonding.

With pair-bonding, for obvious genetic reasons, there developed strong emotional ties within the direct family. This introduced the family group as another layer of evolutionary competition between the individual and the band. Despite this, hominids still retained a chimpanzee-like ability to cooperate with all those in the wider band who they consider as family. For hominids, therefore, there would have been four levels of evolutionary competition: genetic, individual, family and band.

Pair-bonding did not mean that humans ceased to mate with other partners. The degree to which monogamy was practised became a cultural meme. Many different tribes over the history of mankind have successfully practiced polygamy. Even today a Maasai herdsman in Kenya will aspire to have several wives.

Another curious thing about the genus *Homo* is that women are more concerned about their appearance than the men. In most of the animal world it is the male who seeks to attract the females by his striking looks. In my view this female concern with appearance also developed with pair-bonding. As with chimpanzees, it is the female that moves between bands. With pair-bonding it is the man and his family's choice as to which woman he asks to be his partner. Each woman will therefore want to make herself as attractive as possible to be selected.

During mamalian birth babies have to pass through a hole between the two halves of the pelvis. To enable babies with bigger brains to be born hominid woman had to develop wider hips. This may also explain the exaggerated curvaceous shape of the female body compared to our ape ancestors. A young man would want to pair with a woman who can bear children successfully, so he would naturally select a woman with wide hips. Large breasts, slim waists and plump thighs accentuate the appearance of breadth in a woman's hips, making her more attractive to men. Wide female hips were a necessary part of hominid evolution; the female curvaceous shape might be caused by sexual selection during the same development.

As the size of the brain increased, new species of hominid emerged. Around 350,000 years ago Neanderthals evolved in Europe and approximately 200,000 years ago *Homo sapiens* evolved in Africa. Both had larger brains than their predecessors; the Neanderthal cranial capacity was, if anything, slightly larger. They were both of equal height, but Neanderthals were brawnier and apparently better adapted to a cold climate. The ratio of height between men and women was similar for both, so presumably pair-bonding between men and women had fully developed in both species.

We have reached the start of the next stage in our story. *Homo heidelbergensis* had evolved into new species. At around 70,000 B.C., a population of *Homo sapiens* was concentrated in Africa and numbered around 100,000, the size of a medium-sized town in Britain today. These humans were hunter–gatherers who made crude stone, wood and bone tools; they also used fire to cook food. They lived in small bands made up of pair-bonded adults and their children. At least one other genus of *Homo* existed, *Homo neanderthalensis*, who lived in Europe. Both species had developed different memes from ape forbears for organisation, food acquisition, battle tactics and tool technology. Otherwise, the success criteria for a human band were identical to those for chimpanzees. Their behavioural characteristics of an omnivorous diet, training the young over a long period, openness to new ideas and aggressive territorial acquisition were also crucial to mankind's development.

It had taken millions of years of genetic and memetic evolution for hominids to develop from apes. But from now on meme development would become the main driver for mankind's progress.

CHAPTER 2:
THE FIRST BREAKTHROUGH (70,000–8500 B.C.)

Around 70,000 B.C., after two million years of relatively modest development, the pace of hominid advance began to quicken. By 10,000 B.C. *Homo sapiens* had become the first animal species to thrive simultaneously in all five continents. How did this happen, in what is a blink of an eye in the evolutionary timeframe? Evolution, the survival of the fittest, was still working its savage logic, developing human genes and memes that best suited the local environment. The rate of human meme development, however, had been dramatically boosted by the increased facility of speech and the escalating level of communal violence.

At the start of the period there was a step change in sophistication in tools, in what is known as the Upper Palaeolithic revolution. Humans still had a hunter–gatherer lifestyle but now they had more effective technologies. Their hunting tools improved with spear throwers, bows and arrows, fish hooks and sharper stone blades; as a result they were able to stalk and kill even the largest of mammals. Using natural fibres, needles made of bone, bark and animal skins they were able to clothe themselves and survive even in the cold of Siberia. Using rope, skins and forest products they could shelter from adverse weather conditions. Elementary dugout canoes enabled them to cross water and reach new continents. Grinding technology allowed them to digest cereals and expand their diet.

These improvements enabled hunter–gatherers to adapt and prosper in conditions as varied as forests, dry savannah and icy tundra. Even today, where hunter–gatherer populations still survive, it is clear how well they are able to exploit local natural resources. The Bushmen in the Kalahari Desert, the Amazonian Indians in the equatorial rainforests and the Inuit in the frozen north are all from the same species. Whereas other animals could only adapt genetically to local environments by evolving into different species, *Homo sapiens* could survive in any environment by developing specialised technology. Even today we can still marvel at the resilience of the Inuit in the extreme cold of North America using local materials to make igloos, kayaks and warm clothing.

However, human success was not due only to these improvements in technology. Just as with chimpanzees, those memes that defined how bands organised, fought and related to each other were important as well. The term meme, as used in this book is a broader concept than that initially described by Dawkins. I think when Dawkins invented the word meme he envisaged a small discrete unit of information that could be passed on across the generations in the same way as a gene. The problem is that there are a mere 20,000–25,000 human genes but there are literally countless ideas that have been exchanged between generations of humans. There are simply too many memes to analyse at the level of an individual sound bite that Dawkins described in his example of the saddleback bird. I think it is more constructive to think of memes as a pool of ideas shared by individuals in a community. These ideas allow people in the community to exhibit common behaviour and deploy specialist skills. Just as humans are survival mechanisms for genes, communities are survival mechanisms for memes.

There are broadly five categories of memes that characterise a community: technological, war-related, cultural, communicative and relational. Technological memes are responsible for humans becoming more efficient and effective. They are concerned with the development of tools, materials, power sources and transportation. The developments in the Upper Palaeolithic revolution mentioned at the start of this chapter are all technological memes.

War-related memes concern military technology and tactics. Throughout most of history, war has been the ultimate determinant of the success and failure of large communities. Until recently, societies that could not defend themselves against military aggression could not flourish.

Cultural memes concern rituals, dress styles, stories, religion, art and beliefs. They have proved a key factor in the cohesion and motivation of communities. I concentrate on the cognitive part of culture, particularly beliefs and religion. Many forms of religion have evolved naturally during human development and some have played a pivotal role in the story of our evolution.

Communicative memes facilitate the speed and accuracy of transmission of information. The only method of meme transmission mentioned by Dawkins was imitation. Humans, unlike other animals, have the ability to communicate memes by speech. Communicative memes include all other non-genetic ways of communicating ideas including writing, printing, arithmetic and education. It is only recently that information could be instantly shared internationally. For hunter–gatherers the maximum speed of communication was only as fast as they could move from place to place. In practice memes took a very long time to spread from band to band and onward across continents. Up until modern times technological memes, involving the development of specific skills, always took time to mature and could take centuries to spread from one civilisation to another. For example, the first book was printed in Europe six centuries after one was produced in China. As the speed and accuracy of communication developed so did the rate of human advance. For the early hunter–gatherers the standard of living changed little over 10,000 years. Around 5000 years elapsed between the development of agriculture and the first written records. Since the Industrial Revolution the pace of communication and hence of human advance has accelerated dramatically.

Relational memes concern the way humans interact with each other in an organisation. In hunter–gatherer society they concern the relationships between the individual, the family and the band. As human development unfolded, higher level organisations such as tribes and states were formed, each with their own arrangements for how leaders and followers relate to each other. The success of these organisations was determined by their memes. Those organisations that thrived were the ones with the most effective combination of technological, relational, war-related and communicative memes. In more modern times, as the other types of memes became increasingly easy to copy, it was the relational memes that increasingly determined the competitive success of organisations. The history of mankind's advance in this book is the story of progressively complicated and effective organisational structures.

Memes are either passed on by education and training from adults to children or by communication and copying within the community as a whole. They are in a continual state of evolution and most result in minor changes in the skills and behaviour of communities. Important memes that improve the survival prospects of a community are spread further afield by diffusion between communities or by the success of the community itself. In hunter–gatherer society successful bands spread their memes by gaining access to larger hunting grounds, growing in size and importance and eventually splitting into several successor bands that were similarly successful.

The conception of some memes, however, was so significant they affected human society as a whole. I have called these breakthrough memes. From the initial idea these breakthrough memes were developed in a myriad of ways to meet specific needs. They spread widely throughout the human world and significantly changed the lives of its adopters. For example, the idea of utilising stone tools involves developing skills for selecting suitable stones, mining, preparing the stone, sharpening and polishing the cutting edge and (possibly) attaching it to a wooden handle. Each stage of these processes was developed and refined over the years, giving rise to a huge variety of specialised tools by the time of the Neolithic period.

At the start of each subsequent chapter in the book I shall introduce the breakthrough memes for that period and highlight them in bold. The rate of the conception of breakthrough memes has increased throughout human history, which in turn, has increased the pace of human advance.

Chimpanzees have only one type of community acting as a survival mechanism for their memes. Hunter-gatherers have three nested communities, the family, the band and the broader community of bands; each has its distinct sets of memes. I shall call the hierarchy of communities that humans inhabit a society. The structure of evolutionary development of the genes and memes of hunter-gatherer society is shown in Figure 2.1.

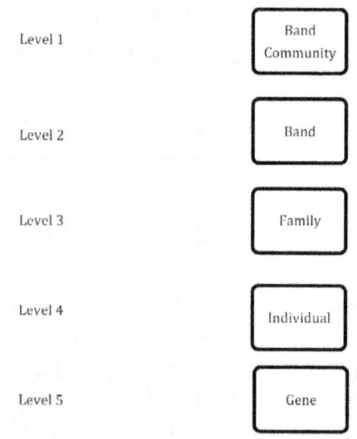

Figure 2.1 Evolutionary structure of hunter-gatherer society

At the lowest level genes are competing and cooperating to produce the greatest number of successful humans. On the next level, individuals compete and cooperate with others to increase their survival prospects and those of their progeny. They can compete in a number of ways. They can use violence or the threat of violence, they can collect allies to support their cause or they can simply act smarter by acquiring special skills or using specialist knowledge. Which combination of these competitive forms of behaviour individuals employ is determined by their genetic make-up, the memes they share with the three organisational levels they belong to and their own ideas and experience. Their own ideas will become memes only if they are adopted by a broader community or taught to their offspring.

On the next level, families of hunter–gathers compete and cooperate with other families. Their purpose is to provide the best support and training for their children. Families are responsible for training most of the band's technological memes. Across the hunter–gatherer world there is a broad standardisation in sexual roles. Women pass on skills involved in gathering food, home making and cooking while men train sons in tool making, hunting and warfare. Unlike other animals some human hunter–gatherers also have developed specialist professional skills. It is from hunter–gatherer times that expert roles such as a herbalist–healer, shaman or midwife began.

On the fourth level bands compete and cooperate with other bands to control the best territory for hunting and gathering. Relational memes determine the way the band organises itself, how it maintains discipline and arranges communal activities. Violent behaviour is quite different when interacting within the band and when encountering foreigners. Whereas aggressive behaviour within a band is disruptive and therefore discouraged, aggressive behaviour is common between competing bands. Skills in warfare are learnt at the band level.

Many of the band's memes are also shared with a wider community, which is at the top level of the evolutionary structure. There is no formal organisation at this level. The band–community all speak the same dialect; this allows community members to share ideas and information. Communities of bands meet regularly for trade, to arrange marriages and to participate at seasonal festivals. Bands also cooperate in armed conflict against other band–communities. This band interaction enables the band–community to share the same cultural memes and superstitious beliefs; they have similar dress, rituals and share stories and myths.

With chimpanzees, it is the success or failure of the band that is the most important factor in the survival prospects of any of their unique memes. With hunter–gatherers it is the success or failure of the family, the band and the band–community as a whole which determines their memes propagation.

There is only one evolutionary structure for chimpanzees. They live in one type of environment, the equatorial rainforest. Just as their genes show very little variation from band to band, so do their meme sets. Hunter-gatherers also have only one evolutionary structure. However, because hunter–gatherers live in a variety of different environments, many different meme sets been formed. The meme sets of Inuit and Aborigines are manifestly different, yet they still have the same evolutionary structure.

Evolutionary structures of societies are arrangements of its constituent communities in a hierarchy of levels above the genetic. These constituent communities demonstrate competitive and/or cooperative behaviour and hold the society's principal meme-sets. This book traces, how, as memes evolved, different types of societies emerged made up of different types of communities linked into new evolutionary structures.

In these early hunter–gatherer times there are no archaeological records for relational and communicative memes, so how do we know about the behaviour of these ancient hunter–gatherers? We can never truly understand those earlier times; their technology was more primitive and their population density much lower. The best anthropologists can achieve is to study modern hunter–gatherer societies. In the early part of the twentieth century there were still pockets of these societies that were not too affected by modern humans; Indians in the Canadian North-West and the Northern aborigine societies in Australia have been the most fruitful areas for research. Anthropologists also studied the Bushmen of Namibia, Inuit tribes in the Arctic and Amazonian Indians.

These remaining hunter–gatherer bands were very egalitarian. They lived in bands with typically 20 to 80 members. There was no single leader of the band; leadership was exercised by consensus. Disputes were resolved as far as possible by discussion. Discipline within the band was maintained by peer pressure, with the threat of ostracism and ultimately expulsion from the band. Expulsion was life threatening, as an individual alone without protection from the band, risked a violent death.

From the Upper Palaeolithic revolution onwards, the level of violence accelerated. The spears and axes were sharper; they had bows and arrows, and shields were used for defence. *Homo sapiens* became the first large animal to kill other members of its species in significant numbers. Hunter–gatherer life expectancy was just over 20 years. It is estimated that 10 to 20% of humans suffered a violent death, compared to less than 1% in the present day. Male violent death rates, at around 25%, were significantly higher than those of the females. In relation to the modern era, this is much greater than young men at the time of the First World War. Violent deaths occurred at all levels of the evolutionary structure. However, violent competition at the community and band levels was most significant for meme development. Humans have the same ability as chimpanzees to cooperate with those they regard as kin and show violence to those they regard as enemies. For the chimpanzees the concept of kinship is extended to include all those in the band. In hunter–gatherer societies, once a stranger is identified as a relative, however distant, this changes the nature of the relationship. Humans have the ability to divide the world into 'us' and 'them.' Hunter–gatherers regarded their own community as 'us' and acted violently against others. There is no group size limit to this human behaviour and in later times dreadful acts of violence against millions of people have been perpetrated once this feeling of communal kinship is aroused in hateful opposition to another group of humans.

Hunter–gatherer bands rarely risked death in open battle. Face-to-face battles between bands or communities did occur, but they were usually symbolic conflicts, consisting of much posturing. The more normal form of warfare was the surprise raid by overwhelming numbers.Warriors from one community would enter another's territory, often at night, and attack isolated pockets of the opposition, taking them unawares. Or perhaps unsuspecting individuals could be lured to a location, to a marriage ceremony for example, and then be ambushed. The aim was to maim or kill as many of the men as possible and capture and rape the women. This way there was minimum risk for maximum gain. Raiding was not continual but would break out sporadically. This could be the result of territorial disputes, sexual adventure, status issues, revenge or sheer belligerence. Once violence started between communities, it usually continued with raid and counter raid until one community gained the upper hand or mediation resulted in a truce.

Genes or memes thrive only if their possessors' rates of survival are greater than the rest of the population. To take a simple example, consider a gene conferring resistance to a disease. If the disease is virulent and leads to an early death, the proportion of the population with the resistant gene will grow rapidly. If, however, the gene is for increased resistance to a disease like the common cold that is not life threatening, the incidence of the gene will remain static. In the case of hunter–gatherers, increased violence acted as an accelerator of evolution because unsuccessful communities were eliminated more rapidly, allowing successful communities and their memes to thrive. Communities with excellent technological, relational, and war-related memes were most likely to prosper and be in a position to win their battles. The vanquished would be pushed out to marginal lands where they either starved or developed new memes appropriate to their new environment. Whatever the result, successful memes continued to operate and unsuccessful memes died out. Just occasionally, however, violence could also be a decelerator of progress; a warlike community could overcome one with more sophisticated technology and organisation. In history, the barbarian has sometimes overcome a civilisation. However, the 'sophisticated' memes were not always lost, because sometimes they were adopted by the victor.

Violence was not the only accelerator of meme development. Probably even more important was the development of language skills. We do not know exactly when *Homo sapiens* was able to express the full range of language with its vocabulary, grammar and syntax. However, there has to be a reason for the step change in the rate of meme development at the start of the Upper Palaeolithic revolution. The development of full language usage is the most likely cause. Language allowed memes to be developed and copied much more rapidly. Hunting and gathering skills improved because animals, food and medicinal crops could be named, and information about them could be shared. Hunter–gatherers became experts on the exploitation of natural resources; their knowledge of the behaviour of wild animals and the habitat of suitable plants for food and medicine was vital to their survival. There are at least 3000 species of fruit found in the Amazonian rainforest, of which 200 are used by the 'civilised' world. However local Indians use over two thousand.

With the power of language new memes could be generated faster and more effectively. The old phrase 'Two heads are better than one' is generally true. For the first time animals could share and develop ideas together, leading to better solutions and a stronger commitment from all members of the community. These memes could also be shared between communities. Whereas chimpanzees could only demonstrate a meme within their own band, now humans could share their ideas across a much broader group. Thus successful memes could migrate faster and become widely adopted.

Speech is of paramount importance in human development. It is the primary advantage that distinguishes humans from other animals. Language itself is a meme. In addition to words for things, expressions were developed for feelings and abstract ideas. Human culture was enriched by poetry, song and storytelling, which enhanced human feelings of bonding.

Two surprising aspects of human behaviour that could not be anticipated from their animal origins were the enjoyment of the arts (pictures, music and dance) and superstition. How did these evolve and what evolutionary benefit did they give? It has been suggested that music and dance developed as a human bonding mechanism even before full language communication was possible. Dawkins suggests in *The God Delusion* that the love of the arts is an unconnected by-product of other necessary genetic traits. It could be, for example, that the genetic disposition for humans to pair-bond and for individual women and men to be sexually attractive to each other has, as a by-product, resulted in a human appreciation of shape and colour in the form of art.

Oscar Wilde famously said that 'All art is quite useless,' implying it had only an aesthetic value. I do not agree with Wilde, as art had three practical uses for hunter–gatherers. The first was communication. When humans painted 2000 figures in the Lascaux caves in France in 15,300 B.C. they were able to communicate about life in those times, just as, in a later era, pictures of ancestors were hung on the walls in stately homes and told of the people and animals alive at that time.

Secondly art was used to define status, to quote Jared Diamond in *The Rise and fall of the Third Chimpanzee*, p. 160:

> [Art] brings indirect benefits to its owner. Art is a quick indicator of status, which – in human as in animal societies – is a key to acquiring food, land, and sexual partners.... [European hunter–gathers] decorated their bodies with bracelets, pendants, and ochre; New Guinea villagers today decorate theirs with shells, fur, and bird-of-paradise plumes.... We know that New Guinea art signals superiority and wealth, because birds of paradise are hard to hunt, beautiful statues take talent to make, and both are very expensive to buy. These badges of distinction are essential to marital sex in New Guinea: brides are bought, and part of the price consists of luxury art.

Art, then as now, is a commodity that can be bought and sold.

Thirdly, sharing arts and crafts could define group identity in clothing design, body decoration, and song and dance. Each community developed their own traditions and aesthetic tastes. Ideas of beauty are relative and time bound. The lip plates worn by the Mursi women in southern Ethiopia are now seen as positively ugly by Westerners. Perhaps the Mursi would feel the same about some of the peculiar fashions of the developed world, such as platform shoes.

Despite the importance of art in communication, defining status and group identity, Wilde is right that art has made a negligible contribution to the overall material welfare of mankind. The development of the arts is best left to other books and will not be considered further here.

The human propensity to develop superstitious and religious practices is, however, a major interest of this book. Throughout history supernatural belief has had a powerful influence on the development of human interaction and organisation. Hunter–gatherers' supernatural beliefs are best characterised by the term shamanism, the belief that the world is populated by good and evil spirits.

How did the belief in spirits evolve and what was its purpose? It goes back to our animal origins. Animals continually search for clues about how events will develop, and react accordingly. For example, a wildebeest will not run away from a lion that has just eaten, because it knows that the lion is unlikely to immediately kill again. This knowledge allows the wildebeest to continue grazing. This ability to relate an effect to a cause is very important in animal evolution. It is thought that humans are genetically predisposed to believe that every event has a cause. Dawkins explains it well in *The God Delusion*, p. 184:

Children, and primitive peoples, impute intentions to the weather, to waves and currents, to falling rocks. All of us are prone to do the same thing with machines, especially when they let us down.... Most of us have been there, at least momentarily, with a computer.... There was a poignant recent report of a man who tripped over his untied shoelace in the Fitzwilliam Museum in Cambridge, fell down the stairs, and smashed three priceless Qing dynasty vases. He landed in the middle of the vases and they splintered into a million pieces. He was still sitting there stunned when staff appeared. Everyone stood around in silence, as if in shock. The man kept pointing to his shoelace, saying, 'There it is; that's the culprit'.

If hunter–gatherers were genetically programmed to assume everything has a cause, it was a short step to believe that spirits were responsible for all events of unknown cause. Spirits could influence animal behaviour or even inanimate things such as sun, rain and disease. Following this logic, humans then tried to influence the spirits in some way, perhaps by giving them food, performing a dance or a chant or by making a sacrifice. In this way, rituals for influencing spirits developed, and myths were created about how spirits thought and acted. Hugh Brody in his book *The Other Side of Eden*, p. 245 explains it as follows:

> The important and distinguishing feature of hunter–gatherer shamanism, however, is its mixture of flexibility, on the one hand, and firm attachment to a particular territory on the other.

> Hunter–gatherer stories reveal an array of spirits that influence what happens and are available to be influenced. These spirits are flexible and ambiguous in character. The supernatural creatures, that have undertaken to make and keep the world as it is, are tricksters as well as godlike. In the myths and histories of the hunter–gatherer world, there is a lack of defining line between good and bad, playful and serious.

Hunter-gatherers were tuned into the uncertainties of nature and attributed these to tricks played by spirits. Hugh Brody explains how the belief in spirits evolved from individual experience in a specific area. Moving to a new territory was quite daunting as this would involve new spirits acting in different ways.

The echoes of this ancient spirit world are still with us in the folk cultures of today. In the West we are still familiar with stories of elves, leprechauns and goblins. Many still believe that houses can be haunted by ghosts and some people are believed to have been possessed by demons. The Catholic Church, if called upon, will exorcise evil spirits.

Although meme development was the most important factor in mankind's advance, there were also genetic changes. As humans spread across the world, the genetic composition of the relatively isolated populations began to change; specific local genetic mutations became prevalent. The peoples of different geographical areas acquired distinguishably different appearances. This was the origin of race. For example, Australian Aborigines can be distinguished from American Inuit both by sight and DNA markers. However, it is important to emphasise that these DNA differences are small; mankind is just one species. The genetic markers that distinguish the Aborigine from the American Inuit represent a very small proportion of the naturally occurring genetic differences between any two non-related humans.

The most important genetic changes were those that reduced the risk of disease. To give an example: in the original Africans, ultra-violet light acting on a receptor in the skin triggered a response that produced a protein, melatonin. Melatonin turns the skin brown and reduces the effect of harmful ultra-violet rays. However a moderate amount of ultra-violet light is necessary to synthesise vitamin D, which is vital for maintaining bone strength and preventing rickets. People living in cold northern climates, who need vitamin D but do not require quite so much protection from ultra-violet light, have evolved lower levels of the skin receptor. Thus the need to avoid rickets has resulted in northern peoples generally having paler skins.

Genetic changes to combat diseases could also be of mixed benefit. Sickle-cell anaemia is a hereditary blood disorder, characterised by red blood cells that assume an abnormal, sickle shape. Sickle cells only survive for 10–20 days as compared to 90–120 days for normal healthy cells. Up until modern times, people with the disease were unlikely to live beyond middle age. Sickle-cell anaemia is common in India, the Middle East and Africa, where malaria was historically endemic. It appears that those with the disease have a better chance of surviving a malaria attack. In an age where overall life expectancy was little over 20 years, the genetic risk of sickle-cell anaemia was lower than the risk of catching malaria. Hence the sickle-cell gene evolved by natural selection.

The general outline of how humans spread out across the globe is known. The first expansion from Africa to the Middle East, occurred about 100,000 years ago and seems to have had limited success. But after the Upper Palaeolithic revolution, mankind proceeded to colonise the world. In broad terms the route of human migration was first from Africa to Asia and from there branching out to Europe, Australia and America.

There are two ways of estimating the dates of arrival in a new continent, by archaeology and genetics. The direct way of calculating human arrival on a continent is by finding remains that are datable. The difficulty with this approach is that human populations in the early years were small and therefore their skeletons are extremely difficult to find. This method only gives a latest date by which humans must have arrived. Measuring the genetic difference between original native populations of two continents gives an indication of how long ago their populations separated. This, too, has its problems, as the first humans to migrate to a new continent may have been superseded by later migrants. Comparing the two approaches, the current consensus is that humans first reached Asia around 100,000 B.C. and Europe about 40,000 B.C. The two most controversial timings are when and how Australia and America were reached.

The first settlement dates of Australia are controversial because it is hard to conceive how humans reached the continent. It is certain that mankind was in Australia by at least 40,000 B.C. and reached Tasmania by a land bridge by 30,000 B.C. The main area of contention is how the Aborigines arrived from Asia. The boat-building technology available to hunter–gatherers at that time can have been no more sophisticated than crudely hollowed-out tree trunks or reed boats. Even though the sea levels were lower at the time, travelling from New Guinea or Timor to Australia would have involved sea journeys out of sight of land. It is hard to understand how the original Aborigines could have made this journey.

For America, the date of the first human settlement remains unclear. Archaeologists have two broad migration models. The first is the short chronology theory with the first movement beyond Alaska into the New World occurring no earlier than 15,000 B.C. The second is the long chronology theory, which proposes that the first group of people entered the American continent at a much earlier date, 20,000–50,000 years ago. Both models assume that humans crossed the Bering straits and then made their way from Alaska downwards. Although at times before 15,000 B.C. the sea level was low enough for there to be a land bridge between the two continents, the problems faced in migrating through a largely ice-bound continent at the time of the ice age appear to be formidable.

We are sure, however, that, by 10,000 B.C. all the major continents were populated by *Homo sapiens*. Only the more remote islands like New Zealand or Iceland, which required longer sea journeys, had not been reached.

It was hunter–gathering humans who first affected the planet's ecology by causing the extinction of large animal species. Earlier hominids had been the first to kill slow–moving large animals (mega-fauna) in large numbers. In Africa and Asia, over time, some of these mega-fauna were driven to extinction by hominids. However, many of the mega-fauna adapted and became sufficiently fearful of humans to survive. However, when *Homo sapiens*, with his advanced tools and hunting techniques honed in Asia and Africa became the first hominid to reach Australia and America, the local mega-fauna were defenceless. A high proportion was driven to extinction from about 50,000 B.C. onwards, in an event known as the Pleistocene or Ice Age extinction. Table 2.1 shows that of 142 mega-fauna over 40 kilos in weight in three continents before the extinction event, only 41 survived. Almost all the mega-fauna in Australia were eliminated, including huge birds, crocodiles and marsupials. South America was also badly affected, notably losing a giant sloth the size of an elephant. In these two continents the timing of these extinctions is coincident with mankind's proven first arrival on the continent. However as the timing for human arrival in North America is uncertain it is not clear whether the disappearance of animals, such as sabre-tooth cats and mastodons, occurred before or after humans arrived.

Continent	No of Mega-Faunas		Examples
	Existing	Extinct	
North America *	45	33	Horses**, camels, lions , sabre tooth catscheetahs, mammoths, masadons, llamas
South America	58	46	Giant sloth, horse, stegomastodon
Australia	16	15	giant kangaroos, rhino-like marsupials , a marsupial 'leopard', a 200 kg bird, a 1 tonne lizard
Europe	23	7	woolly mammoth, woolly rhinoceros, cave lion, cave bear
Total	142	101	

Table 2.1 The Ice Age extinction

* Only 15 of the extinctions occurred after 15,000 B.C.

**Horses were reintroduced by the Spanish in the sixteenth century

There are two possible reasons for extinction. The suspicion that humans are involved derives from the coincidence in the timing with man's first settlement of the continent. The only direct proof of human involvement is in Europe where mammoth bones have been found pierced by arrow heads.

Climate change is the other possible cause, as the end of this period coincided with the end of the ice age. Opponents of this view point out that ice ages had waxed and waned previously without the same rate of extinction of mega-fauna. The general view is that mankind is implicated in many of the extinctions without necessarily being the cause in all cases.

One, especially significant mega–fauna, Neanderthal man, was driven to extinction in Europe. Given *Homo sapiens'* competitive, aggressive nature, it was impossible for the two species to occupy the same hunter–gathering niche. Evidently *Homo sapiens* won out. However, it appears that some cross-breeding took place, as Neanderthal genes have been identified in the human genome. Perhaps the same aggressive raids occurred as among *Homo sapiens*, which resulted in killing the Neanderthal men and capturing and raping the women.

The last mega-fauna to become extinct was the mammoth. Mammoths were living in 3,000 B.C. in northern Russia, before being finally eliminated.

The period after the start of the Upper Palaeolithic revolution lasted around 60,000 years before the next major human advance. This is a short period of time by evolutionary standards but a very long period compared with later developments. Soon new memes would accelerate mankind's progress at an even faster rate.

By 8500 B.C. the population of the world had reached about a million people. To understand the scale of this, imagine a world that contains no more than the human population of Birmingham. Robin Dunbar speculates in *A Pelican Introduction to Human Evolution*, p. 294 that band–communities contain about 150 people. This would mean there were only about 7000 different band–communities in the whole world.

Hunter–gatherers may have reached all the continents of the world, but overall the population was still small and fairly stable. With the extinction of many mega-fauna, the human population might even have passed its peak. Certainly, in some places, the population limits that a hunter–gatherer lifestyle can support were beginning to be tested. Home sapiens began to experiment with new food sources, which was to lead to a further advance in mankind.

CHAPTER 3:
FARMERS AND PASTORALISTS (8500–3500 B.C.)

Biologists have identified two broad categories of survival and reproduction, known as the r and K strategies. r strategies are generally adopted by small animals such as insects, fish and small mammals. These animals live in a precarious equilibrium with the environment. Life for them is a lottery, with a high risk of death. Their strategy for reproduction is to have a short gestation period and large litters with short intervals between births. Their population will vary considerably from year to year depending on the environmental conditions.

K strategies are practiced by larger animals. Their population tends towards a stable equilibrium with the environment. They live longer, have smaller litters, invest more time in rearing their offspring and have longer intervals between births. There are none of the surges and troughs of population experienced with r strategy animals. *Homo sapiens* adopted a K strategy. However, human population has never reached stability but continued to grow as humans found ever more ingenious ways of adapting to the environment.

There was a time around 8500 B.C. in the Middle East when the wild sources of food were over-exploited and the hunter–gatherer lifestyle could no longer support the indigenous population. The local people were driven to augment their food supply by cultivating wild plants and domesticating animals. The same process happened later and independently in different parts of the world, involving different plants and animals. This spread of **plant cultivation** and **animal domestication**, taken together with the manufacture of **pottery** and **textiles**, constituted the Neolithic revolution.

This was the start of the next fundamental change of human lifestyle after the Upper Palaeolithic revolution. Farming could support a greater density of population than hunter–gathering and introduced a new human role, the peasant farmer. Born out of necessity for survival, the peasant farmer's lifestyle was always one of hard graft, heavily dependent on the vagaries of the weather and the vicissitudes of nature. Whereas the peasant farmer himself was not better off, a new alpha male, the big man or chief, was able to exploit the situation to his benefit and achieve new levels of wealth and power. Chiefs led larger organisations which we know as tribes. Successful leaders were venerated and **ancestor worship** developed.

Pottery manufacture came first. Humans were making pots long before they cultivated plants and domesticated animals. Indeed, the availability of pottery was a precondition for consuming cereals; rat-proof containers were necessary for storing grain. Pottery developed independently in many parts of the world, as a natural experimental by-product of the use of fire. The Chinese claim the earliest date, the eighteenth millennium B.C.; Japanese pottery goes back to the eleventh millennium and pottery existed in the Middle East from the tenth millennium. Fortunately for archaeologists, each culture developed its own distinctive style of pots and this allowed them to trace how the Neolithic lifestyle spread across the Earth.

Cultivating plants from the wild is not straightforward. The seed of wild plants needs to be carefully selected to be suitable for crops. For example, naturally occurring peas explode from their pods and scatter their seeds across a wide area in order to germinate. As only peas that did not explode would be available for humans to harvest and plant for the following year, this mutation was selected by necessity. Over time, all domesticated peas came to have non-exploding pods. Another example is cereals. Ears of wild wheat and barley seed grow at the top of a stalk. When they are ripe the ears spontaneously shatter, dropping the seeds to the ground where they can germinate. A single-gene mutation stops the ears from shattering. In the wild, this mutation would be lethal to the plant. But Neolithic humans encouraged this mutation by choosing it for next year's growth.

Legumes and cereals were the first crops to be cultivated. Fruit cultivation took place later in the Middle East around 4000 B.C. and included olives, figs, dates and grapes. These fruit trees and bushes had the drawback of not yielding fruit for at least 3 years after planting and not reaching full production until as much as 10 years later. Fruit was, therefore, only available to people with a settled village life.

Few wild animals were suitable for domestication. Domesticated animals have to be fast growing, able to breed in captivity and to live in herds, as well as being relatively docile to handle and not easily panicked. Diamond estimates that, of the 148 mammals that could have been candidates for domestication, only 14 proved suitable to Neolithic man. These were the major five: sheep, goat, cattle, pig and horse, plus the remaining nine: the Arabian and Bactrian camel, llama/alpaca, donkey, reindeer, water buffalo, yak, banteng (south-east Asian cattle) and mithan (an Indian variety of cattle).

Of the big five, the sheep and the goat were first domesticated in the Middle East and the pig in China. The ancestor of Western domesticated cattle was the auroch, which was common in forests in North Africa and Eurasia, but is now extinct. Cattle were independently domesticated in South-East Asia and India. As with plants, selective breeding was also important for animals, both to enhance yields and to improve their ability to be handled. For example, the original wild horse was 1.3 m tall and was useful only as a pack animal. It was only after centuries of breeding to develop their strength that horses could be ridden.

Tables 3.1 and 3.2 are adapted from *Guns, Germs and Steel* by Jared Diamond and show the principal areas of the world where the Neolithic revolution occurred and the different plants and animals that were domesticated. Table 3.1 shows the major Eurasian sites. The Neolithic revolution first occurred in the Middle East and China. Thereafter, the revolution spread to West Europe, the Indus valley (in modern Pakistan) and Egypt, where other new local plants and animals were farmed. These five areas became the cradles of future civilisations.

Area	Plants	Animals	Earliest Attested Date
Middle East*	Wheat, pea , olive , flax	Sheet, goat	8500 BC
China*	Rice, millet	Pig, silkworm	7500 BC
Indus valley+	Sesame, eggplant, cotton	Humped cattle	7000 BC
Egypt+	Sycamore fig, chufa	Donkey, cat	6000 BC
Western Europe+	Poppy, oat		6000-3500 BC

Table 3.1 Major Eurasian areas of plant cultivation and animal domestication

* Independent areas of domestication

+Local Domestication, following arrival of founder plants from elsewhere

Neolithic humans also developed the ability to create cloth by spinning and weaving natural fibres from domesticated plants and animals. Different centres specialised in different textiles. Flax for weaving linen was one of the earliest crops to be domesticated in the Middle East. Cotton was being used in the Indus valley civilisation in the fourth to the fifth millennium B.C. Sheep with a woolly fleece were first domesticated around 3000 B.C. Silk from cultivated silkworms was produced surprisingly early in China around 2700 B.C.

With animal domestication there also came an improved supply of animal hides. Raw hide is stiff and it requires treatment by tanning to become leather. In ancient times tanning involved soaking the hides with various combinations of animal brains, forest products, dung and urine. This combination of smells later ensured that tanneries were well separated from populated areas. The oldest known piece of leather footwear, dated to around 3,500 B.C., was found in Armenia.

The Neolithic revolution also occurred independently in at least eight other areas (Table 3.2). Although locally important, none of these centres went on to have a major influence on memetic development outside their own continent. Only two of these locations, the Andes and Mesoamerica developed into major centres of civilisation. Culturally some elements of the old American ways have survived and been absorbed into the Christian tradition, for example the Day of the Dead in Mexico became identified with All Saints Day. However, the native evolution of American civilisation was cut short by the 'discovery' of the New World by Europeans in the fifteenth to sixteenth century.

Area	Plants	Animals	Earliest Attested Date
Mesoamerica	Maize, Beans, Squash	turkey	3500BC
Andes and Amazonia	Potato,Manioc	Llama, guinea pig	3500BC
Eastern USA	Sunflower, Goosefoot		2500BC
Sahel	Sorghum, African Rice	Guinea fowl	5000BC
Tropical West Africa	African Yams, Oil Palm		3000BC
Ethiopia	Coffee, Teff		?
New Guinea	Sugar Cane, Banana		7000BC?

Table 3.2 Other areas of plant cultivation and animal domestication

There is some indication that the lifestyle of Neolithic humans was less healthy than that of hunter–gatherers. Certainly, by living so close to animals, they became exposed to more diseases. Many animal pathogens mutated, crossed species and affected humans. Measles and smallpox mutated from cattle. Flu and pertussis came from pigs and ducks. Over the course of time people living in Eurasia developed some resistance to these diseases. However, when Europeans arrived in the New World these Western diseases had a catastrophic impact on Native Americans.

One favourable genetic change that developed at this time was the emergence of lactose tolerance, allowing adults to drink milk. Before animal husbandry became common, milk could be consumed by humans only in the first four years of life. This is because the natural production of the enzyme, lactase, which is essential to allow humans to digest the lactose in milk, ceases after four years. A genetic mutation among an early cattle-raising people who lived in north-central Europe around 3000 to 5000 B.C. allowed lactase production to continue into adulthood. Today the lactase-persistence gene is found in more than 90% of Europeans. However, most adult Africans and Asians who never had milk as a major component of their diets remain lactose intolerant.

Despite the higher risk of disease the farming lifestyle supported higher population levels. This in turn changed the evolutionary structure. The same area that once supported one hunter–gatherer band now supported a few hundred people, clustered together for protection in villages. Larger communities of these farmers, bound by kinship ties, are known as tribes. There is no reason to believe that Neolithic society was any less violent than that of the hunter–gatherers. Farmers were highly vulnerable to attack; an enemy stealing grain or livestock could condemn a tribe to a year of starvation.

Raiding was still the predominant form of warfare. As with hunter–gatherers the tactic was to kill the men and rape the women. Acquiring livestock became an additional prize. An example is given in the Bible. In Numbers 31 after beating the Midianites, the Israelites 'took all their cattle and all their flocks' and they were instructed by Moses to keep all the virgins 'alive for themselves.'

With the higher populations involved, the consensus approach to leadership of hunter–gathers was not practical. Selecting an individual leader became essential for a tribe to defend itself and to be successful in conflict. Anthropological studies of tribal societies in Polynesia, New Guinea and Africa provide an indication of how human society changed from the egalitarianism of the hunter–gatherers towards central leadership, firstly by choosing a 'big man' and then a chief. A big man was chosen by the tribal elders. He had no legal power of authority and he operated just on the strength of his personality. Successful big men would receive presents from other members of the tribe, gain more than their share of cattle from raiding and could support more than one wife.

Over time, the big men in successful tribes acquired a retinue of young warriors eager to share the spoils of war. These young warriors could bully others in the tribe, giving the big man power to command and collect tributes. Big men became chiefs when they no longer ruled by consensus but could command obedience. Neighbouring tribes were cajoled into sending gifts. A hierarchy of power was exercised through subordinate chiefs and village headmen. Although kinship and tribal fraternity remained important, the chief and his warriors became an elite group within the tribe. These more complex tribal organisations are known as tribal chiefdoms.

In his book *War in Human Civilisation*, p.223 Azar Gat gives the classic example of tribal behaviour in modern times by considering the lifestyle of Highland Scottish chiefdoms in the period before the battle of Culloden in 1746 A.D.

the chief played the role of father and patron to his clan. Although the clansmen owed him tribute, mainly in the form of foodstuff, drink, and cloth, a fraction of their contribution was returned to them, as the chief ... demonstrated his generosity in helping out clients in need and during hard times. A household retinue of armed men and henchmen was kept on the tribute, dining on the chief's table in feasts that, on suitable occasions, when hosting other chiefs or dignitaries, turned into lavish displays of power and wealth. Personal ornaments, weapons, and other prestige goods were exchanged as gifts and commodities among chiefs, and between them and the outside world..., many Scottish chiefs had two or three wives during their lifetime ... siring many children, with marriages partly calculated to cement alliances. Feuding with and raiding on neighbouring clans were incessant, with cattle stolen and corn stores set on fire.

Chiefdoms were part of a broader group of tribes that shared the same ethnic origins, with a common language and culture. The Gaelic-speaking Scottish chiefdoms mentioned above form one example of an ethnic group. Another example is the 40 or so tribes of Maoris who lived in New Zealand before the British conquest.

Chiefs were often polygamous and sought to establish direct family relationships with key individuals to maintain their authority. For chiefs, the concept of family broadened to become an extended family or clan, containing all those on whom he could make the claim of family loyalty. Succession on the death of a chief could be a problem if different clans supported rival candidates for chief.

We are now in a position to build the evolutionary structure for tribal society and compare it with Table 2.1, the evolutionary structure for hunter–gatherers. The first two levels are the same, representing genes and individuals. On the third level I have shown two organisations; the village and the family/clan. Villages were the prime structure of a farming community. They were the focus for important relational memes as there were critical issues to be decided on land ownership, farming policy and discipline. Villages were largely self-administering. A group of elders decided village issues sometimes led by a village headman. Farmers were also expected to be able to fight, both in defence of their village and to participate in raids on other tribes. Violent conflict between villages within a tribe could still occur but most inter-village disputes were resolved without excessive violence.

Like hunter–gatherers, families were largely responsible for imparting technical skills from generation to generation. Although the skill sets were quite different from hunter–gatherers, the number of specialist roles in the village had not increased. Villages still needed the services of a midwife, herbalist–healer and shaman–priest.

Family relationships that developed beyond the village were also important, particularly if they included the ruling elites. These extended families, or clans, became an important factor in determining the success of a leadership bid and the way a tribal chiefdom was administered. The family–clan is shown as a parallel entity to the village on level three of the evolutionary structure. Farmers participated in village life and would support their clan in intra-tribal conflict.

Tribes were the main competitive unit in Neolithic society and were the focus for warfare activity. In successful tribes their leaders were supported by a retinue of warriors. The prevalent form of warfare between tribes was still raids. Where livestock existed, it was almost always the principle prize. The possession of a large herd was a key indicator of social standing. Rape and capture of women was also common. The taking of male prisoners, however, was rare; slavery, though it existed in some tribes, was unusual.

Ethnic groups of tribes were at the highest level of the evolutionary structure. There was no formal organisation linking the tribes. However, similar to the top–level of the hunter–gatherer evolutionary structure its constituent tribes shared the same language and cultural memes and cooperated in warfare when confronted by other ethnic groups.

The resultant evolutionary structure for tribal society is shown in Figure 3.1. It is very similar in form to that of hunter–gatherers, containing just one more constituent community group. There was only one evolutionary structure for tribes and it appears to have naturally evolved in all continents in which the Neolithic revolution occurred.

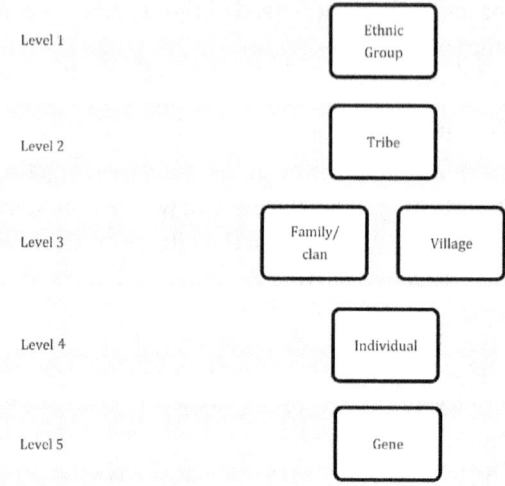

Figure 3.1 Evolutionary structure for tribal society

Hunter–gatherers could not prevent Neolithic farmers invading their territory. In the second millennium A.D. this happened in southern Africa, when Bantu farmers pushing south invaded the homelands of the Bushmen in South Africa. It is evident that the hunter–gatherers hit back with cattle raids but they could not overcome the loss of the wild environment as it was converted to arable land. Conflict between farmers and hunter–gatherers appears to have been a slow process of gradual conversion of the habitat, interspersed with stock and crop raids. Across the world hunter–gatherer bands all lost out in the long run as Neolithic farmers spread out and occupied their lands. This can be shown genetically as explained in the book by Luigi Cavalli-Sforza, *Genes, Peoples and Languages*. The markers left by a spread of agriculture from the Middle East into Europe coincided with the genetic markers left by people who were migrating along the Mediterranean from East to West.

The spread of farming across the Middle East, Europe, China and India took place at the expense of woodland. After the Pleistocene extinction of mega-fauna, man's second major change to Earth's environment was underway, eliminating deciduous forests and converting it to arable land. The UK countryside that we admire today is not natural but was initially created by Neolithic farmers. At first their technique was simple slash and burn farming. This in its turn was overtaken by other more sophisticated forms of agriculture, involving crop rotation, irrigation and, much later, the use of the plough.

Two types of Neolithic lifestyle emerged from the fifth and fourth millennia B.C. onwards, according to whether the main sources of food were crops or animals. The farming lifestyle was relatively fixed to one parcel of land. Farmers worked the land while keeping a few animals. An alternative pastoralist lifestyle emerged, particularly in the Middle East, North Africa and in the steppe lands of East Europe and modern Russia. This lifestyle relied on animal husbandry, managing flocks or herds grazing on land not suitable for cultivation. Pastoralists subsisted largely on dairy products and animal blood, rather than meat. The evolutionary stucture of pastoralist societies was the same as that of farmers. Like farmers their chiefs were also likely to be polygamous. Polygamy left a large number of unattached young men keen to enhance their status by gaining the prizes of conflict. The most famous example of these pastoralist groups existing today are the Maasai in Kenya. Even today a successful Maasai man will have several wives and the bride price (or bridewealth) paid to the woman's parents is in cattle.

The greater mobility of pastoralists gave them a military advantage over the farmers. In the next chapters we shall see how pastoralists threatened states and empires; the threat to farming tribes at the time can be seen from African history. In East Africa in the second millennium A.D. pastoralists speaking Nilotic languages spread south from southern Sudan into Ethiopia, Kenya, Tanzania, Uganda, Rwanda and Burundi, sometimes displacing and sometimes dominating the local Bantu-speaking farmers. Thus the Maasai came to Kenya and the Tutsi came to rule over the Bantu Hutu people in Rwanda and Burundi.

Religious culture also changed in Neolithic times. Tribes formalised traditional hunter–gatherer shamanism so that each ethnic group had its own religion, with its own myths, doctrines, places of worship, rituals and taboos. These religions were as complicated and many-layered as any modern day religion and as strange and broad as the scope of human imagination. The Buganda creation story mentioned in the introduction is a typical example. The spirit of disease and death, Walumbe, is alerted to Kinto and his wife's presence on Earth when they return to heaven to collect millet for their favourite chicken! The Garden of Eden story with snakes and apples is no less bizarre to modern ears.

All these religions had ceremonies for key events such as rites of passage, successful harvests and war preparations. Ceremonies included singing, dancing and drumming and were emotional experiences that acted as bonding activities for the tribe's members.

One new characteristic of tribal religions was ancestor worship. This seems to have grown with the importance of the leadership role. Successful leaders were venerated and on their death their spirit was perceived to live on. Inspiration and hope was sought by the living from the past successes of the dead. Each tribe worshipped its own ancestors, going back at least four or five generations. Ancestor worship consisted of rites carried out by descendants, usually involving offerings of food and drink at sacred sites. To justify their position it was important for chiefs, in particular, and all aspiring leaders to establish their relationship to their ancestors. In Scotland tribes were named after the clan of the first leader. In Gaelic 'mac' means 'son of'; thus being a MacDonald implies the tribesman was led by the clan whose founder was Donald. It was very important for a potential chief to also show that Donald was among his ancestors.

Chiefs now began to play an important role in the tribe's culture. Tribal rituals, especially religious ones, reinforced the role of leader. In key ceremonies the chief often led the tribe in asking the gods to act favourably towards it. This religious role enhanced the chief's status. He could be seen as a person acting between the tribe and the gods, giving him semi-divine authority. We can perhaps glimpse one of the ideas behind this tradition by considering the role of the emperor in ancestor worship in China at the time of the Zhou dynasty (1046–256 B.C.). I quote here from *Confucianism* pp. 21–2 by Joseph and Dorothy Hoobler:

> The reigning emperor, the Son of heaven, carried out certain religious rites that ensured Heaven's continued blessings on China and its people. The emperor's right to rule was called the mandate of Heaven. Sometimes Heaven could make known its displeasure with the ruler through natural disasters such as floods or earthquakes. Rebels who sought to overthrow the dynasty often pointed to such events as signs that the emperor had lost the mandate of Heaven. If rebels succeeded in defeating the emperor's forces, it signalled that the mandate of Heaven has shifted to a new family.

As the Chinese emperor represented his people, so the chief represented his tribe in communications with the gods. If he did his job well the gods were deemed to be supportive; if not his position was vulnerable.

Summary: situation at 3500 B.C.

The Neolithic revolution had established itself in north China, India, the Middle East and Europe, where woodlands were being replaced by arable land. In these areas, hunter–gatherers were being pushed into the steppes, deserts and deep forests. In America the Neolithic revolution was just starting; Africa, South-East Asia and Oceania still remained the preserve of hunter–gatherers.

Farming enabled the Earth to support higher population densities than hunter–gathering. The resultant tribal society had a different evolutionary structure from that of hunter-gatherers. The main difference was the emergence of a new alpha male role, the chief or big man who coordinated military activity within the tribe. The chief was wealthier than others in the tribe, looked to his clan for support and rewarded them for services rendered. Successful chiefs also maintained a retinue of warriors. Shamans became priests as religions based on ancestor worship developed. Otherwise there were no additional specialist roles in the tribe.

By 3500 B.C. human population was ten times greater than the start of the Neolithic period. Nevertheless, humans still lived in relatively sparsely populated geographic areas. The population of the whole world was only about 10 million people, roughly the population of the Netherlands today, and growing slowly at about 5% every 100 years.

While Neolithic technology, organisation, religion and lifestyle had changed significantly, this was still a Stone Age. Tools were fabricated from flints and other hard stones. For humans to advance further they would have to have new tools made of metal.

CHAPTER 4:
THE DAWN OF CIVILISATION (3500–500 B.C.)

The Victorians thought that humans were civilised whereas animals were driven purely by instinct and were shocked by Darwin's view that humans evolved from apes. However civilisation, as defined in the Oxford Dictionary as 'an advanced system of human development,' has only been present for a few thousand of the 200,000 years of human existence. Up to 3500 B.C. most human activity, like that of any animal, involved collecting food for survival. We have also seen that life expectancy was short and death by violent means was common. Little time was available for that artistic or intellectual activity that we associate with civilisation. It took the creation of a new type of community, the state, to allow civilisation to develop. In states military power was centralised under the control of one leader. Despite the new **use of metals** in warfare, this centralisation of power reduced the overall levels of violence and created a new privileged elite. This elite had the necessary time and means to commission those works of art, scholarship and engineering that distinguish a civilisation. In advanced states **writing** skills developed, firstly as a means to collect taxes and then as a method of communication. Writing enabled some to indulge in intellectual activities; formalised systems of **arithmetic** were developed, **time was measured** in standardised units and the rich began to be **formally educated**. The elite now included priests who administered the **state-sponsored religion**. In these more settled societies trade grew. This was made easier by the use of **money** as a medium of exchange and by improved transportation using **ships, camels** and **horses**. However, only the elite and a few city dwellers were able to experience civilisation; most continued in their subsistence lifestyle.

I shall be considering three types of state; tribal, regional and city, which had separate development paths and different impacts on human development. However, before the evolution of states is discussed there were further developments in tribal society that need to be reviewed.

The use of metals and its effect on tribal society

In Neolithic times, humans had discovered that some ores when heated would release metal in a process known as smelting. Archaeological artefacts made of gold, silver, copper and lead have all been dated to before 4000 B.C. These metals were soft and they could be used as ornaments or vessels, but not for fighting. This changed when it was discovered that bronze (an alloy of copper and tin) could be formed into blades with a sharp, tough cutting edge. This immediately increased the ability of farmers to cut down trees, shape wood and hoe the land. It resulted in a huge change in farming effectiveness and led to a large growth in population. However it also increased the ability for humans to injure each other; with bronze daggers, swords and axes, human conflict entered a more lethal phase.

We know bronze was made by the Sumerians in Iraq and the Chinese in Gansu province around 3000 B.C. and it became widely used in Eurasia during the second millennium. The biggest problem in bronze manufacture was acquiring the ore. Copper ores were scarce and tin ores were even rarer. We know of only five major sources of tin in Europe and the Middle East and all, like Cornwall and Brittany, are remote from the centres of civilisation.

The Hittites, a people who created an empire in Anatolia, were the first to use iron from the fourteenth century B.C. onwards. Iron manufacture was also difficult. Although iron ore is relatively plentiful, it is much more difficult to smelt than copper. Copper is easy to separate from the ore as the molten metal forms a liquid at the bottom of the furnace. However, iron did not melt at the temperatures that the early, primitive furnaces could achieve. The ore changed to iron in the solid phase. To extract the iron it was necessary to repeatedly heat and hammer the ore to force out the liquid slag. This labour-intensive preparation meant that its first use was confined to fairly small objects. To produce larger pieces required more heating and hammering and only the best smiths were able to achieve uniform quality. Smithing was thus a highly skilled trade and it is from this period that legends begin to feature magic swords produced by smiths who were also wizards. It was not until the first millennium that iron-working spread from its Middle Eastern base, reaching Italy in 800 B.C. and China in 600 B.C.

Given the difficulties involved in producing metal weaponry, tribes who could support specialist armed warriors had a significant competitive advantage. The scale of tribal fighting intensified and the warrior retinues of successful tribal chiefs could be numbered in the hundreds. Raiding parties could go deeper into enemy territory. Pitched battles were fought for the first time; in which the main military tactic was to overwhelm the enemy with an aggressive charge. These battles involved hundreds or even thousands of combatants. Looking back over the centuries with our compressed view of time, this is an age when tribal ethnic groups apparently swept across continents leaving a trail of buried artefacts to be discovered later by archaeologists.

It turns out that we can trace the existence of earlier successful ethnic groups by studying the roots of languages. Languages are memes: words pass from generation to generation with the occasional mutation as pronunciation changes over time. Consonants could make minor changes in articulation, the sound of 'f' might change to 'v '; 'd' to 't '; vowel sounds could become longer or more clipped. By comparing two similar languages, we can map the changes that have occurred and deduce their common language ancestor. For example, we know today's English evolved from ancient Anglo-Saxon after the Angle and Saxon tribes invaded Britain from Germany in the fifth and sixth centuries A.D. Anglo-Saxon is related to Dutch, Swedish and German. We can deduce that these languages shared a common proto-Germanic root language. This root language is no longer spoken, but linguists have reasoned that it was spoken in the area around the Baltic in the early centuries B.C. Proto-German in turn is linked to a much larger family of languages called Indo-European.

Table 4.1 shows a comparison of simple words across many of the branches of this huge family, including: a Celtic language, Old Irish; a Baltic language, Lithuanian; and Sanskrit, which is the ancestor of Hindi and other North Indian languages. The similarity between them is clear and linguists can deduce the original language spoken by the founding Indo-European tribes.

English	one	two	three	mother	brother	sister
German	ein	zwei	drei	mutter	bruder	schwester
French	un	deux	trois	mère	frère	soeur
Latin	unus	duo	tres	mater	frater	soror
Russian	odin	dva	tri	mat	brat	sestra
Old Irish	oen	do	tri	mathir	brathir	siur
Lithuanian	vienas	du	trys	motina	brolis	seser
Sanskrit	eka	duva	trayas	matar	bhrater	svasar
Deduced Indo-European	oynos	dwa	treyes	mater	bhrater	suesor

Table 4.1 A comparison of some words in the Indo-European language family

Across the whole world there is only a limited number of these ancient language families. The major ones in use today are:

Indo-European: the largest group, including many of the most widely spoken languages of the world: English, Spanish, French, Portuguese, German, Russian, Persian, Hindi and Bengali

Sino-Tibetan: including the main Chinese language Putonghua (Mandarin) as well as Wu, spoken around Shanghai, Yue spoken around Guangzhou, Tibetan and Burmese

Afro-Asiatic: covering Semitic languages, such as Arabic and Hebrew,

Berber languages spoken in North Africa and Cushite languages spoken in Ethiopia, Somalia and Kenya

Niger-Congo: a very numerous family of languages, the dominant language group of sub-Saharan Africa, it includes Yoruba from Nigeria, Swahili (the lingua franca in East Africa) and Bantu languages such as Zulu in the South

Nilo-Saharan: a group of languages in Central Africa, perhaps the most well-known of which are the Luo and the Maasai languages from Kenya

Dravidian: the languages of South India, such as Malayalam, spoken in Kerala, and Tamil, spoken in Tamil Nadu

Austronesian: includes Malay and all the Polynesian languages

Altaic: the language originating from the east Siberian steppes, including Turkish and Uighur, which is spoken in China

Uralic: languages originating from western Siberia, including Hungarian and Finnish

Kadai: South-East Asian languages, of which the most important is Thai

Austroasiatic: another group of South-East Asian languages, including Khmer from Cambodia and possibly Vietnamese.

Of the major languages of the world, only Korean and Japanese remain outside these groupings, indicating that both languages must have had many millennia of independent development. As these 11 language groups are spoken by 95% of the world's population, most of us can trace the origins of the languages we speak to just 11 tribal ethnic groups that lived several millennia ago. For most of the groups listed above we can identify the original homeland from which pastoralist or farming ethnic groups spread. We cannot be sure which specific competitive advantage all the original ethnic groups possessed. But as an illustration, a brief summary is given below of the memes that allowed the success of the original speakers of three of these language groups: Indo-European, Austronesian and Bantu (a major branch of the Niger-Congo family).

There is fairly strong evidence that the original Indo-Europeans were pastoralists from the Russian steppe. This is corroborated genetically. In *Genes, Peoples and Languages* Cavalli-Sforza notes there is a clear indication of the genetic spread into Europe of a people from the presumed Indo-European homeland north of the Black Sea. A key part of Indo-European success was due to the military use of the first horse-drawn war chariot. The horse was first domesticated in the fourth millennium B.C. on the Russian steppe. At first these horses were small and could only be used as pack animals. Gradually, by selective breeding, the animal's size was increased and by 2000 B.C. the horse was capable of pulling a light spoke-wheeled chariot. This became an important military vehicle, permitting the more rapid deployment of forces and improved manoeuvrability in battle. Using the horse in this way would have given pastoralists a huge advantage in mobility over static farming communities. In open countryside they would have been able to raid at will, with little chance of retaliation. In written history we find successor ethnic groups speaking a derivative of Indo-European. These include Celts, who conquered Britain and France and once sacked Rome; the Persians, who conquered all of the Middle East and formed the first great empire; the Aryans, who conquered the city-states of India and set up the first great Indian empires and the Hittites, who fought against the Pharaohs of Egypt and first made iron.

Tribes needed just one sustainable technical advantage to be successful. For the Austronesians it was the outrigger. Simple dugout canoes can be stabilised by using an outrigger, two smaller logs held parallel on either side of the main canoe. The outrigger might seem a minor technical achievement in our story of the advance of humans but it enabled Austronesian speakers to become the first great ocean sailors of the world. The language family is spread across three continents, covering half the globe, and accounts for about a fifth of the world's known languages.

From reconstructing proto-Austronesian it is apparent that the original Austronesians were fisherman from Taiwan. From about 3500 B.C. they moved south from Taiwan to the Philippines, Indonesia and Malaysia, where they settled and became the dominant culture, replacing hunter–gatherer groups. The number of Austronesian languages started to multiply as individual islands developed their own dialects.

Ocean travel always remained part of their culture. Their ability to navigate the oceans increased; travelling from island to island they voyaged further eastward into the Pacific, reaching Samoa and New Caledonia, deep in the Pacific, in 1200 B.C. Then in the first millennium A.D., centuries before the Portuguese began their great ocean journeys, they astonishingly reached three places thousands of miles apart in dramatically opposite directions: Hawaii (north Pacific), which was not reached again by other ethnic groups until the fateful voyage of Captain Cook in 1779; Easter Island (east Pacific), which is one of the most remote places on Earth, being thousands of miles away from a major landmass; and Madagascar (the Indian Ocean). This latter island, just 250 miles off the coast of Africa was apparently colonised in by a tribal group of Austronesian language speakers from Borneo! Finally, in the second millennium, around 1250 A.D. they reached New Zealand in the South Pacific, the largest temperate island untouched by hunter–gatherers.

An example of tribal success that occurred during recorded history is the Bantu expansion into central and southern Africa from 3000 B.C. to 1700 A.D. Bantu languages are part of the Niger-Congo group, and are associated with the spread of an indigenous West African farming culture, growing African rice, yams, oil palm and kola nut. The people who spoke proto-Bantu adapted their agriculture to forest environments from about 3000 B.C. Gradually they extended eastwards into Uganda and Kenya and found their crops could thrive in wet areas unsuitable for other local crops. At that time the whole of central and southern Africa was occupied by peoples known as the Khoisan, commonly known as Bushmen on the savannah and Pygmies in the forests. In central and southern Africa the local flora and fauna had not, so far, proved suitable for cultivation and as a result the Khoisan lifestyle remained that of Stone Age hunter–gathering. The Bantu acquired iron technology around 1000 B.C.; this gave them a huge advantage over the Stone Age Khoisan. Bantu tribes gradually migrated south, decimating the Khoisan population. Eventually only small bands of Khoisan remained in deep forests and arid areas.

One interesting postscript remains; when the Dutch arrived in South Africa in 1642, the local population was still Khoisan. The Dutch were able to establish their own colony with little resistance. Only in 1702, after the Dutch met the Bantu Xhosa people moving south, did they encounter fierce native reaction. It is interesting to speculate what would have happened if the Bantu had reached the Cape before the Dutch. Cape Town would probably have been just a trading post, like other similar European establishments at that time down the West Coast of Africa. No white colony would have ever been established and Apartheid might never have happened.

The development of tribal states

The most famous Bantu tribal ethnic group is the Zulus, who created a tribal state in South Africa in the nineteenth century. The creation of this Zulu kingdom is an example in modern times of how states were formed from tribes. A group of tribes became a state when one person attained sufficient power to command all the tribesmen. This leader was able to exert central control of the armed forces. Services to the chief, which had been voluntary, now became compulsory with corvée (forced) labour and military call-up; optional gifts were converted into regular taxes. Subordinate tribal chiefs lost their independence; they either married into the leader's clan or were replaced by the leader's nominee. The spoils of war were retained by the leader, further enhancing his power. When all this occurred the leader was generally referred to as a king.

All these changes were made by the Zulus when their kingdom was formed from the Nguni-speaking Bantu who practised cattle-raising and shifting agriculture. At the beginning of the nineteenth century one chief, Dingiswayo, established overlordship over the other tribes; he replaced the subjugated tribal chiefs with leaders from his own clan, disbanded the tribal militias and created a central military unit reporting to him. In 1817 Dingiswayo was killed and was succeeded by Shaka of the Zulu clan, which gave the new state its name. Shaka introduced new battle tactics: he forced his warriors to use a new thrusting spear in face-to-face fighting rather than the traditional spear-throwing battles. His raids became increasingly successful until Shaka's state extended to an area the size of England with a population measured in hundreds of thousands. To consolidate his hold on power, he instituted communal rituals to glorify his successes. State consolidation continued under one of Shaka's successors, Mpande. He turned tribal domains into state territorial administrative districts and placed sons from polygamous marriages in important administrative positions. At the same time, he married his daughters to distinguished followers, further extending his ruling clan network. All this forged a strong sense of Zulu identity that still exists today.

The Zulu state came to an end when, not long after Mpande's death in 1872, the fearsome Zulu mass charges were famously broken by the firepower of the British Empire. Nevertheless, a British official described the Zulu state as 'a collection of tribes, more or less autonomous, and more or less discontented; a rope of sand whose only cohesive property was furnished by the presence of the Zulu ruling family and its command of a standing army.'

The Zulu language was not a written one. There are other tribal states in history without writing skills, such as the Inca empire which left behind substantial buildings, and the African kingdom of Buganda (in present day Uganda), which had a population of hundreds of thousands. However, if the state lacks a centralised administration with written records, it can never achieve the full benefits of civilisation. This requires laws, tax records, and written instructions so that the government can better exercise its control. This is necessary not only for reasons of operational efficiency but also for the continuation of the state after a change of ruler.

The development of writing and the emergence of regional states

Writing developed from records of taxes and trades. Around 4000 B.C. in the Middle East tokens were produced in baked clay representing different numbers of animals: 1 sheep, 10 sheep, 50 goats and so on. These were strung together to give a record of the tax transaction. Beginning about 3500 B.C. in Sumeria, in southern Iraq, the tokens were replaced by a symbol for a number and a picture of the commodity, called a pictograph. These were inscribed in clay and then baked to preserve the accounting record. The use of these pictographs was gradually extended to represent words not related to tax. For example, a loaf plus a head combined might mean eat. In order to express the full grammar of language, written conventions developed in a number of haphazard ways. To enable nouns to be used in their abstract form a symbol might be added after the word. For example, to write the abstract word life, which has the same sound as arrow in Sumerian, a picture of an arrow was shown plus a symbol (called a determinant) to characterise the type of noun. Words could also be made up by breaking down longer words into syllables, as in the game charades. Conventions also had to be established as to how to add signs for plurals of nouns and tenses of verbs.

In Sumeria the pictograms evolved into logograms as they lost their pictorial form and became an abstract representation of the original concept. Written language was a complex collection of logograms, determinants, syllables and grammatical elements. It could be written and read only by highly trained scribes. Cuneiform was the first written logogram script, consisting of wedge-shaped marks in clay. The first language to use the script was Sumerian and, since texts were baked on clay tablets, several texts have survived that can be read today.

Conventional wisdom is that the concept of writing spread to Egypt in about 3200 B.C. The Egyptians, however, developed their own pictograms. Because they wrote with ink on papyrus (a thick, paper-like material produced from the pith of the papyrus plant) their hieroglyphs maintained a more pictorial form, even though the script developed the same degree of complexity. The Chinese writing system was in place by 1300 B.C. It is thought that most early Chinese writing was done with ink on bamboo. Like Sumerian, it developed into an abstract logogram form. Over time different Chinese states developed different scripts, even though they had the same style of brushed ink marks.

The Indus valley civilisation also developed a script but this has never been deciphered. It is significant that the four oldest centres of agriculture all developed writing, as each developed the institution of statehood.

This underlines the point that writing systems developed naturally in an agricultural society in which tax and other records were important.

There were two evolutionary paths towards statehood. One that we have already seen was the accumulation of central military power over a number of tribes, as in the example of the Zulu state. When these states acquired writing skills they are called regional states. The other route to statehood was for the village organisation to increase in size and wealth and grow to dominate the surrounding areas. These became city-states and are considered separately in the next section.

Three famous regional states that developed in this period were Egypt, China and the Hebrew state of Israel. In Egypt during the fourth millennium some 40 petty tribal states coalesced in two areas: one in Upper and one in Lower Egypt. According to tradition, King Menes, a sort of King Arthur figure, first established overall control over the whole country. However the first king of record was Narmer, who reigned around 2900 B.C. Once united, Egypt's geographical isolation allowed the state to establish itself as one coherent entity relatively free from invasion by neighbours. In China state development did not start until the second millennium B.C. as the clan of Shang established overlordship over the Yellow River basin in north China. It was a loose structure with regional clans tied to the royal clan by frequent intermarriage. In Israel, King David, one of the heroes of the Old Testament, is thought to have ruled over the combined state of Israel in about 1000 B.C.

The organisation of society in a regional state had a different evolutionary structure from that of a tribe as shown in Figure 4.1.

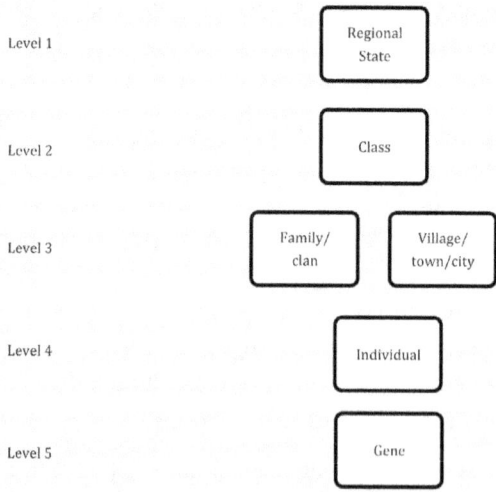

Figure 4.1 Evolutionary structure of regional state societies

At the top level is the state itself. The state had been formed by establishing one centre of power over tribes within an ethnic group. The state had common communicative memes; one language was spoken and one writing system used. As with tribes, all its population shared the same religious culture. This now began to be more formalised; for the first time we have written records that indicate religious practice at the time. In tribal religion, ancestor worship was the dominant form of religious practice. It remained so in China where rulers clung to their role as the intermediaries to the gods. Elsewhere a separate class of priests emerged to fulfil this task. Each state had a specific god who acted as the state patron. Examples of patron gods are Marduk for Babylon, Assur for Assyria, and Yahweh for Israel. Battles between states were perceived as battles between sponsoring gods. This is an extract from the Bible; Jeremiah, 46: 1–4, 10:

> The word of the Lord which came to Jeremiah, the prophet against the Gentiles; Against Egypt, against the army of Pharaoh-Necho King of Egypt, which was by the Euphrates in Carchemesh ...
>
> Order ye the buckler and shield, and draw near into battle.
>
> Harness the horses; and get up, ye horsemen, and stand forthwith your helmets; furbish the spears and put on your brigandines ...
>
> For this is the day of the Lord God of hosts, a day of vengeance, that he may avenge him of his adversaries; and the sword shall devour, and it shall be satiate and made drunk with their blood; for the Lord God of hosts hath a sacrifice in the north country by the river Euphrates.

As with the gods of all states, the God of the Hebrews was expected to support his side and to arrange for the slaughter of their enemies. Priests conducted rituals which often involved animal sacrifice, to seek help from their gods. Leviticus 9, 2, 8–10 in the Bible gives a typical description:

> And he [Moses] said unto Aaron, Take thee a young calf for a sin offering and a ram for a burnt offering, without blemish and offer them before the Lord ...
>
> Aaron therefore went unto the altar and slew the calf of the sin offering, which was for himself.
>
> And the sons of Aaron brought the blood unto him; and he dipped his finger in the blood, and put it on the horns of the altar, and poured out blood at the bottom of the altar;
>
> But the fat and the kidneys, and the caulk above the liver of the sin offering, he burnt on the altar as the Lord commanded Moses.

The book of Leviticus is full of tedious descriptions of rituals for festivals, cures for disease and other favours asked of God. To placate or thank the gods, substantial temples were built, of which one famous example is the temple complex at Luxor in Egypt, first constructed in the second millennium. The temple-building tradition, however, goes back to the beginning of civilisation. This is a description of the old temple of Uruk in Sumeria, built in the middle of the fourth millennium B.C.

> a huge building, larger than the Parthenon in Athens.... The shrine was even more remarkable for the fact its ground plan almost exactly anticipated, by 3000 years, the layout of early Christian churches. There was a central nave, a crossways transept, a ... lobby and an apse at one end.... A magnificent walkway leading to a wide public terrace ran alongside. The huge embedded pillars of the colonnade, 2 metres in diameter and built of sun-dried bricks ... were protected from the sun ... The labour that went into these buildings was immense: many millions of work hours. (*Babylon*, p. 40, by Paul Kriwaczek)

In Egypt the priests invented a powerful hook to draw people in and to emphasise their own importance. Priests claimed to know in detail what would happen after death; there would be a blissful afterlife for those who followed god's rulings and a savage penalty for non-conformity. The passage into the afterlife could only be eased with the help of prayers and offerings from the priests. In Egypt, the Book of the Dead is a canon of mantras aimed at ensuring a pleasant afterlife for the deceased. Among the perils after death to be overcome was the weighing of the heart ritual. In the presence of the god Osiris the dead person was asked to swear they had not committed any of a list of 42 sins. The dead person's heart was then weighed on scales representing truth and justice – if the scales balanced this meant the deceased person had led a good life and he would have a pleasant afterlife. If the heart was out of balance then a fearsome beast Ammit (the devourer) would eat the dead person bringing his afterlife to an unpleasant end. This idea of reward for the good and punishment for the wicked in the afterlife has been carried forward by most modern religions.

The state organised the collection of taxes and administered justice. Scribes were the civil servants of the time; they organised tax gathering, wrote laws, codified religious ceremonies and documented trade agreements. They enabled instructions to be sent reliably over a distance from a king to a lord and from a lord to his agents. Scribes became very important people, and were relied upon by kings to oversee the administration of their kingdom. In China they became scholars and constituted a separate prestigious class to rank just below the aristocracy.

Regional states were organisations that were created by elites for their own benefit. As former tribal chiefs had grown in power they had acquired more land in their own name. The possession of land, rather than herds of animals, was now the measure of wealth. Slaves were captured in battle and used by these chiefs to work their lands. Chiefs became kings and subordinate chiefs became aristocrats, intent on maintaining their position of power and authority. In tribal societies farmers had been freemen who worked their fields and paid taxes to their chief. With the coming of writing, all their tax, dues and duties were now fully documented with a bureaucrat to administer them. Now the chief could be imposed on them, their kinship ties to those in authority were lost and they were considered just a source of income. Independent farmers became peasants bound to the land.

Thus, in regional states, a hierarchy of king, aristocrat and peasant emerged. Jared Diamond calls the kings and aristocrats kleptocrats – a term for rulers who use their powers to steal from their own country. The purpose of regional states was to enhance the power and wealth of the ruling king and aristocracy. They achieved this by controlling and taxing the rest of the population and by defending their state in war against other states.

Kings, aristocrats and their dependent warriors constituted the main military arm of the state. The peasantry could still be called upon to fight, but now were much less motivated as a fighting force, having lost their kinship ties to those in authority. Recorded history, written for the kleptocrats, largely supports their view of the world as a succession of military adventures masterminded by brave and resourceful kings. For example, in the battle of Kadesh in 1274 B.C. between the Egyptians led by Ramesses II and the Hittites, Ramesses II seems to have blundered into an ambush by the Hittite chariots but was able to rally his troops and escape back to Egypt. The best result Ramesses could justifiably claim was a draw. However, in Egypt there are inscriptions and reliefs declaring a victory in five separate places, including Karnak, Luxor and Abu Simbal. The inscription in Luxor reads as if Ramesses won the battle almost singlehandedly:

> His majesty slaughtered the armed forces of the Hittites in their entirety, their great rulers and all their brothers … their infantry and chariot troops fell prostrate, one on top of the other. His majesty killed them … and they lay stretched out in front of their horses. But his majesty was alone, nobody accompanied him.

Kings and aristocrats coveted status goods to display and enjoy their wealth. They built palaces, temples and mausoleums, clothed themselves in richly dyed clothes and acquired valuable crafted goods made of fine pottery, precious metals and jewels. To do this they needed the assistance of merchants and artisans to acquire and make the requisite items. It is in this period that new trades developed with specific skill sets such as carpenters, masons and weavers.

States became stratified by class and this is shown on the second level of the evolutionary structure. There were eight possible classes: kings, aristocrats, warriors, priests, scribes, merchants, artisans and peasant farmers. Slaves had no status at all. Each individual state gave different relative importance to each of these classes depending on the organisation of the religion, the importance of trade and the warlike nature of the state. Each class attempted to increase its status; in this competition balance of power might change but the peasant generally lost out.

The importance of class structure is inherent in Vedic religion, the antecedent of Hinduism. This was brought to India by the Aryan tribes from the steppes of south Russia who destroyed the Indus Valley civilisation. The Aryan form of religion was characteristic of a tribal state with a fully developed priesthood. It was very different in character from the Hindu religion of today. This Vedic religion had an impersonal pantheon of gods, including Agni the fire god and Surya the sun god. The Rig Veda, which contains hymns and mantras to these gods, was compiled between 1700 and 1100 B.C.

The *Rig Veda* and other *Vedas* were composed in Sanskrit, an Indo-European language, and memorised by Brahmin priests. As an Indian alphabet had not developed until around 300 B.C., Sanskrit was already a dead language by the time the *Vedas* came to be written down. A famous verse in the *Rig Veda* describes how society was to be divided.

> When they divided the Man, into how many parts did they apportion him?
>
> What do they call his mouth, his two arms and thighs and feet?
>
> His mouth became the Brahmin; his arms were made into the warrior, his thighs the people and from his feet the servants were born.

This is the origin of the varna – the four classes of Indian society: brahmins (the performers of rituals and experts on the religious texts), kshastriyas (warriors and kings), vaishas (independent merchants, farmers, cattle-herders and artisans) and sudras (servants). Hindu theology teaches that by birth you are born into one of these varnas and your duty (dharma) is to fulfil the role allotted to you. This division is thought to have developed after the invasion of India by the Aryan pastoralists. The sudras are thought to be the original inhabitants of India, speaking a Dravidian language. Sudras were required to live separately from the three top classes, presumed to be Aryans, and were excluded from their socio-religious life. Below the sudras there was a fifth class, the original hunter–gatherer tribes, who became the untouchables.

The Vedic religion formalised a class society in India, keeping the defeated people and the peasants at the bottom and reserving the upper classes for warriors, priests and merchants. It goes further in making it difficult to scale the class ladder. Even in the upper ranks of society it was difficult to switch from merchant to soldier or soldier to priest. What is extraordinary is that in Vedic religion the premier class is not the aristocrats but the priests; brahmins place themselves above kings in status.

On the third level of the hierarchy villages remained important administrative units and a rich source of cultural tradition. However, villages had lost their independence to compete directly and had become pawns in the aristocratic power game.

As the state increased in complexity, artisans and merchants congregated in newly formed towns. We know little of how these towns were organised. As we shall see with city states, towns were less easy for an aristocrat to control directly. It is likely some form of self-government emerged. In medieval times trades congregated together and federations of artisans were formed. These federations, known as guilds, formed the basis of a system of local government.

Clans also remained present on the third level. Kings and aristocrats still relied on family relations to occupy key positions of power. Nepotism was still an important way of trying to ensure the allegiance of the state organisation to the king.

Individuals were still responsible for training others in technical skills. As in tribes, peasants trained their sons and daughters in farming skills. Now, though, there were many specialist trades such as blacksmiths, weavers and tanners whose skills also needed to be passed on from generation to generation. These became family-run businesses and, if they thrived, the family would train new staff from outside the family to become the world's first employed workers.

With the formation of states human society had become much more complex than the simple structure of hunter–gatherer bands. However, successful meme propagation, particularly of cultural and relational memes, still largely depended on the ability of the state to control a large territory and defend it against its enemies. Regional states with excellent technological, relational, and war-related memes were most likely to prosper and be in a position to win their battles. The propagation of cultural, relational and communicative memes was largely dependent on the success of the state in war. We have seen in the first part of this chapter how languages were spread by the success of tribes. Success in war would also determine the spread of religions and the written scripts of states. In contrast war-related memes were often spread by diffusion between states and tribes; the absolute necessity of a state to defend itself meant that the acquisition of the latest war tactics and technologies was of vital importance. Technology transfer between states was slow, hindered by relatively low population densities, language and distance barriers. As we have already seen, it took several centuries for iron-making skills to travel from the Middle East to China and India.

The development of city-states

City-states evolved from prosperous villages that had increased their population and grown into towns and cities. Initially, they developed only in areas of highly concentrated agricultural production where land was irrigated. Farmers attended their fields by day and returned to the city for protection at night. City-states evolved in clusters as each village in an area of high agricultural productivity prospered. The oldest city-states and the birth of modern civilisation occurred in the fourth and third millennium B.C. in the Fertile Crescent, an area of land stretching from Lebanon through Syria to southern Iraq.

Some of the most famous city-states are Ur, Uruk, Akkad, Kadesh, Ashur and Babylon. Although writing began here, we have only scraps of knowledge about how their organisation developed. We know the name of the first recorded person in history: Gilgamesh, the legendary ruler of Uruk. But in truth, all we know of this early period are legends and the boasts of kleptocrats. In these accounts, history is seen as war; one state struggled for dominance over another in order to loot treasure, rape and enslave women, enslave men and extract tribute from the losing state.

Later city-states grew up around centres of trade, based in the Mediterranean. Increased levels of trade were a consequence of the spread of civilisation. We have seen how the demand for tin and copper ore led to the need to trade goods over hundreds of miles. Trading in these ores must have involved many contacts between diverse tribes all the way from the source to the market. It will have spread wealth outside the centres of civilisation and made citizens aware of the many peoples and geographies outside their local experience. Copper, tin and iron and their ores were not the most important traded goods in the Bronze and early Iron Age. In a period of increasing prosperity, the ruling classes required many goods to display their wealth: timber to build their palaces, gold, silver and ivory for ornamentation, fine pottery and glass for dining, textiles coloured using specialist dyes, writing materials and so on. For the first time mining, manufacturing and commerce became a significant part of some societies' economies.

There were no paved roads, and the carriage of heavy goods over land was time-consuming and expensive. Horses and donkeys were used as pack animals and camels were domesticated to carry goods over deserts. Nevertheless, land transport was slow, with limited capacity. Journeys by sea and river were much more practical. The use of bronze tools allowed larger ships to be made from wooden planks, laid side by side and joined by a small square mortise. The joins were filled with a mixture of resins to produce a seaworthy ship. These ships could be rowed, but also sailed. The Mediterranean Sea became a centre for commerce. The Minoans from Crete were the first known specialist trading community and thrived from about 1900 to 1400 B.C. They had an expertise in ship-building, sailing and sea-borne warfare. Traders needed records and writing to register contracts and trades. The Minoans had their own (undeciphered) logogram script – Linear A. The Minoans were followed by the Mycenaean Greeks, who adapted their script to be suitable for the Greek language in a form called Linear B.

After the Mycenaeans came the Phoenicians, a Semitic people whose ancestors were thought to be the Canaanites of the Bible. They developed a number of city-states on the west coast of the Fertile Crescent. The most famous cities are Tyre, Sidon and Byblos. Phoenician trade was founded on the export of purple Tyrian dye produced from a particular sea shell. Trade grew to include Lebanese cedar, glass, pottery and silver from their mines in Spain. At the height of their powers from 1200–800 B.C. they established trading posts along the North African coast and in Cyprus, Sicily and Spain.

The successors to the Phoenicians were the Greeks. From 800 B.C. they gained predominance in the eastern and northern Mediterranean. Greek colonies were set up in Italy, the south of France, Asia Minor and on the Black Sea. Later the city-states of Rome and Ashur grew into regional powers and established great empires.

Over history, from the Phoenicians to the British in the modern era, the states that rely on income from commerce have contributed a large proportion of mankind's most significant memes. This may be because by its very nature trade is competitive, new ideas bring monetary rewards and failure brings bankruptcy. These early trading city-states were particularly innovative, leading to a number of significant advances for human society; they can be credited with the invention of the alphabet, the first use of credit and money, the world's first democracy and the development of mathematics.

Society was structured differently in a city-state from a regional state (Figure 4.2).

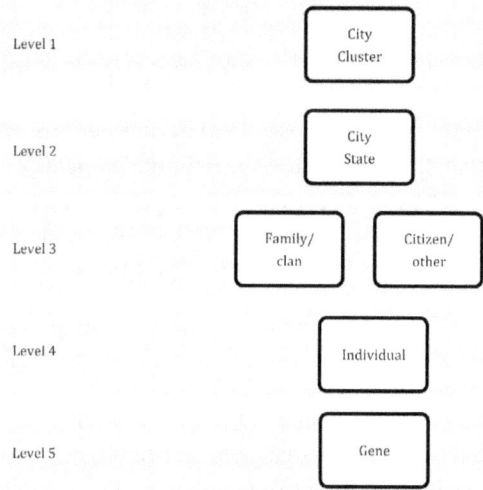

Figure 4.2 Evolutionary structure of society in city-states

At the top level there was a cluster of city-states that shared the same language and writing system. The cluster of Phoenician city states developed their writing system to become the first alphabet. The first known inscription, dated to 1200 B.C., is on the sarcophagus of King Ahriam and can be seen today in the National Museum of Beirut. The evolution of the alphabet can be traced back to Egyptian hieroglyphs. The Egyptians had 24 signs for 24 words beginning with different consonants: aleph = ox, beth= house, gimel = camel, daleth = door and so on. The inventors of the alphabet took these 24 hieroglyphs to represent their first letter. Thus the first letter of the alphabet, aleph, used the Egyptian hieroglyph for ox, the second letter, beth, that for house and so on. Over time the hieroglyph was simplified to a few strokes of the quill and duly became the symbols for 24 consonants. There was no sign for the vowels; these had to be deduced, and this is still true for all Semitic languages, such as Arabic and Hebrew.

The alphabet was invented only once in history; and all common alphabets of today derive from that of the Phoenicians. The Greeks added vowels in the eighth century B.C. and their alphabet was adapted by the

Etruscans and later the Romans. The Roman alphabet is the base for the one that is used across the world today for most European languages. Importantly, the alphabet was also copied in India; the Brahmi script was first recorded in India in the third century B.C. and is the origin of the many scripts now used in India.

The use of alphabets allowed common people to communicate in written form. No longer was writing just the work of specially trained scribes and accessed only by the privileged classes. The Chinese never made this transition to alphabetical writing systems; in China reading and writing was always a highly skilled process, not easily mastered by the ordinary person. This led to scholars becoming highly valued in Chinese society, taking on much of the execution of state administration. This simple fact had great implications for the future directions of East and West. In the West reading and writing skills could be acquired widely across all groups of society. This higher level of literacy in the West is one reason why the West developed into a more dynamic society after cheaper printed books became available from the fifteenth century onwards.

In the first millennium clusters of city-states also began to accept trade tokens or coinage made from gold, silver and bronze. Coinage eliminated the inefficiencies of barter. It simplified calculation of profit and loss and allowed wealth to be stored in a portable manner. In the Middle East this seemingly abstract concept is generally believed to have started in Lydia in modern Turkey in 650 B.C. The Babylonians had already established silver, measured by weight, as a common unit of trade and tax as far back as the second millennium B.C. The Lydians developed a more convenient system of exchange, consisting of standardised weights of precious metal each stamped with the image of an authority figure to guarantee their quality.

The growth of the power of Athens was built on the export of silver coinage from their mine in Southern Attica. By later Roman times the monetary system had standardised on the aureus (gold), the denarius (silver) and the sesterius (bronze), all bearing the head of the emperor of one side. This set of gold, silver and bronze coins, set the standard pattern in Europe right up to the early twentieth century, when paper money first began to be adopted.

Regional states also developed coinage systems around the same time. The Chinese independently invented bronze coinage. Cast bronze replicas of cowry shells were in use as money tokens before 1000 B.C. The first Chinese manufactured coins were made between 700 and 500

B.C. and were cast bronze discs with holes in the centre allowing them to be strung together. India, too, had its own coins in the form of punched metal disks from around 700–500 B.C.

At the next level of the evolutionary structure was the city-state itself. We only have information dating from the first millennium B.C. to understand something of how city-states were formed. They developed from tribes or tribal states in which there was a sufficient concentration of ordinary citizens in one place to allow them to maintain their freedom. Citizens were capable of banding together and forming their own military force to oppose the warrior elite, who in the first millennium were mounted on horses. The leader of this citizen army was called a tyrant. In Greece, this citizen army, properly trained and disciplined, developed into the phalanx, an infantry formation in which soldiers were massed together, defended by shields and armed with spears and short swords. We see glimpses in Roman and Greek ancient history of tyrants, aided by citizen armies, overcoming aristocratic leaders. Two types of authority seem to have developed in city-states; dictatorship by a tyrant or an oligarchy. A tyrant ruled with his family clan supported by a citizen army. In an oligarchy rule was shared by a number of aristocratic families bound by a constitutional arrangement.

Competition between city-states now involved battles between phalanxes. War in its modern form became more recognisable, with separate organised groups of infantry, archers and, eventually, cavalry.

Battles between city-states were thought of as battles between the sponsoring gods. An example of this is seen in the Bible. When the Babylonians conquered Judah in 586 B.C. they destroyed the temple in Jerusalem. This was considered a triumph for their god, Marduk, over the Jewish god, Yahweh.

On the third level, class did not have the same significance as for regional states. This is because the role of the farmer and trader remained independent, not controlled by the aristocracy. The key distinction between different inhabitants was citizenship status; citizens were accorded rights to have a say in state affairs. The prime example was Athens, which is credited with having the first democracy. According to legend, Athens was formerly ruled by kings selected from the four tribes who dominated the area. The kings were overthrown by tyrants. Then around 500 B.C., a complicated constitutional structure, which is known as Athenian democracy, was put in place. The city was divided into 10 phyles, each of which elected members to a boule or council that governed Athens on a day-to-day basis.

Central to Athenian democracy were the meetings of the Assembly, which had the function of a legislature and a supreme court. All adult male citizens who had undergone military training were eligible to attend. This was about 20% of the population. The standard form of decision-making was a debate with speakers making speeches for and against a position, followed by a general vote with a show of hands.

This system was remarkably stable, being abandoned only when Athens lost its independence to the Macedonians in 338 B.C. It would be almost two millennia before any such devolved constitutional structures were to be seen again.

In city-states families or clans were extremely important. Many city-states were oligarchies. Rome was a later example of an oligarchic city-state. It was supposedly founded in 753 B.C. and around 500 B.C. an oligarchy managed to throw out the ruling tyrant and form the Roman republic. This developed a complicated constitutional organisation. Decisions were taken in a council of oligarchs known as the Senate. Very nervous of devolving power, the Senate gave time-limited authority to consuls who, elected for a year's term, governed the state. As a large city it was also wary of the power of its citizens; the post of Tribune was created to represent the land-owning citizens (plebeians). They were elected by the citizens and had the power to veto actions taken by the consuls. The Roman system of government, with checks on the power of its administrators and scope for wealthy citizens to influence government, was very successful; Rome grew to dominate first Italy and then the western Mediterranean area.

For most ordinary citizens on the fourth level, life remained a precarious matter that appeared to be dependent on the will of the gods. The state god was usually part of a pantheon of gods, who were the basis for a whole series of myths and legends about creation, the origin of the state and the cause of human foibles. The gods that were worshipped were not moral creatures; they were capricious and vengeful like humans. For example, in the Sumerian version of the biblical story of Noah and the flood, the flood is caused, not by the transgression of man, but because there were too many people making too much noise and giving the god insomnia! The ordinary man loved stories about the gods and ancient heroes. Stories like the wanderings of Odysseus in Greek legend were told by travelling storytellers with phenomenal memories. Even in modern times storytellers in India can recite by heart all the classic adventures of gods and mythical figures: the *Ramayana* (24,000 stanzas), the *Mahabharata* (100,000 stanzas) and the *Bhagavata* (18,000 stanzas).

Gods were asked for favours and looked to for favourable portents or to curse others. For the common man religion was a straightforward deal with the gods; if he worshipped and gave offerings to a particular god, he expected favours in return. If that did not work, he either concluded that the god was angry with him and needed more attention, or he chose another god who might be more favourably disposed.

With the administrative organisation of states, expanding levels of trade and the ability to communicate by writing, all the ingredients were now in place for humans to develop civilisation. Civilisation at this time had three major characteristics. Firstly, there was a rich elite who had leisure time to enjoy themselves and display their wealth in art, architecture and entertaining. Secondly, there were cities with artisans, artists and artistes able to support the elite. Thirdly, there were administrators who collected taxes, organised government and administered the law. For the first time some people had sufficient free time to reflect on the world around them. Humans started to understand the movements of the planets and the seasons, to speculate on nature and to contemplate numbers and shapes. Home sapiens began to develop and display those wholly new and unprecedented powers of reasoning, which eventually led to the development of scientific knowledge.

The human ability for abstract reasoning appears to be based on a combination of two brain functions naturally occurring in higher animals. We have already discussed their ability to associate cause with effect. Higher animals also appear to have the ability to consider alternative courses of action, imagine their consequences and choose the best action to take. Humans, with the power of speech, could propose different scenarios and discuss the resulting theoretical consequences. Thus, the possible causes of an effect could be debated and the most likely cause determined. In this way, over the millennia, humans developed that ability for logical reasoning which distinguishes man from all other animals.

This ability was demonstrated at the start of civilisation. The reason why there are 60 minutes in an hour, or 360 degrees in a circle, are truly ancient. The Sumerians and their successors the Akkadians used the sexagesimal numbering system (with a base of 60). In their system ⟅ represented one, ⟅ followed by a space represented 60, and ⟅ followed by two spaces, represented 3600. They never developed a symbol for zero, but the positional numbering system allowed them to develop the use of that great aid to arithmetic, the abacus, and hence to be the first to undertake complicated arithmetical calculations.

From 1800 B.C. onwards the Babylonians studied the periodic movements of the sun, planets and stars and were the first to use mathematics to predict events in the heavens. They also had water clocks and standardised the calendar, with 24 hours in a day and seven days in a week. This time recording system was exported from Babylonia and used throughout the Mediterranean by the Greeks, Romans and Egyptians.

Thus far the development of reasoning in the human animal was astonishing enough, but the advance in scholarship by the Greeks was truly amazing. The Greeks were the first to develop formal reasoning. They applied logic to deduce proofs in mathematics, sought reasons for natural phenomena and made speculations on the nature of mankind. It is their philosophy that is mostly commented on these days, but it was in the field of mathematics that the Greeks had the most profound effect on human civilisation. They invented the idea of mathematical proof and their geometric theorems are still taught today. They also contributed important ideas on number theory and mathematical analysis. Mathematics went on to become the language of the physical sciences, leading directly to Newton and on to Einstein and the other great mathematical physicists. Not all Greek theorising was successful. Their medicinal theories have underpinned generations of bogus medical practice up to modern times. Nonetheless, it was the Greeks who first showed that humans could explain natural phenomena using human logic without resorting to explanations of divine influence from gods and spirits.

The ancient Greeks realised the commercial value of literacy to a state whose wealth came from trade and, the Greek states were the first in which primary education was generally available to citizens. Education included military skills as well as academic subjects. Each city had its own educational curriculum and all recognised that education was important not only for commercial and military reasons but also to train children in the culture and values of the state. In Athens citizens who could afford it sent their children to elementary school at approximately 7 years of age; children coming from poorer families would only have been educated informally by their parents and relations. Elementary-school children were taught how to read, write, count and draw. Students would write using a stylus to etch symbols onto a wax-covered board. Physical training began during their elementary education but was developed further when the boys left elementary school and attended the gymnasium. Physical training was seen as necessary for improving one's appearance, for preparation for war, and to ensure good health at an old age. Traditionally, attendance at the gymnasium completed most of the post-elementary education in Athens.

After turning 14 years old, boys from the wealthiest families had the option of secondary education. This might take place in a permanent school or be received from travelling teachers. The Greeks placed great value on knowledge and education, which they advanced to a new level. Secondary education included natural science (biology and chemistry), rhetoric (the art of speaking or writing effectively), geometry, astronomy and meteorology. Academic accomplishments gained in secondary school would help an individual to gain the respect of his peers. There was an intellectual element in Athenian society that encouraged academic speculation that did not have to have a practical outlet. Greek mathematics was for a while an evolutionary cul-de-sac, preserved in written records and translated into Arabic and Latin; only centuries later did it provide the basis for modern science.

In China, in contrast, education was restricted to boys of the aristocratic classes. To become a perfect gentleman during the Zhou dynasty (1126–256 B.C.), candidates had to master the six arts of rites, music, archery, charioteering, calligraphy and mathematics. Only the wealthy could afford education. However, as in Greece, a class of scholars emerged who acted as teachers and mentors to the young. We shall meet the most famous of these, Confucius, in the next chapter.

Over time city-states either expanded, like the Romans, and became regional states themselves or were absorbed into larger empires as they were not militarily strong enough to survive on their own. In the early part of the first millennium B.C. the most successful city-state that became a regional power was Ashur in modern-day Iraq. Today this state is known as the Assyrian empire. This empire was initially a war machine with a large standing army, paid for by looting and extorting tribute from the surrounding countries. In 670 B.C. Ashur became the first state to control the whole Fertile Crescent and large parts of the Middle East. Once the neighbouring states had been looted and subjugated, the Assyrians attempted to consolidate their territory into one regional organisation. All people in the Assyrian empire were considered Assyrians, no matter what city or tribe they belonged to; this included Jews from Israel. All were subject to the same taxation and risk of conscription. There was one leader, one dominant god and one common language of administration. The dominant god was Ashur who had just a single temple in Ashur; his earthly representative was the Assyrian emperor. The common spoken language was Aramaic. For this reason, Aramaic was the spoken language of Jesus: by the time of Jesus, Hebrew had become a dead language preserved only through written texts. Assyria was a portent of what was to come. We shall see in the next chapter how huge empires much bigger than Assyria would emerge to control the civilised world.

Summary: situation at 500 B.C.

By 500 B.C. the human world had become more complex with five types of society coexisting, hunter–gatherers, tribes, tribal, regional, and city states. Tribes had pushed out the hunter–gatherers from the arable land in most of Europe, Middle East, India and North Africa and large parts of China. Tribes themselves had been replaced by states in the Fertile Crescent and along the shores of the Mediterranean and independently throughout northern China.

States were created to support a privileged elite and for the first time large differences in wealth between rich and poor emerged. The rich used their wealth to commission elaborate buildings, beautiful artefacts and clothing. New materials and items of status were widely traded and specialist artisanal skills developed. Large cities appeared for the first time; the largest in 500 B.C. was Babylon, with around 150,000 people and the second largest was Luoyang in China, with 80,000.

Regional and city states developed writing skills. The West used alphabetical systems of writing, while the Chinese still wrote in logograms. Formal education systems for the wealthy evolved to train numeracy and literacy skills in addition to existing fitness and war-related training. Some teachers became scholars and the first major works in astronomy, mathematics, philosophy and natural sciences date from this period.

The use of metals had made farming more efficient and effective; world population expanded dramatically from 10 million in 3500 B.C. to reach 150 million people. Although the Fertile Crescent was at the forefront of meme development, the opportunities for population expansion in that area were limited. Indeed, there was already some depletion of the landscape due to the loss of woodland for smelting ores. In contrast, Indian and Chinese populations were able to grow by exploiting the potential of another major fertile river system. In India, the Aryans moved on from the original Dravidian centres of civilisation on the Indus to set up chiefdoms and tribal states along the Ganges. In China the civilisation spread from the Huang He (Yellow River) in the north to the Yangtze basin, where rice cultures were productive. The stage was already set for China and India to be the most populous areas of the world, as they already accounted for about half of mankind.

The use of metals, the development of infantry formations and use of the chariot had transformed warfare. Military conflict now involved pitched battles between armies. Militarily neither tribes nor states had established an ascendancy over one another. In the Indian subcontinent, the Indus Valley civilisation had been destroyed by Aryan tribes and writing skills would not be acquired there for another 200 years.

CHAPTER 5:
THE AGE OF EMPIRES (500 B.C.–400 A.D.)

The ability of regional powers like Assyria to support large standing armies changed the pattern of state development. Big states swallowed smaller states, only to be swallowed again by still bigger states with even bigger standing armies until huge empires were formed. In Europe, the Middle East, India and China, in the first millennium B.C. there emerged empires of huge size, absorbing most of the surrounding centres of civilisation. These empires developed top-down organisations, creating administrative provinces, formalising tax systems, standardising currency and encouraging economic development. Literacy skills were acquired by nobles, administrators and tradesmen and a common language developed so that records, instructions and ideas could be understood by all. Trade blossomed in this relatively peaceful environment.

The whole ethos of government changed. Empires were less intent on the rape and pillage of neighbouring states. The large empires grew rich by absorbing other states. Defeated kings could simply be replaced by governors while the local bureaucrats were maintained in their post to ensure that taxes kept flowing in. By this means the invading empires were guaranteed a continual income for all their years in power. For the majority, the peasant farmers, life continued as a hard daily grind. Nobles now focused not just on the glories of war but were also interested in a lucrative posting to a position of power in the empire. The wealthy were now extremely wealthy; they could build their own palaces and villas and support their own churches and temples in a way that was possible only for kings in the past. As the economy grew, towns blossomed, complete with shops and artisan workshops. There was also scope for scholars, artists and sculptors to make a living if they could gain the favour of a rich benefactor.

A defeated state might survive, if it agreed to pay tribute, but once its leadership was overthrown and replaced by centrally appointed governors there was little chance of it regaining its independence. States had been created by kings and aristocrats for their own benefit. Once the aristocracy lost their power and their gods were seen to have been defeated, the old state was absorbed into a province of the empire. The conquered peasantry was still taxed and was indifferent to the leadership as long as they were left to get on with their life without additional impositions. Great states of the past like Egypt, Babylon, Ashur, Ur and Athens could no longer provide a military threat; empires had standing armies that could overwhelm internal rebellion. The danger to centrally run empires now came from tribal communities outside their boundaries; as a result the majority of their army units were stationed at the edges of the empire.

The Chinese and Middle Eastern empires had similar basic structures. Each empire was divided into self-governing provinces led by centrally appointed officials. They had a common language of administration and a standard system of writing. They improved communication by investing in **paved roads** and **canals**. This was primarily to facilitate the process of tax collection and the movement of military troops, but it also had the benefit of improving trade and commerce. Trade was also facilitated by having one common currency in use throughout the empire.

Empires were huge; it could take months to travel from one end of the empire to the other. Good communication systems were vital. The further development of horse breeding and riding tack (saddles and horseshoes) allowed horses to be ridden. This allowed the emperor's messages to be delivered to the outposts of his empire in a few days. Writing was the communication tool that held the empire together. In China this was helped by the invention of **paper**. The Indians also invented **Arabic numerals** to simplify writing arithmetic.

Stable empires lacked the competitive forces necessary for rapid meme development. This was not an age of great technical breakthrough, instead it was an age in which organisation, engineering and infrastructure developed. The most important new technical meme was the **waterwheel**, which was independently developed in Europe and China.

In warfare the war chariot was now replaced by **cavalry**. However cavalry did not reach its full potential until the invention of the stirrup as we shall see in the next chapter. The arrival of the **mounted horse**, though, had implications for society as a whole – to quote Azar Ghat in *War* p. 329:

> The introduction of the horse added a new dimension to elite supremacy. It should be noted that in sedentary societies the horse possessed little economic utility value. Carts and ploughs were driven by oxen until the breast-and shoulder horse harnesses spread through Eurasia during the first millennium A.D.... On the other hand, the horse required specialised and expensive feeding. It follows that in sedentary societies the horse was the possession of the elite – because it was expensive and luxurious rather than utilitarian.

Thus, the horse became the practical symbol of the difference in power between aristocrat and peasant.

It was in the area of religion and culture that new memes had the most significant impact. The identity of a religion with the state had been broken. There were as many sets of gods as there were conquered states.

These gods had been discredited and there emerged new types of religion and philosophy aimed at the poorer people. Itinerant religious teachers promoted their own religions and philosophies of life, independent of the state. In former times such activity would have been stopped by local rulers, but in India, the Middle East and Europe, the empires were often tolerant of local religions and were indifferent as to which god the locals worshipped. There began to be free competition between religious beliefs for the first time in history. The development of **Zoroastrianism, Buddhism, Christianity, Brahmanism** and **Rabbinical Judaism** dates to this period. In addition a new philosophy of life, **Confucianism** developed in China. Religions and philosophies of life became a separate competitive organisation in the evolutionary structure of empire societies.

The evolutionary structure of empires evolved from those of regional states (Figure 5.1). City-states did not have the requisite size to survive the military power of large regional states; they either changed their structure to become regional states or were conquered. There were two new communities in empire society compared to regional states, religion and state institutions. State institutions included the army, the administration of the law and provincial government. They competed for attention from the emperor and for an increased share of state resources. They were managed by aristocrats and administered by scribes. They lessened the competitive importance of family and clan controlling the empire. However clans, a relic of tribal times, remained very important in China.

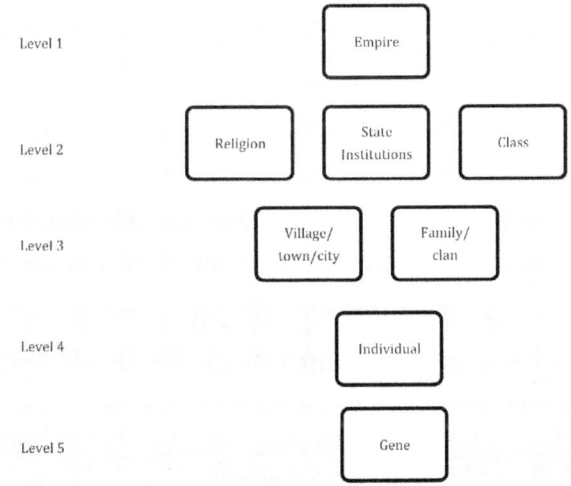

Figure 5.1 Evolutionary structure of empires

China

In China in 771 B.C. there were 148 states and for the next half millennium these states battled it out in a vicious period of state consolidation. After 300 years the number of states had been reduced to 14 and then, in 221 B.C., to one, the Qin. The winning emperor called himself Qin Shi Huang (the first August Qin emperor). He controlled a huge territory stretching from the Great Wall of China in the north to modern day Hunan Province in the south, which incorporated the two great rivers: the Huang He (Yellow River) and the Yangtze.

After his victory, in the remaining 10 years of his reign, together with his minister Li Si, he undertook a number of administrative reforms that defined the organisational structure of imperial China for the next two millennia. The old warring states were disbanded and a new administrative unit, a commandery, was created. The leaders of the commanderies were appointed centrally and leadership was no longer inherited. Perhaps most importantly, the script used by the Qin was adopted as the standard for the whole empire. All the regional scripts were abandoned and the emperor created one common communication system. He also unified China economically by standardising the currency, units of measurement and the width of cart axles to improve road transport. Communications were improved by building an extensive network of roads and canals.

To the north the Chinese were continually plagued by warring pastoralist tribes. The Qin dynasty consolidated several early protecting walls, creating what is now referred to as the Great Wall of China. The Qin wall was a series of fortifications made of stone, brick, tamped earth, wood and other materials. Little of the original wall now remains. The current wall was reconstructed during the Ming dynasty (1368–1644) and stretches from the sea for 8,850 km into Mongolia.

The huge changes made by Qin Shi Huang in his incredible reign could be achieved only by assuming absolute dictatorial powers. In the end, he proved himself to be a maniacal monster. He built an incredible army of life-sized terracotta warriors to protect him in the afterlife. Any books and scholars he disapproved of were liable to be burnt and he abused conscripted labour to complete his enormous building works. The human cost of the construction of the Great Wall alone has been estimated by some authors to be in the hundreds of thousands.

On his death his dynasty's grip on the empire quickly dissolved, but the succeeding Han dynasty (206 B.C.– 220 A.D.) kept most of his administrative reforms and established the long-lived Chinese empire on a firm footing, extending its boundaries to cover all of south China, including the important provinces of Guangdong, Hainan and Fujian.

China became a centrally managed empire controlled by a huge bureaucracy, capable of reading and writing the Chinese script. By 5 B.C., when the population of China was 60 million, there were about 130,000 bureaucrats serving in the capital and provinces. With one standard script all regions of China could read the imperial instructions, even if they spoke a different dialect or language. There was no incentive for scholars to move towards an alphabetical system as this would have further complicated communication.

The major building material of the Han Chinese was timber, and because of this no Han palace or house survives today. The Han architecture that still remains was built largely for the purpose of fortification or transportation. The ruins of the rammed-earth walls that once surrounded the capitals Chang'an (modern-day Xi'an) and Luoyang still stand, along with their drainage systems of brick arches, ditches and ceramic water pipes. The Hong Gou (Hong Canal) was China's earliest man-made canal linking the Huang He River and the capital Luoyang to the coast. The construction of the canal began in 361 B.C. and was completed during the Han dynasty.

It was wholly appropriate that the Chinese Empire with its many scholars should be the first to use paper for writing. Early in the Han dynasty coarse paper had been produced for wrapping. The manufacture of writing paper is the first breakthrough meme for which we have the inventor's name and date, Cai Lun (a court eunuch) in 105 A.D. His process was complicated: mulberry tree bark, hemp, old linen and fish nets were boiled together to make a pulp that was pounded, stirred in water and then drained in a wooden sieve. The resulting sheet was dried and bleached before becoming usable paper. Paper production was initially a laborious process. However the use of watermills in paper production allowed paper to become China's chief writing medium by the third century A.D.

Water mills were, almost certainly, independently invented in China; we know that in the first century A.D. mills were being used not only to crush grain but for important industrial processes. For example, in China, watermills powered bellows to force air into blast furnaces, improving the efficiency of iron manufacture. Europeans did not acquire this technology until the Middle Ages.

The culture, both political and religious, that developed during the Han period was to endure for centuries. While ancestor worship remained the dominant religion of ordinary Chinese people, Confucianism became the underlying philosophy of government from the time of the Emperor

Wudi (141–87 B.C.). Confucius (551–479 B.C.) had been a teacher of the nobility at the time of the warring states. Like Buddha and Socrates, none of his original writings survive. His philosophy has been preserved in the *Lun yu (Analects)*, compiled by his followers. It is explained in outline in *Confucianism* p.29 by Thomas and Dorothy Hobbler as follows:

> At the heart of Confucius' teaching is the idea that society works well (harmoniously) when each person understands his proper role and acts accordingly.... As Confucius said, 'Let the ruler be a ruler and the subject be a subject. Let the father be a father and the son a son.' In this way, there would be harmony.

> Confucius taught that harmony begins in the basic unit of society, the family. Sons must show respect for the father. Younger brothers must respect older brothers. Wives must respect husbands. The father must guide through example. Then the family would be harmonious. The family was a microcosm of society itself. In the same way, subjects must show respect for the ruler. However the ruler too had his duty – to treat his subjects as a loving father would. By doing so, he would ensure harmony.

This was an appealing philosophy for emperors, emphasising the duty of the subjects to obey. It was also the ideal philosophy for institutional bureaucrats, emphasising structure and authority; challenging the status quo was not encouraged. In 124 B.C. Wudi established an imperial University in Chang'an, the state capital, to train budding scholars in Confucian doctrine. Initial enrolment was only 50 students, but student numbers had grown to 3000 by 8 B.C. and to 30,000 by the second century A.D.

Traditionally, selection of civil servants had been based on blood ties, but during the Han dynasty the Chinese government developed a system of selecting bureaucrats for office based on merit. Beginning in 165 B.C., senior officials throughout China had to nominate young men of distinction for service in the bureaucracy. From 124 A.D. the nominated pupils travelled to the capital to take an exam; the best ones went on for another year's training before they were tested again for entry into higher government. In this way the Chinese established civil service selection based on merit 1800 years before the West. Over time the examination system would become the main road to advancement of a separate scholar class. Now Confucian teachings could be used as the master had hoped, as a philosophy of government in which subjects and rulers alike would be encouraged to act according to the Confucian creed.

The Han dynasty had a huge state bureaucracy. The three highest officials were the chancellor, who prepared the budget, the counsellor, who was in charge of public works and the discipline of state officials and the grand marshal, in charge of the armed forces. Ranked below these were nine ministers; some of whom would be recognised today as holding important state posts such as the minister of finance, the minister of justice and the minister herald, responsible for receiving honoured guests, but most posts directly supported the emperor. There was a separate minister for the emperor's ceremonies, security, stables, clan, coaches and entertainment. The size of the Han bureaucracy can be judged by the fact that, within the Ministry of Ceremonies, the director of prayer alone had a staff of 35, and the director of music controlled 380 musicians.

In Han times some territories within the commanderies were directly ruled by members of the emperor's clan and other aristocrats. Elsewhere, all the most important administrators were centrally appointed. The most critical posts were the administrators of the commanderies, who were responsible for local tax collection, and below these, the magistrates who maintained law and order. Aristocrats proved too independent and rebellious; over time, their local administrative power was removed and they became passive recipients of a state pension. China became centrally governed with little scope for individual territories to exercise their own initiative. All power was centralised, with no local accountability. The law was administered autocratically by local magistrates, hearing both criminal and civil cases. Appeal against a verdict could only be made by petition to the emperor and was beyond the means of most families. Such magistrates had the power of summary execution. The position of magistrate was also lucrative as sentences of imprisonment and corporal punishment could be commuted into cash payments.

Chinese society in this period was broadly structured like the West with emperors, kings and aristocrats living off the taxation of the peasantry. However, its class structure was more formalised and complex. The emperor was at the apex, and ranked immediately below him were the members of the emperor's clan. The rest of society, from nobles to commoners but excluding slaves, belonged to one of 20 ranks. The scholar bureaucrats were ranked just below the nobles in social prestige. This group began to identify themselves as members of a larger, nationwide class of gentry with shared values and a commitment to mainstream scholarship.

Thus by 200 A.D., at the end of the Han period, we have five broad classes of society: the emperor and his clan, the nobles, the gentlemen scholars, merchants/artisans and peasants. In addition, as in the West, there were slaves.

Unlike India and the West, the Chinese government also tried to maintain central control over religion. There was one senior god whom the Chinese emperor alone was allowed to worship. Common people still venerated their ancestors and local gods. Also, as in the other empires, itinerant preachers spread a gospel of hope and moral support to the poor. Over time the teachings of two of these teachers Laozi (from sixth century B.C.) and Zhuangxi (365–280 B.C.) was condensed into a collection of ideas and superstitions called Daoism, which around 100 A.D. emerged as a distinctive socio-religious phenomenon. Dao is usually translated as the 'way' or 'path'. Daoist ethics emphasised compassion, moderation and humility. Daoist philosophy included concepts such as yin-yang, the five elements (earth, fire, wood, metal and water), alchemy, and herbal medicine. Hope for the poor was provided by the mythical Queen Mother of the West, who was believed to have the power to make people immortal and to provide her followers with children and wealth. The health and longevity of the individual, not of family or society, were Daoism's primary concerns. The philosophy promoted the idea that harmony between mankind and nature led to better health. It could be considered a philosophy of life, a religion or a Chinese folk religion, depending on the view of its adherents. As a popular belief equivalent to Buddhism in India at this time, Daoism was never formally accepted by the state as a religion until the Tang dynasty (618–907).

India

The first great Indian empire was the Mauryan empire which, at its peak, covered the entire Indian subcontinent except for the southernmost tip. Its great leaders were Chandragupta (322–298 B.C.) and Ashoka (273–232 B.C.). Our knowledge of the Mauryan empire is poor, as remaining written records are sparse. Indians did not use parchment because the cow was sacred; instead they wrote on birch bark and palm leaf, which rapidly deteriorated. We have no court records remaining from the time of Ashoka. Ashoka was not even recognised as a great king by later generations of Indians. It was only in 1915 when the Brahmi script on a large number of rock inscriptions was deciphered that archaeologists pieced together the extent of his conquests.

Unlike the other empires in China and the Middle East, the local organisation of the defeated states was never eliminated and the empire did not survive Ashoka's death. From this point on, China and India, the two great population centres of the world, took different courses; the route to the strong centrally managed state in China was never copied in India. Another native Indian empire did not emerge for another 500 years until the Gupta dynasty (320–550 A.D.). However, neither this nor the early Mauryan empire ever achieved the ruthless bureaucratic control necessary for the establishment of a permanent centrally managed empire.

While never being as politically powerful as China, India played a pivotal role in the development of religious thought and ideas. At around 500 B.C. there were two main religious movements. We have already met the Vedic religion with its Brahmin priests and their Vedic scriptures and traditional Aryan gods. However, there was also a group of itinerant religious leaders called samanas. They came from a different religious background to the Aryan invaders, maybe from the Dravidians or even the original hunter-gatherer populations. They rejected the Vedic scriptures and the division of society into varnas. They ignored social norms and expectations; they were usually celibate and in spiritual matters gave preference to experience rather than scriptural authority. Samanas studied the effects on the mind and spirit of meditation, yogic breathing, fasts and sensory deprivation. When, as a result of such practices, individuals had some kind of mystical experience, they might, in a fashion similar to gurus in modern-day India, attract disciples and found a sect. In such a manner Siddhartha Gotama founded Buddhism.

Siddhartha Gotama (who became known as the Buddha) was born in the sixth century B.C. among a people known as the Sakyas who lived just inside the borders of modern day Nepal. He became a samana and claimed to have discovered enlightenment through meditation. His thesis was that all beings (humans, animals and gods) were subject to an endless cycle of birth, death and rebirth. To escape this cycle and achieve nirvana, the state of highest happiness, it was necessary to achieve enlightenment. This was possible only for those who were both virtuous and wise. Virtue was achieved by doing good deeds. Wisdom in Buddhism means a profound philosophical understanding of the human condition. It requires insight into the nature of human existence which comes only after long reflection and deep thought. Buddha did not say that enlightenment is achieved through faith and worship in any god. The Buddha had little time for theological speculation. This is illustrated in the passage below from *The Case for God* pps31/32 by Karen Anderson:

> One of his monks was a philosopher manqué and, instead of getting on with his yoga, constantly pestered the Buddha about metaphysical questions: Was there a god? Had the world been created in time or had it always existed. The Buddha told him that he was like a man who had been shot with a poisoned arrow and refused medical treatment until he had discovered the name of his assailant and what village he came from. He would die before he got this perfectly useless

information. What difference would it make to discover that a god had created the world? Pain, hatred, grief and sorrow would still exist. These issues were fascinating, but the Buddha refused to discuss them because they were irrelevant: 'My disciples, they will not help you, they are not useful in the quest for holiness; they do not lead to peace and the direct knowledge of Nirvana'.

Buddhism, at least at its founding, was not a religion. Religion is defined in the *Concise Oxford Dictionary* as: 'The belief in and worship of a superhuman power, especially a personal god or gods'. This definition includes three elements typical of all religions: a deep-felt conviction in the existence of a superhuman power or god, a method of worshipping the god by prayers and offerings, and chanting phrases, and a belief that the god will take an interest in human activity, listen and, on occasions, respond. In India at this time all people believed in the existence of gods. However, these gods were considered by the Buddha as merely other beings and played no role in the achievement of nirvana. 'Cease to do evil, learn to do good, purify the heart' is one Buddhist summary of their way of life. This was not a top-down religion, buttressed by rituals that supported aristocrats and the state. This was a way of life designed to reduce the sorrow and distress of the poor and provide hope and purpose to their lives. It was the first known philosophy of life that was devised for ordinary people.

Buddhism requires intense meditative discipline of the sort that can be achieved only by monks. In practice, Buddhism evolved to be a partnership between a community and a monastery. Lay people provided food and other donations such as clothing. In return Buddhist monks, forbidden to work for gain, grow food and handle money, acted as the community's teachers and its focus for ritual observance.

We know little of the pace of development of Buddhism, but from the outset it was a missionary organisation. Buddhists were specifically charged to: 'Go, monks, and wander for the good and welfare of the multitudes'. Two hundred years later, by 300 B.C., Buddhism had grown so dramatically that it would become the state religion of the Mauryan empire. Its most famous emperor, Ashoka, appears to be one of the few great rulers in history with a conscience. Apart from Buddhist texts, which were written down later, the only documentation of Ashoka's empire is recorded on the pillars of Ashoka and various rock carvings in the Brahmi script. There are 33 of these inscriptions, known as edicts. In edict 13 he recorded his sorrow after his conquest of Kalinga on the east coast of India which caused the deaths of more than 100,000 soldiers and civilians. Edict 13 records his words as:

What have I done? If this is a victory, what's a defeat then? Is this a victory or a defeat? Is this justice or injustice? Is it gallantry or a rout? Is it valour to kill innocent children and women? Did I do it to widen the empire and for prosperity or to destroy the other's kingdom and splendour? One has lost her husband, someone else a father, someone a child, someone an unborn infant.... What's this debris of the corpses? Are these marks of victory or defeat? Are these vultures, crows, eagles the messengers of death or evil?

His reaction to the brutality of the conquest led him to adopt Buddhism and place great emphasis on piety in his edicts. Buddhism became a state-sponsored philosophy around 260 B.C. and was propagated and preached both within the empire and worldwide.

However the Mauryan empire did not last long enough for Buddhism to totally replace the Vedic religion in India. Poor literacy levels and therefore a weak administration could be one of the reasons for the empire's collapse. The first Indian alphabet (Brahmi) was a derivative of the Aramaic alphabet and was developed only around the time of the Mauryan empire. As the name of the script implies, Brahmins would have acted as administrators of the empire and developed the script. It is thought that the proud oral tradition of these priests, which involved memorising all the Vedas, may have led them to be less enthusiastic about developing written texts. It is certainly extraordinary that, except in rock inscriptions, no written source from the Vedic tradition exists describing life in the Mauryan empire. The rise of Buddhism with its egalitarian, anti-varna message would have been threatening to the status of brahmins; they could therefore be seen as the natural enemies of the Buddhist state that Ashoka created.

The Vedic religion changed in response to Buddhist ideas. In this new form (Brahmanism) animal sacrifice was abandoned and brahmins became vegetarians. The hitherto distant, impersonal gods like Surya the sun god were downgraded in importance. Instead, in the same way that the figure of the Buddha gave inspiration to the Buddhists, Brahmanism emphasised new gods who had human forms, in particular, the three gods of the *trimurti*: Brahma, the creator, Vishnu, the preserver and Shiva, the destroyer. These latter two gods came to be the most important deities when Hinduism finally developed from Brahmanism.

Brahmins also codified many of the prevailing customs in a document called the *Manu Smriti*. This emphasised the status of brahmins as the premier class and is an unpleasant documentation of class prejudice and misogynistic attitudes. The duties of the varna were again emphasised. A brahmin's duty was to study and teach, to sacrifice and give and receive gifts. A kshatriya's role was to protect the people and a vaishya's was to breed cattle, till the earth, to pursue trade and to lend money. Both these varnas should maintain their religious beliefs through sacrifice and study. The sudra's task was to serve the other three higher classes. The fifth class, the chandalas (untouchables) were banned from owning property and had to live outside the village. They were referred to as dog-eaters to emphasise their lowly status and they had to perform the most demeaning tasks (sweeping, working with leather and removing excrement). To touch such a person was polluting to one of the higher groups and required ritual purification.

As to women, this is a quote from Kim Knott in *Hinduism – a Very Brief Introduction*, p. 81:

> Also polluting was a menstruating woman. Her touch would require a Brahmin to bathe. High caste women, *Manu* said, should be under the protection of fathers, husbands and then sons. They should never be independent, owing to their weak, fickle nature and the social consequences of allowing women to act outside male authority. A man must honour his wife, however, though he must also control her, by force if necessary, and keep her focused on her domestic duties. Bearing children – particularly sons – was her virtue. A good wife should serve even a bad husband as god. She should not leave him and once widowed she should not remarry.

Society and all religions at this time were dominated by male authority, so perhaps we should not be so shocked by this uncompromising statement of male dominance. From hunter–gatherer times when women concentrated on gathering, cooking and child rearing, all the developments in society seemed to have pushed women further into the background. The man–woman partnership of the time of the empires in which the woman manages the domestic environment and supports the man in an outwardly subservient role appeared to have evolved naturally. Brahmins, very conscious of their own power and status, seemed to regard any challenge by women to male authority, and by implication their priestly role, as very threatening.

Over time it was this new Brahman religion, rather than Buddhism, that gained support from the ruling classes. In particular the Gupta dynasty which dominated north India in the fourth and fifth centuries A.D. were staunch supporters of Brahmanism and its inherent class divisions.

Before leaving this section on India it is appropriate to acknowledge their great contribution to mathematics by the creation of a symbol for zero. We have seen that the Babylonians had developed a place system for counting but had no symbol for zero. It was in India that the full Arabic decimal system developed. We do not know who was responsible. We know the Gupta dynasty supported a number of important mathematicians and astronomers, but the earliest Indian inscription we have is dated 876 A.D., long after the dynasty dissolved. Already in 825 A.D. the notation had been introduced from India to the Arab Caliphate by a Persian scientist, al-Khwarizmi.

The Middle East and Europe

After the Assyrian empire fell, an even greater empire emerged in the Middle East. The Persian Achaemenid dynasty (550–336 B.C.) ruled over an empire that included at its height all the ancient centres of civilisation in the Fertile Crescent and Egypt, plus modern-day Turkey, Iran, Afghanistan and parts of Pakistan. As in China the central administration was greatly improved. States were reorganised into satrapies. A satrap (governor) was the vassal king who administered the region. He was assisted by a state-appointed military leader who separately supervised military recruitment and ensured order, and a state secretary who kept the official records. During the reign of its greatest ruler, Darius the Great, the economy was stabilised by standardising the silver and gold coinage. Darius introduced a regulated and sustainable tax system with each satrapy supplying food, gold or silver according to its agricultural and commercial development. The legal system was also codified and standardised.

The Persian empire was the world's first superpower and set a blueprint for all empires in the future. The Persians ultimately lost to the Macedonian Greeks led by Alexander the Great. On Alexander's untimely death at the age of 33 his empire was divided between his generals, who set up their own dynasties and established Greek cultural leadership in the eastern Mediterranean and Middle East. Across their conquered territories, the Greeks anticipated Roman practice by building new cities as centres for Greek administration and culture. Two of these cities became great centres of learning and trade: Antioch in southern Turkey and Alexandria in Egypt. Greek ascendancy lasted 200 years and then in turn they lost out to the greatest of all empires, the Roman empire.

The Roman empire was a long time in a gestation that covered four distinct phases. The first phase was an extended period of consolidation of power in Italy. It ended with the Romans finally overcoming the Sicilian Greeks in 281 B.C. This period was crucial to its future success. Roman policy was to include other parts of Italy as full members of the Roman Republic and accord its population the status of citizenship. As a result,

Rome was able to raise larger standing armies than its enemies. The second phase consolidated power in the western Mediterranean, including the defeat of Carthage and the incorporation of Spain and North Africa into the empire. Gat estimates that during the war against Carthage (218–202 B.C.), Rome was able to mobilise 250,000 soldiers and sailors, half of whom were Roman citizens. In the later part of the second century B.C. the Romans overwhelmed the Greeks in Macedonia and Syria. Finally the basic outline of the Roman empire was completed with the conquest of Gaul, Britain and Egypt from 60 B.C. to 70 A.D.

The remarkable feature of the early Roman expansion was its complicated republican organisation; leadership of all other empires was based on inherited succession. However, the republican Senate was unable to hold onto power as the empire increased in size and generals became increasingly important. When the Senate ordered Julius Caesar to disband his victorious army in Gaul and return to Rome, he initiated a civil war which ended the republican era. After Julius Caesar's death in 44 B.C., his nephew, Augustus, became the first in a line of emperors.

Succession at first was based on a combination of family ties and the outgoing emperor's recommendation. This worked reasonably well until the middle of the third century A.D. Thereafter, the Praetorian (palace) guards, being the only legion in Rome, came to realise they could determine who would succeed. Succession became increasingly problematic, based on bribery and military might; as a result 23 emperors were proclaimed between 235 A.D. and 285 A.D.

In the end Rome lost its position at the heart of the empire and the empire divided. The western part only survived until around 400 A.D., whereas the eastern part continued, albeit with gradually decreasing power, until the fifteenth century.

Unlike the Han Chinese, the Romans built in stone and brick. As a result we can see the remains of many of their buildings. Some of these were constructed on such a colossal scale that they still astound us today. The temple site at Baalbek in modern Lebanon is a wonderful example, with its great court, enormous pillars in the temple of Jupiter and wonderful frieze on the temple of Bacchus. Construction is said to have begun in 60 B.C. and it was built and enlarged over a period of two centuries. Over 100,000 slaves are thought to have worked on the site.

The Romans developed the use of cement, which had been invented by the Greeks. During the Roman Empire, cement was made from quicklime, pozzolana (a volcanic ash deposit) and an aggregate of pumice (a volcanic rock). Concrete made from cement had widespread use in many Roman structures, freeing architecture from the limitations of building in stone and brick and allowing revolutionary new structures to be built with soaring arches and large domes. Some still exist, such as the Pantheon and the huge Baths of Caracalla in Rome. The Pantheon is still the largest unreinforced solid concrete dome in the world. The Baths of Caracalla have walls still standing 37 m high and could accommodate 5000 bathers at a time. However, it is what happened underground that gives a true picture of Roman architectural achievement. In order to heat the baths a network of 3 km of triple-tiered tunnels was constructed. Each of the tunnels was 6 m in diameter and received water via an aqueduct from a source 100 km away. Below ground an army of hundreds of slaves laboured feeding 50 ovens with tonnes of wood a day to heat the water.

It is instructive to note the huge numbers of slaves mentioned in these accounts. Rome was powered by slaves: the total number of slaves in the Roman Empire has been put at 10% of the population. The highest concentration was in Italy, with 2.5 million slaves accounting for an astounding 35% of the inhabitants.

Roads were vital to the maintenance and development of the Roman state, and were built from about 500 B.C. onwards. There were several grades of roads, from small local roads to broad, long-distance highways built to connect cities, major towns and military bases. The major roads were often stone-paved and were cambered for drainage. At the peak of Rome's development, no fewer than 29 great military highways radiated from the capital, and the empire's 113 provinces were interconnected by 372 great road links, of which over 80,500 km were stone-paved. Roman bridges were the first large and lasting bridges built. Causeways were built over marshy ground. Along the roads, there were posting stations and travellers hostels every 30 to 50 km. The emperor's mail could be delivered by a relay of riders on horseback who could cover several hundred kilometres in 24 hours.

As the Roman empire expanded into Europe and North Africa it brought civilisation. Roman cities were built in brick, mortar and stone on a rectangular grid plan. They were surrounded by walls and approached through gates. The most important towns had forums, temples, amphitheatres and baths. Some of these cities, like Leptis Magna in Libya or Jerash in Jordon, still have a substantial amount of their structure standing after two millennia.

Watermills were the first naturally powered machines. The Greeks invented the two key components: the waterwheel itself and toothed gearing. The history of watermills in the West goes back to the third century

B.C. and the technology was widely adopted by the Romans both to grind grain and for industrial processes such as sawing wood or shaping metals. The most impressive site is at Barbegal near Arles in southern France, which has been referred to as the greatest concentration of power in the ancient world. The site consists of two parallel sets of eight wheels fed by two aqueducts, powering a flour mill that was capable of producing an astonishing 4500 kg of flour a day.

We have seen the growth of popular religion in India and China. Other new religious developments also took place in the Middle East, beginning in Persia. Aryan and Persian forebears both came from the Russian steppe. We know the Aryans had their own ideas of the afterlife, as the early hymns of the Rig Veda state that the virtuous will go to the World of Fathers while the wicked will be condemned to dwell indefinitely in the House of Clay, These Aryan ideas of good and evil appear to have been developed by the prophet Zoroaster in the pre-empire period. He inspired Zoroastrianism, which became the state religion of the Persian empire.

In Zoroastrianism, there is a creator, Ahura Mazda, who is all good and there is a separate source of evil who is trying to destroy the creation of Mazda. We know from brief references and quotations in later works that this was developed into a complete religious canon, Unfortunately, the most important texts of the religion have been lost and only the liturgies have survived.

However these ideas of heaven and hell and good and evil must have influenced other religious developments in the Middle East; the two most important are Rabbinic Judaism and Christianity.

Rabbinic or modern Judaism developed following the second destruction of the temple in Jerusalem by the Romans in 70 A.D. and the subsequent wanderings of the Jewish people away from Palestine. With the temple destroyed, whole chunks of Biblical teachings that insisted on ritual sacrifice at the temple in Jerusalem, became unworkable. The emerging system of Judaism would claim that it was possible to serve God not only through sacrifice but also through study of the Torah (the first five books of the Old Testament). Rabbis produced commentary on the text of the Torah, reinterpreting it for Jews to establish a modified set of beliefs and methods of worship suitable for life away from Jerusalem. It was further codified into a list of proscriptions known as the *Halakhah*. This is a system of 613 rules or *mizvot*, of which 248 are positive injunctions and 365 are prohibitions. To be a good Jew meant that the individual had to observe these rules, which include at a practical level, male circumcision, eating kosher food, separating meat from dairy during cooking and at a religious level celebrating festivals in certain prescribed ways. Some of these laws are outwardly arbitrary; for example it is not immediately obvious that that one sentence in *Deuteronomy*: 'You shall not boil a kid in its mother's milk,' should mean that housewives have to prepare meat and dairy in separate dishes.

Observance of these *mizvot* bound the community together in ceremonies of mutual religious observance and cultural practices. Jews who did not observe the cultural norms were subject to peer pressure and risk of ostracism from the Jewish community. The *herem* was a 30-day renewable ban of excommunication. Excommunicated Jews were cut off from their community and as they did not belong anywhere else there was no way for them to earn a living. Normally, émigré communities will become absorbed into the mainstream after a few generations. Maintaining their religious beliefs and cultural practices ensured that, in the centuries before the Enlightenment, Jews always maintained their status as a separate sub-community. Exclusion and resulting persecution only strengthened the need for Jews to bind together and conform to their own ideas and beliefs.

Following the destruction of the Temple in Jerusalem, the main religious competitor to Rabbinic Judaism for expatriate Jewish communities was Christianity. As in Buddhism, a considerable time elapsed between the life of the founder and the first written record of his life and teachings. Our knowledge of Jesus is restricted to what the early church wanted us to know about the life of Christ. The gospel (good news) of Mark was written 30–50 years after Jesus' death and the remaining three gospels 20–40 years later. There is no independent secular source. However it is clear that Jesus, like Buddha, was from a tradition of successful wandering spiritual teachers who attracted a school of adherents. It is also clear that the message of Christianity is aimed at the poor. To quote a couple of examples:

From Mark, 10, 17–23:

> And when he was gone forth into the way, there came one running, and kneeled to him, Good Master, what shall I do that I may inherit eternal life? And Jesus said unto him, Why callest thou me good? there is none good but one, that is God. Thou knowest the commandments, Do not commit adultery, Do not bear false witness, Do not steal, Honour thy father and mother. And he answered and said unto him, Master, all these I observed from my youth. Then Jesus beholding him loved him, and said unto him, One thing thou lackest: go thy way, sell whatsoever thou hast, and give to the poor, and thou shalt have treasure in heaven: and come, take up the cross and follow me. And he was sad at the saying, and went away grieved: for he had great possessions. And Jesus looked round about and saith unto

his disciples, How hardly shall they that have riches enter into the kingdom of God! And the disciples were astonished at his words. But Jesus answereth again, and saith unto them, Children, how hard is it for them that trust in riches to enter into the kingdom of God.! It is easier for a camel to go through the eye of a needle, than for a rich man to enter into the kingdom of God.

The Beatitudes from Matthew, 5, 3–12:

Blessed are the poor in spirit,
  for theirs is the kingdom of heaven

Blessed are those that mourn,
  for they will be comforted.

Blessed are the meek,
  for they will inherit the earth.

Blessed are those who hunger and thirst for righteousness,
  for they will be filled.

Blessed are the merciful,
  for they will receive mercy.

Blessed are the pure in heart,
  for they will see God.

Blessed are the peacemakers,
  for they will be called children of God.

Blessed are those who are persecuted for righteousness' sake,
  for theirs is the kingdom of heaven.

Blessed are you when people revile you and persecute you and utter all kinds of evil against you falsely on my account. Rejoice and be glad, for your reward is great in heaven, for in the same way they persecuted the prophets who were before you.

This was clearly a popular religious message aimed at ordinary Jews, and it was equally threatening to the establishment and the rabbis. The message was apparently simple: be a good person, give due homage to God and you will be rewarded in the afterlife. In the end, Rabbinic Judaism, not Christianity, was adopted by the Jewish communities. However, Jewish converts to Christianity found their message had a strong appeal to gentiles. With the work of Paul (circa 5–67 A.D.) and others, it spread to non-Jews in the towns and villages of, firstly Anatolia, and later the whole Roman Empire.

The Romans had a long history of adopting and adapting cults from other countries. Around 200 B.C. in Rome, Cybele (a goddess originally from Anatolia) was thought by superstitious Romans to be a key reason for their success against Carthage during the Punic Wars. In the first century A.D. the cult of Isis (the wife of the ancient Egyptian god Osiris) was promoted by the mad emperor Caligula. From the first to the fourth century A.D. the cult of Mithras was the major competitor to Christianity. Mithras was an ancient Persian god whose cult was adapted by the Greeks and then by the Romans.

Adapting a Jewish religion to suit Roman ideas would therefore have had several precedents. What was unusual about Christianity was its discipline and organisation. No other popular religion of its time achieved such centralised control of its communities without the assistance of the state. Jesus was originally biased against authority. *According to Christianity – A Very Short Introduction*, p. 48 by Linda Woodhead:

> It is quite possible that Jesus envisaged a fully egalitarian society whose members share table fellowship, teach and minister to one another, and refuse to acknowledge any authority except that of a God of Love.

It appears that the church's hierarchical power structure began with Paul, the architect of the Christian mission to the gentiles. From his Epistles, it is clear that he and all those who had directly encountered Jesus, in the flesh or arisen, claimed to have special authority over others. When Jesus' contemporaries died they passed this special authority onto their chosen successors. The idea of apostolic succession was born. In a special ceremony involving the laying on of hands, authority that could be directly linked to the apostles was passed onto new priests. The traditional function of a priest was to perform sacrifices to the gods. The most important ceremony in Christianity was the Eucharist, which celebrated the sacrifice of Jesus' life by consecrating bread and wine to represent Jesus' body and blood. Christian priests claimed, by virtue of apostolic succession, the sole right to conduct this ritual. These ordained priests helped to maintain a standard orthodox set of beliefs. Religious community leaders who might have experienced direct contact with God through the Holy Spirit were compelled to follow the orthodox religious view rather than develop their own ideas of the faith. By the second century the full church hierarchy had established itself. Bishops appointed priests, who had authority over the deacons who were responsible for pastoral care. Belief systems came to be standardised, being termed catholic or orthodox. Discipline was enforced and a hierarchical administrative system was introduced with all Christian communities linked under Jesus Christ, who was represented on earth by the bishop. Christianity developed into an orderly, conservative society. As such it was particularly appealing to urban Romans, (in much the same way as Puritanism would prove to appeal to the urban middle class at the time of the European Reformation).

Nevertheless Christianity remained a minority religion, always liable to persecution, until the battle of Milvian Bridge in 310 A.D. There, the Emperor Constantine was inspired in a vision that a Christian God would promise him victory if he marked the sign of the cross on the shields of his soldiers. After this victory Christianity gradually increased in importance until it became the state religion of the Roman Empire. The Christian Church immediately saw the advantages of state sponsorship. Constantine, described as 'the deputy of Christ,' was eager to insist that the alliance of church and empire was part of God's providential plan for the world.

With state sponsorship, Christianity was widely adopted across the Roman Empire. Pagan temples were transformed into Christian basilicas. Christianity became the religion of the rich and the poor, benefitting the state by giving it a new unity of purpose. The bishops acquired wealth and privilege. The emperor became known as a 'friend of God' and was thus clearly able to enter heaven despite the difficulty of getting camels through needles.

Trade between the empires

Memes developed mostly within empires. There was little direct contact between the civilisations on different continents, although some valuable goods were traded intercontinentally for the first time. With the conquest of south China by the Han dynasty there are written records of maritime trade in South-East Asia and the Indian Ocean. Merchants brought lapis lazuli, pearls, jade and glassware to the Han empire in sea-going junks.

Trade from the Middle East to India was helped by the discovery of seasonal monsoons blowing between the exit from the Red Sea and the west coast of India. The non-stop crossing made possible by these winds led to greatly decreased costs and increased traffic. The major goods supplied were ivory, ebony and spices, of which the most important Indian spice was pepper.

The fabled Silk Road extending 4000 miles across Eurasia first opened up during the Han dynasty. The staple of the trans-Asian route, Chinese silk, had been an instant success in the West despite its high price; and the restrictive regulations of disapproving Roman emperors had been ineffective in an aristocratic society geared to conspicuous consumption. The Chinese monopoly on silk production was defended by an imperial decree, condemning to death anyone attempting to export silkworms or their eggs. Along the Silk Road up to the frontiers of the Roman empire, silk became a monetary standard for estimating the value of different products. It was not until 552 A.D. that the Roman Emperor, Justinian, obtained the first silkworm eggs. He had sent two Nestorian monks to Central Asia and they were able to smuggle out silkworm eggs hidden in rods of bamboo. The Church was thus able to make silk fabric for the emperor and was eventually able to develop a silk industry in the eastern Roman empire.

By controlling the land-based trade routes, a number of independent states were able to thrive by taxing goods in transit. The city-state of Palmyra in Syria on the Silk Road and the Nabataean kingdom based in Petra in Jordan lived from the profits of the spice trade. Both these fully exploited the domestication of the camel, the 'ship of the desert,' as load-bearing transport for these relatively light but valued goods.

Now that trade linked the three main centres of population in Eurasia, diseases incubated on one continent were able to reach another. We do not know specifically what diseases caused the terrible epidemics that recurred every generation or so from 170 A.D. onwards. In Eurasia this was a time when the death due to disease was the equivalent of the devastation of the Black Death in the Middle Ages. In the West the worst years were 251–266 A.D., when for a while 5000 people died each day in the city of Rome. Between 200 and 400 A.D. the population of the Roman empire is estimated to have fallen by 20%. In the sixth and seventh centuries the eastern Mediterranean saw the first major outbreak of bubonic plague. Estimates have been made that as much as 40% of the population of Constantinople died from the plague and that half of Europe's population was wiped out before the plague disappeared in the 700s.

In China the situation was worse; the epidemics came on top of famines, floods and rebellions. This resulted in a fall in the Chinese population by an astonishing 40 million people, or two-thirds of the total, between 150 A.D. and 280 A.D.

Summary: situation in 400 A.D.

At the end of the age of empires there had been a massive consolidation of power and culture. Whereas once there was a myriad of states and chiefdoms, each with its own religious culture, there were now, effectively, only four such cultures dominating the whole civilised world excluding

China: Christianity, Zoroastrianism, Brahmanism and Buddhism. These politically supported religious cultures were shared between rulers and subjects alike; they supported a shared belief in the rights, customs and responsibilities of each part of society.

The Roman empire dominated Europe and the western Mediterranean. The population of Rome at its peak had been around a million people, but was now about half the size. However, it was still the largest city in the world. Constantinople, the new capital of the eastern Roman empire, had about 300,000 inhabitants. Alexandria and Antioch had over 100,000 people.

The Sassanid Persian dynasty (successor to the Achaemenids) ruled in the eastern Middle East and Iran. Here, Zoroastrianism was still the state religion, although it faced stiff competition from Christianity. The capital Ctesiphon (near Bagdad), had about 250,000 inhabitants.

The Gupta dynasty controlled north and central India. They had adopted Brahmanism and the decline in Buddhist support in India had begun. Pataliputra, the capital of the Gupta empire, had about 150,000 inhabitants.

In China, political control had been strong and no popular religion had been given state support. The emperor still was the people's official representative to God. Confucianism imposed a common political culture. After the collapse of the long-lived Han dynasty there was some political disunity. Even then, the whole area of China was controlled by only three or four states. The largest cities were Nanjing (330,000) and Luoyang (250,000) with about five other cities of over 100,000 people. Civilisation and the Confucianism philosophy had spread from China to Japan, Korea and Vietnam.

World population had reached around 250 million people, about the same as the inhabitants of Indonesia today. In the new order established by the empires the proportion of the population suffering violent death had decreased dramatically and was now less than 5% compared to 10–20% in tribal societies. However, as noted above, death due to disease had increased and life expectancy remained low, at around 22 years.

Outside the empires, tribes and tribal states dominated the rest of the world, although they had lost territory to the empires over this period. The area controlled by hunter–gatherer societies decreased further. However, war still determined the future of tribes, states and empires. The balance of military power was about to make a huge shift towards tribal societies and cause a crisis in civilisation.

CHAPTER 6:
THE DECLINE AND RECOVERY OF CIVILISATION (400–1450)

All the cultural developments we associate with civilisation: art, architecture, scholarship and the more ordinary artisanal goods, required people to live in an urban environment where these creative specialties could be learned and developed. Every important civilisation contained towns and cities where artisans, merchants and scholars mingled with aristocrats, kings and emperors. The number of people working in specialised trades in cities is an indicator of wealth generation. Indeed, a crude measure of the degree of civilisation is the proportion of people living in cities compared to those working in the fields. With the lack of any detailed population censuses in ancient history such a measure is clearly not available to us. However, we can gain a sense of it by comparing the estimated proportion of the world's population living in the largest 10 cities over time. Table 6.1 shows that the top 10 cities accounted for 0.8% of the population in 100 A.D.; this percentage declined to 0.5% by 1000; it did not begin to recover until the eighteenth century and would not achieve higher levels until the nineteenth century. This is broadly in line with agreed knowledge. In the West the level of intellectual activity achieved by the Greeks and Romans was not reached again until the Renaissance in the fifteenth century and not surpassed until the Enlightenment in the eighteenth. On a material level the network of roads established by the Romans was not matched until the eighteenth and nineteenth centuries in Europe. Across Eurasia, it appears that city life and hence civilisation in the thousand years of the Middle Ages from 400 A.D. to 1450 A.D. was less advanced than in the age of empires.

The reason for the decline in the levels of civilisation was the military advantage attained by the peoples of the Eurasian steppe. These tribes from the Uralic and Altaic language families developed a totally mobile pastoralist lifestyle, mounted on horses, living off herds of animals and able to survive the harsh winter environments. The earlier Indo-European tribes of the western steppe had been semi-nomadic, engaged in a combination of herding and farming, based in the river valleys. They had used heavy ox-drawn carts to move their families around. In contrast, the extremely hardy Siberian peoples were fully mobile horse riders who had light tents or yurts that could be rapidly moved. The invention of the stirrup gave them a firm base to fight effectively while remaining mounted. Their principal weapon was a bow and arrow which they were able to shoot at full gallop in attack or retreat. Their armies could travel fast to either strike an enemy off guard

Rank	Year 100 Name	Population(K)	Year 1000 Name	Population(K)	Year 1500 Name	Population(K)	Year 1800 Name	Population(K)	Year 1900 Name	Population(K)
1	Rome	450	Cordova, Spain	450	Beijing, China	672	Beijing, China	1,100	London, UK	6,480
2	Luoyang, China	420	Kaifeng, China	400	Vijayanagar, India	400	London, UK	861	New York, USA	4,242
3	Seleucia, Iraq	250	Constantinople (Istanbul), Turkey	300	Cairo, Egypt	400	Guangzhou, China	800	Paris, France	3,330
4	Alexandria, Egypt	250	Angkor, Cambodia	200	Hangzhou, China	250	Tokyo, Japan	685	Berlin, Germany	2,707
5	Antioch, Turkey	150	Kyoto, Japan	175	Tabriz, Iran	250	Istanbul, Turkey	570	Chicago, United States	1,717
6	Anuradhapura, Sri Lanka	130	Cairo, Egypt	135	Istanbul, Turkey	200	Paris, France	547	Vienna, Austria	1,698
7	Peshawar, Pakistan	120	Baghdad, Iraq	125	Gaur, India	200	Naples, Italy	430	Tokyo, Japan	1,497
8	Carthage, Tunisia	100	Nishapur, Iran	125	Paris, France	185	Hangzhou, China	387	St. Petersburg, Russia	1,439
9	Suzhou, China	95	Al-Hasa, Saudi Arabia	110	Guangzhou, China	150	Osaka, Japan	383	Manchester, UK	1,435
10	Smyrna, Turkey	90	Patan, India	100	Nanjing, China	147	Kyoto, Japan	377	Philadelphia, USA	1,418
Total Top 10 cities		2,055		2,120		2,954		6,140		25,963
World Population(M)		250		400		578		954		1,634
Percentage		0.8		0.5		0.5		0.6		1.6

Table 6.1 Top 10 cities in the world as a percentage of world population 100–1900

Source: Tertius Chandler: Four Thousand Years of Urban Growth, An Historical Census.

or to avoid battle. They had no base which they had to stand and defend; hence there was no target that their more sedentary neighbours could attack in reprisal. Before long they began to threaten their neighbours and exhort protection money. The level of violence escalated, as there was no way of ensuring continued payment except by terror. They looked down on the more effete property-based lifestyle of their neighbours and attacked them with a level of brutality that was some of the worst in history. Genghis Khan (1162–1227), the great and notorious leader of the Mongols, is reputed to have said:

> The greatest joy a man can know is to conquer his enemies and drive them before him. To ride their horses and take away their possessions. To see the faces of those who were dear to them bedewed with tears, and to clasp their wives and daughters in his arms.

It clearly was not good to be counted as one of Genghis Khan's enemies. The effect can be seen from this report by Giovanni de Plano Carpini, the pope's envoy to the Mongol Great Khan, as he travelled through Kiev in February 1246 and wrote:

> They [the Mongols] attacked Rus, where they made great havoc, destroying cities and fortresses and slaughtering men; and they laid siege to Kiev, the capital of Rus; after they had besieged the city for a long time, they took it and put the inhabitants to death. When we were journeying through that land we came across countless skulls and bones of dead men lying about on the ground. Kiev had been a very large and heavily populated town, but now it has been reduced almost to nothing, for there are at the present time scarce two hundred houses there and the inhabitants are kept in complete slavery.

The Mongols were the last of a series of Eurasian tribal ethnic groups that devastated civilisation. Their trail of destruction started in 200 A.D. and lasted until around 1300 A.D. They came in waves, as every hundred or so years a new ethnic group consolidated its position on the steppe, expanded its territory, terrorised civilisation and then collapsed in disunity. It seems to have begun with the Huns and finished with the most destructive of all, the Mongols.

The story of this chapter is how the four different areas of the civilised world coped with and recovered from the threat of the Asian pastoralists. Except in China the great empires of the past were destroyed. In Europe local communities were forced to defend themselves and a new **feudal structure** of society emerged centred on a local baron. In the face of adversity, civilisation survived, preserved in the religious customs of the time. Christianity and the new religion of **Islam** came to dominate the lives of those in Europe and the Middle East.

It is also the story of how Chinese civilisation thrived in these difficult times. Their technological developments, **printing, gunpowder** and **the compass** were eventually to cross to the West and initiate the next stage in human development.

I start with the most distant area from Siberian pastoralists – Europe.

Europe

The Huns were a pastoral people who settled around the Caucasus in about 150 A.D. When the Huns migrated into the Hungarian plain in about 370 A.D. they began raiding the neighbouring countries. German tribes were the first to feel the effect of the terror. At the time the Roman Empire was divided into two, with the eastern part centred on Constantinople and the western part containing most of what we now consider Western Europe. German tribes in the west, harried by the Huns, crossed the Rhine and attacked the Romans. The western Roman Empire was unable to stop the advance and by 470 A.D., it had collapsed. The next five hundred years was a chaotic and miserable time in Europe as Asiatic invaders struck from the east, the Vikings attacked from the north and Islamic invaders hit from the south.

Somehow, the German tribes clung onto power, but it was not until the eleventh century that recovery started. There were two key factors enabling this recovery: feudalism and religion. Feudalism enabled local barons and their knights to respond more rapidly to enemy raids while the literacy retained in the Catholic Church allowed civilisation to re-establish itself.

The large central armies of the old Roman Empire had proved they could not cope with the raiding tactics of its new enemies. Due to the newfound mobility of the enemy, big armies, which by their nature were ponderous and slow to react, were no longer militarily effective. A new, more flexible solution had to be found, based on a more reactive local military organisation. Regional lords (barons) took on the responsibility of maintaining security in their territory with a local military force. The use of the stirrup, which was copied from the invading pastoralists, gave mounted cavalry the opportunity of wielding swords and lances from the saddle. Heavily armed mounted knights operating from castles and walled towns proved the most effective way of confronting the invaders.

To pay for this level of defence, the local peasantry were coerced into working as serfs on the baron's lands, thus creating feudal states. In feudal states power was effectively decentralised, as barons had powers both to raise taxes and provide their own military force. Peasants were tied to the land and at the mercy of the barons. While barons had a duty to respect the king's wishes, kings, not having a large central standing army, had limited ability to enforce their wishes. Europe became a set of loosely connected statelets, controlled by barons. The barons had their own private armies, their own castles and their own source of income. They could impose their own will on the countryside around. Feudalism succeeded in bringing back a degree of order to the countryside but at the price of the fragmentation of central power.

That central organisation in Europe was retained at all is thanks to the Catholic Church. When Roman power collapsed and the pagan tribes became rulers, only the priests remained literate. Invading tribes wanted to attain the wealth and status of previous Roman elites and sought to achieve it by gaining literacy skills only available to Christians. Kings needed clerics as state bureaucrats to administer their laws and try to keep their barons under control. Rulers gave bishops territories to manage outright and granted huge tracts of land to monastic foundations, sometimes as a reward for services rendered but also to offset the power of the barons. Taxes or (tithes) to support the church were paid directly to the religious authorities.

At a time when Europe was imploding, the boundaries of Christianity actually increased. Several new Christian kingdoms had established themselves by the end of the eleventh century. Tribal Vikings converted to Christianity and formed the kingdoms of Norway, Sweden and Denmark. The Christian Slavic kingdoms of Poland, Croatia and Bohemia gave some political stability to the east. Even the terrifying Magyars from the Eurasian steppe settled down to form the Christian kingdom of Hungary.

It was during the second part of the first millennium that Christianity formally divided into two; with the Catholic Church based in Rome and the Orthodox Church based in Constantinople. The Orthodox Church also had success in bringing civilisation to the tribes of East Europe. Missionaries invented the Cyrillic script to record the different sounds of the Slavic languages and it was used in new kingdoms in Serbia, Bulgaria and Russia.

Notwithstanding the even larger distances involved and the difficulty of travel and communication, the Catholic Church remained centrally managed from Rome by the pope. In Rome the pope was the sovereign of the city and the surrounding countryside and was in a unique position of being both the religious and temporal ruler. This gave him freedom to act independently from other kings and princes. Although his military might was small, the pope had the status of a super-sovereign; he was in many ways the most powerful man in Europe.

The Catholic Church persuaded the barons to fight to extend its boundaries further. The heathen Baltic states were invaded by a specially created military and religious order called the Teutonic Knights. The kingdoms of Castile, Aragon and Portugal slowly pushed the Muslims from Spain. Most astonishingly, against the interests of leaders in Europe, the Catholic Church persuaded them to participate in a series of fruitless attempts to recover Jerusalem for Christianity, known as the Crusades.

The Crusades were a depressing exercise in human duplicity, avarice and cruelty. However, they opened the eyes of the largely ignorant West to the superior technology and science of the East. One idea they may have brought back was the windmill. The Persians seem to have been the first to develop windmills (between 500 and 900 A.D.). The sails in their mills rotated horizontally around a vertical axis, spinning a shaft directly connected to a grindstone or pump. Windmills first appeared in Europe about the time of the Crusades (1096–1270 A.D.) but had a different structure called a post-mill, with a horizontally mounted axis. The name is derived from the fact that the main structure sits on a post that allows the entire windmill to turn to face the wind. Power was transferred to a grindstone, pump or hammer by a system of cogs and gears.

As the invasions from Vikings, Arabs and pastoralists lessened, the process of state consolidation began. Both regional and city states were formed. In the important regional kingdoms, like England, France and Castile, the kings managed to wrest control back from the barons and establish a strong central administration. In city-states like Venice and Florence, where trade was important, oligarchies established control over government. However, in places like Germany, where central control was never established, many small baronies and cities became states in their own right. Europe settled down to a tedious period of interstate wars and territory consolidations, all for the benefit and glory of kings and knights.

Nevertheless, Europe increased its wealth and population. A European population of about 40 million in 400 A.D. fell to about 20 million by 600 A.D., then recovered and grew to about 75 million, before the Black Death struck in the 1340s. It was clear that Europe had survived the impact of the pastoralist invaders and was recovering some aspects of civilisation. However, few would have predicted that the remaining relatively small European states would go on to become such a dynamic world force in the next period. However, the very fact that no one state had established predominance meant that competition between the states was strong. The evolutionary rules of survival of the fittest were working and encouraging the development of innovative memes. When printing enabled the dissemination of new memes in Europe in the next period, state rulers were keen to develop and exploit the resulting opportunities.

Middle East, Iran and North Africa

While Christianity gained ground in Europe it lost out in the Middle East and North Africa to an extraordinary new religious phenomenon, Islam. In the seventh century A.D., the Byzantine Empire (as the successor of the Roman Empire came to be called) was the centre of Orthodox Christianity and controlled North Africa, the Balkans and the Eastern Mediterranean. The Byzantines fought with the Sassanian Persian Empire for control of the Middle East. All this changed when tribal groups of pastoralists, now known as Arabs, drastically reduced one empire and obliterated the other in a phenomenal surge of conquest which left them holding a vast area of land that stretched from Spain to Pakistan.

The original Arabs were made up of two distinct communities. In the north, there were pastoralist tribes in the borderlands between the Byzantine and Persian empires. These tribes were mutually antagonistic and warlike, hiring their services out to the Byzantines or the Persians for border protection. To the south along the trade route from India there were a number of towns, including Mecca and Medina. There were no states; the area was entirely tribal. Each tribe had its own brand of religion, including elements of Baal worship, Judaism, Christianity and Zoroastrianism. The Arab empire was largely the result of the inspiration of one man, the prophet Mohammed (570–632 A.D.). Mohammed considered himself a prophet of God in the Christian-Judaic tradition. He had the same outline message: be a good person, pay due homage to God and you will be rewarded in the afterlife. He insisted on the uniqueness of a God, harking back to the central tenet of Judaism, and claimed to have received the direct words of God dictated to him and written down in the Qur'an.

Mohammed was both a religious and temporal leader. From a base in Medina he persuaded, cajoled and forced Arab tribes to follow his religious guidelines and unite into one community or ummah. By the time of his death most of the Arabian peninsula had united under his temporal rule and followed the Islamic faith.

Under his three immediate successors, raids on the Sassanian and Roman empires rapidly turned into conquest. All of North Africa, Spain and the whole of the Middle East except for Anatolia fell into their hands. In the east the Sassanian empire was crushed and became part of the new Arab state.

By and large the natives of the territories conquered were Christian or Zoroastrian. No forced conversion was attempted; Mohammed always had respect for 'people of the book', that is Jews and Christians, whom he regarded as kindred spirits. At the beginning Islam was seen as a religion for Arabs only. Non-Muslims were taxed more heavily than Muslims but left to pursue their own faith. Over time the economic and political incentive to change faith to that of the ruling elite meant that non-Arab citizens converted to Islam without any coercion. Arabic became first the lingua franca and then the native language of North Africa and the Fertile Crescent.

Central political control was initially in the hands of a caliph who, like Mohammed, exercised military as well as religious leadership. The new Arab empire was known as the Caliphate and lasted about 250 years. From 850 A.D. onwards civil wars and leadership struggles limited the caliph's power, and eventually central political control over all Islamic lands ceased to exist. However, like Christianity in Europe, the power of Islam as a culture remained. All Islamic states shared the same cultural values. Islam, like Christianity in Europe, had become deeply embedded in the culture of the region and would survive whatever the nature of the ruling states.

A Muslim is someone who declares the shahada, 'there is no God but God and Muhammad is his prophet'. This absolute insistence of the uniqueness of God meant that its adherents had to cast aside their old gods and unite under one divine law in dar al Islam – the world of Islam. All Muslims hold to the five pillars of Islam which, in addition to the shahada, are praying five times a day, fasting during the month of Ramadan, a pilgrimage to Mecca once in a lifetime and making charitable donations.

When central military authority ceased, the only source of religious authority was the written word in the form of the Qur'an and the hadiths (a compilation of the acts and sayings of Muhammad collected after his death). Local interpretation of these documents gradually gave way to a more formal legal approach, now known as Shari'a law. These laws, which govern the minutiae of a good Muslim's life, are common throughout dar al Islam in the same way as the laws of the halakhah govern that the life of a Jew. These rules notably include inheritance laws, dietary regimes, a dress code and fasting at Ramadan. Religious scholars or imams studied the Qur'an and were experts on Shari'a law. Like the Christians in Europe, their literacy assisted kings or sultans to administer their states.

In 749 A.D. disputes over the rights of the caliph's nephew Ali to succeed to power led to a significant split in the Muslim world between the Shi'a, who recognised the successors to Ali as their Imam or religious leader, and the Sunni, who continued with the elected caliph. Other sects splintered from the Shi'ite line such as the Druze from Lebanon and the Alawis from Syria. However, the Shi'ites and all the other minority sects comprise no more than 15% of the Muslim faithful; it is the majority Sunni believers who dominate the Muslim world.

Sunni Islam does not have the same hierarchical structure as the Christian Church; the imams are considered religious scholars rather than priests. Sunni Islam became a community-based religion. Religious leadership was exercised locally by the ulama, a class of religious scholars, whose authority was based on their knowledge of the scripture. The result was a largely traditional society but one in which a large number of varieties were tolerated under Islam's broad banner.

In contrast, Shi'a clergy exercise power more in the Catholic tradition. Shi'a imams can independently interpret Shari'a law without being bound by tradition and have financial independence as the trustees of several shrines and the recipients of religious taxes and properties. This gives them much more authority in relation to the state. They are also more organised, with entrance examinations and with a body of senior clergy called ayatollahs. The sort of theocratic state we have in Shi'ite Iran today is unlikely to exist in a Sunni country.

The period from about 750 to 1250 A.D. is considered the Islamic golden age. Influenced by hadiths such as 'the ink of a scholar is more holy than the blood of a martyr', the Muslim world became an intellectual centre for science, philosophy, medicine and education. In the caliphate's capital Baghdad, a House of Wisdom was established where Muslim and non-Muslim scholars were brought together to compile works of scholarship from across the world and translate them into Arabic. Many classic works of antiquity that would otherwise have been lost were translated into Arabic and Persian and later, in turn, translated into Turkish, Hebrew and Latin. During this period the Muslim world was a cauldron of cultures which collected, synthesised and advanced the knowledge gained from ancient civilisations. Significant advances were made in mathematics. The Persian al-Khwarizmi (780–850 A.D.) published the first book on the systematic solution of linear and quadratic equations; it is from a term in this book that the word algebra is derived. Al-Uqlidis 'the Euclidian' (920–980 A.D.) summarised the advances made with Hindu-Arabic numerals and was the first to use decimal notation in place of fractions.

Thus far the Middle East had been relatively untroubled by Asian pastoralists. All this changed in the eleventh century when the Seljuq Turks swept through Iran, Iraq and Syria and advanced into Anatolia; at the battle of Manzikert they destroyed the army of the Byzantine Empire, reducing it to a small area around Constantinople. The Seljuk Turks, who came from an area north of the Caspian Sea, had already converted to Islam and absorbed Persian culture. Thus, their invasion did not shatter the fundamental structure of existing Islamic society. Their Turkish successors, imbued in Persian culture and the Islamic faith, formed an elite that governed the Middle East and Iran for most of the succeeding millennium.

The Muslim Seljuq Turks were followed by the pagan Mongols who, in 1258, destroyed Baghdad in bloody Mongol style. Grand buildings that had been the work of generations were burnt to the ground. Mosques, palaces, libraries and hospitals were all destroyed. The countless precious historical documents and books on subjects ranging from medicine to astronomy in the Grand Library of Baghdad were lost forever. Survivors said that the waters of the Tigris ran black with ink from the enormous quantities of books flung into the river and red from the blood of the scientists and philosophers killed. The caliph was captured and forced to watch as his citizens were murdered and his treasury plundered. According to most accounts, the caliph was killed by trampling. As they believed that the earth was offended if touched by royal blood; the Mongols rolled the caliph in a rug before they rode over him. All but one of his sons was also killed. Citizens who attempted to flee were intercepted and dispatched by Mongol soldiers. The total death toll was estimated to be from 90,000 to a million people; Baghdad was ruined and depopulated. Arab culture never reached the same heights again. Gradually, over the centuries, the Turkish elites regained control. Their glory days were still to come, but the great native civilisations of the Middle East had ceased to exist. What was left was Islam, and that would grow and expand as a world force.

India

Invasion by the Hunas tribes (relatives of the Huns who invaded Europe at the same time) from 450 A.D. eventually put paid to the Gupta Empire. India reverted to its old structure of warring states. Developments in religion were significant. Buddhism split into a number of different schools. The Theravada Buddhists remained closest to the original concepts of Buddha. Of the other schools only the Mahayana school is now known. In Mahayana Buddhism the Buddha is thought of as a semi-divine being. Followers of the Mahayana reasoned that the Buddha, being so compassionate, would never cut himself off from others; they believed he was still out there somewhere actively working for the welfare of human beings. Further, adherents to this new sect believed that Buddha had supernatural assistants known as bodhisattvas. In some branches of the Mahayana tradition a number of bodhisattvas were almost indistinguishable from Buddha himself and became part of a pantheon of Buddha-like super-beings. In line with this belief, devotional cults sprang up in which reverence and homage were offered and intercessions sought. Mahayana Buddhism became a religion rather than a philosophy of life.

In Tamil Nadu, in the south of India, the idea of a compassionate and supportive god was developed in the Vedic/Brahmin tradition as well. They developed the idea of personal devotion or Bhakti to a particular god. Popular heroes from folk stories, like Rama and Krishna, gained devotees and became gods. As no-one was excluded from the Bhakti cult on caste grounds, it seriously rivalled Buddhism as a refuge from caste prejudice and the power of the Brahmins. The cult took hold in the south and gradually spread north.

As had happened before in India, a synthesis occurred between old and new. Brahmaic doctrine merged with these new devotional forms of the religion. The result was the emergence of Hinduism that we know today. They copied the Mahayana Buddhist practice of showing the Buddha and bodhisattvas as icons and presented their gods in human form. However, these icons were considered symbols. The statues often had many limbs, each bearing a representation of an attribute of the god. There was an attempt to consolidate the traditions of thousands of local gods into the overall structure of the religion. By identifying regional gods with any of the three trimuthi, Brahma, Vishnu or Shiva, regional stories of the gods and their images could be integrated into the whole. Vishnu became associated with his supposed avatars (or manifestations on earth), including Krishna and Rama, and Buddha himself. Shiva became identified with the lingam or phallus which had been worshipped in many temples since pre-Vedic times.

In this new form Hinduism triumphed in India and Buddhism declined. By the second millennium, Buddhism was confined to Sri Lanka in the Indian subcontinent and had to look to countries outside of India, Buddha's birthplace, for its future development.

India has always been vulnerable to invasion through Afghanistan. Turkish Muslim raids from this direction started in the eleventh century. Local Hindu rulers could not match the superior tactics, power and religious motivation of the invader. Idol worship is anathema to all Muslims and while Mohammed was tolerant of all peoples of the book, this tolerance was not extended to any other religions. In the Middle East, Christians had been accorded respect after their conquest. In contrast, in India, as well as the traditional rape and pillage, local Indian culture was not respected; all significant Hindu temples were destroyed. This set up a bitter clash between Muslims and Hindus, which still divides the Indian subcontinent today. By

1200 a Muslim sultanate based in Delhi dominated north India. The centre of development of native Indian culture switched to the south, which remains to this day the only part of India where you can see old Hindu temples undefiled by Muslim invaders.

China

As early as 350 B.C. the Xiongnu, sometimes considered the eastern wing of the Huns, were reported as attacking the state of Qin. Early in the fourth century, after the Han dynasty fell apart, the Xiongnu invaded northern China and sacked the key capital cities of Loyang and Chang'an. There followed a period of disunity in China in which various pastoralist tribes set up a series of states in north China. This was not a complete negation of civilisation, like the collapse of the Roman Empire in Europe. The states were still large; the pastoralists used the existing bureaucracy to raise taxes. Over time the pastoralist invaders absorbed Chinese culture and tradition. In the south various native Chinese dynasties controlled the land below the Yangzi.

This period is important in China for the growth of Buddhism. Introduced to China via the Silk Road, Buddhism was adopted by leaders of north Chinese states and then established a foothold in the south. This is an extract from the *Rough Guide Chronicle to China* p. 122:

> By 600 A.D. an estimated 90% of China's population subscribed to Buddhist beliefs and practices. Such a statistic, however, even if accurate, requires qualification. The advent of Buddhism scarcely led to an abandonment of existing beliefs. Rather it became part of a syncretic outlook that continued to develop Daoist and Confucian concepts. For most Chinese, there was, and is, no contradiction in following different paths simultaneously. Nor are those paths wholly distinct. If sometimes acrimonious competition arose between exclusively dedicated Buddhist and Daoist sects, the two owed much to each other.

The early spread of Buddhism in China was sponsored by northern pastoralist rulers who already had had contact with Buddhism as it spread northwards from India via Tibet. This is illustrated by the career of the monk Kumarajiva (344–413 A.D.). His father was a monk from Kashmir who moved to a Mahayana Buddhist community in the independent town of Kucha in modern north-west China. In accordance with tradition Kumarajiva joined the local monastery at the age of 7. There he showed outstanding language skills, studying ancient Buddhist and Hindu texts and learning

Chinese from passing merchants. After a period studying in Kashmir he returned to Kucha and gained a reputation for scholarship and learning. He was kidnapped by the ruler of the small kingdom of Liangzhou, and later in 401 he was brought to Chang'an, at the request of Emperor Yao Xing. There he assembled a team of scholars and was given the title of national teacher. Between them they translated 39 Buddhist texts into Chinese, including the *Diamond Sutra* and the *Lotus Sutra.* Kumarajiva developed a specialist Chinese vocabulary for Buddhist concepts, which was used by all successive writers. Although later translations might have been more accurate, his translations had a smooth literary style and have remained the most popular.

The initial success of Buddhism arose partly because sponsoring a local monastery was a way of avoiding tax. Buddhist monasteries became very wealthy. However, Buddhism in its turn gave something back to the community. Local monasteries stimulated communal self-help, which was otherwise lacking in China; monastic wealth provided capital for local bridges, roads and mills.

In 589 China was reunited again and this ushered in a golden era of Chinese civilisation: the Sui dynasty (581–618 A.D.), followed by the Tang (618–907 A.D.) and then by the Song dynasty (960–1279 A.D.). In this period the technique of printing was developed, the greatest contribution of the Chinese to human civilisation. The earliest example of printing is of a woodblock print on silk at the time of the Han dynasty. However, it took some time before the woodblock was used for printing text. The earliest known example of printed text was produced around 750 A.D. and the earliest book was a *Diamond Sutra* scroll printed in 868 A.D. The printing process was laborious, requiring a separately carved woodblock for each page of the book.

The movable type system was invented around 1040 A.D. with characters made of ceramics. However, in China, movable type printing had huge logistical problems. A typical movable type set requires 100,000 character blocks, including 20 or more copies of the characters in common usage. This was a substantial investment. Typesetting was also difficult; one printer used revolving tables about 2 m in diameter. The characters were all numbered and one man holding the list called out the number to another who would fetch the type. The first known book produced with a wooden movable type was produced in the mid-twelfth century. At the same time government officials used a copper block embedded with bronze movable type to print paper money and formal official documents.

This system worked well when the print run was large. But for the smaller runs typical of the time, movable type was not practical. Reprints would have required resetting and re-proofreading, unlike the wooden block system in which blocks were stored and reused. For these reasons woodblock printing remained the most common form of printing in China. Printing technology did not reach Europe until the last part of the fifteenth century.

After paper and printing, the third of the four great inventions of China was gunpowder, a mixture of sulphur, charcoal and potassium nitrate (saltpetre). Its invention in the ninth century is attributed to Daoist alchemists looking for the elixir of life. Its first military use in the tenth century was to make a slow fuse for a flame thrower. It was then discovered that by enclosing the mixture, a greater explosive effect could be achieved. During the eleventh century the Chinese rapidly developed a new arsenal of weapons: bombs, grenades, exploding arrows, rockets and mines and, later, guns and mortars. Chinese military forces used gunpowder-based weapons against the Mongols. The Mongols in turn used them in the Middle East and from there the use of gunpowder spread to Europe and eventually changed the nature of war.

The last of the four great Chinese inventions was the compass. It is not clear exactly when the compass was invented. Descriptions of magnetism go back to the fourth century B.C. The first incontestable reference to a magnetised needle is in 1088 A.D. with a detailed description of rubbing the tip of a needle with lodestone, suspending the needle with a strand of silk and observing its north–south orientation. It is likely that the compass was first used for navigational purposes between the ninth and eleventh centuries A.D.

The great building feat of the period was the creation of the Grand Canal, linking the Yangzi with the Huang He and on into Beijing. This was the most incredible work of its age; it was 1200 miles long and 130 feet wide and could have only be conceived of and built by a powerful and ruthless emperor. It was built by corvée labour in six years of furious construction between 605 and 610 A.D. Over 5 million men and women were recorded as working on the first section of the canal alone. Originally intended for military use, it functioned as the Mediterranean Sea did for the Romans, and within a generation it had become China's economic artery, carrying rice from the south to feed the cities of the north. Running alongside and parallel to the canal was an imperial roadway and post offices supporting a courier system.

Buddhist monasteries continued to acquire wealth and prestige during the early Tang period. However they grew be too powerful and became a threat to the state administration and its Confucian scholars. In a memorandum to the emperor in 819 A.D. Han Yu (768–824AD) said, 'the

Buddha was merely a barbarian from the western kingdoms who recognised neither the loyalty that binds a subject to his ruler, nor the obedience due from a son to a father'. This is a reference to sons being committed at the age of nine or earlier to life in a monastery, cutting across the whole Confucian concept of family order. Calling Buddhism 'a cult of barbarian peoples', Han Yu energised a Confucian backlash. In 842 the Emperor cracked down on Buddhism and all foreign religions in China, including Nestorian Christianity and Islam. He ordered monks and nuns to pay taxes or return to their villages. Non-compliance meant defrocking monks, closing monasteries and plundering their assets. Like in sixteenth century England when Henry the Eighth closed the monasteries, an enormous amount of wealth was released; as much as 40% of China's wealth had been controlled by Buddhist institutions. Unlike the Chinese Christian and Muslim communities however, Buddhism survived the Emperor's purge, but it was never the same force again and was never again a challenge to the scholar class.

Confucianism, as a philosophy of life developed into its final form in the late Tang and Song Dynasty. Better described as neo-Confucianism, it integrated traditional Confucian teaching with Daoist and Buddhist ideas. Its great exponent was Zhu Xi (1130–1200 A.D.). Zhu Xi replaced ancient rites and supernatural forces in ancient Confucian texts with an abstract moral standard which he called the 'Supreme Ultimate'. In doing so, he created a more rational and secular form of Confucianism. The subtleties of the philosophy are difficult to explain succinctly. Zhu Xi believed that through investigation the scholar should come to understand the 'li', or the principle of the world and how to act correctly. His favourite text was from the *Great Learning* which describes how correct learning leads to a stable state and society.

> The extension of knowledge consists in the investigation of things. When things are investigated, knowledge is extended, the will becomes sincere; when the will is sincere, the mind is rectified; when the mind is rectified, personal life is cultivated; when personal life is cultivated, the family will be regulated; when the family will be regulated, the state will be in order; when the state is in order, there will be peace in the world.

The practical aspect of neo-Confucianism resulted in scholars concentrating on just four Confucian texts rather than the 13 previously recognised classics. Zhu Xi also wrote a manual for ritual and behaviour that became a popular classic used by ordinary people. It was as influential as any Confucian work ever published and centuries later would be found in homes throughout China.

Under the Song dynasty the Chinese Empire reached a new stage of wealth and culture. Underlying these changes was the increased fertility of the paddy fields of the Yangtze. The bounty which had been increasing for some 400 years soared again after 1012 when a new variety of early-ripening rice was created that allowed two harvests a year. This surplus allowed the population to grow from 60 million to 100 million over the next 500 years.

The economic advance of the Chinese over this period is explained in this excerpt from *Atlas of the Year 1000*, p. 88 by John Man:

> Reforms tackled economic and social problems, setting up orphanages, hospitals, schools and grain reserves. All this revolutionised Chinese society. Self supporting, patriarchal aristocratic estates continued to decline in importance. Ordinary people started tilling land for themselves, especially in the rich Yangtze valley. Taxes, once based on the number of people of working age, were now related to agricultural production. A new class of tenant farmers emerged.

The changes led to social and economic advances that have a remarkably modern ring. Peasants fled the countryside for the cities, seeking work in new industries – mining, metallurgy, ceramics, paper, printing and saltworks.

China was already more efficient at producing iron with its use of the blast furnace. Around 1000 A.D., due to the demand for charcoal in iron-making, the hills around the main production site, Kaifeng, began to become deforested. Through necessity the Chinese discovered how to use coal instead of charcoal in processing iron. Iron production soared to about 125,000 tons – almost as much as the whole of Europe would produce in 1700. Coal production grew to match that of iron and by 1050 coal was commonly used to heat houses, centuries before it would become available in Europe.

Trade blossomed and as southern ports expanded so too did China's merchant navy. Chinese shipbuilders learned from Arab and other visiting seafarers and added their own innovations: stern rudders, water-tight compartments and multi-masted ships with fore-rigging and aft-rigging. Their largest vessels were able to carry 600 tonnes and had crews of several hundred sailors.

The volume of coinage issued could not keep place with the expansion of trade. Traditional coinage was a metal disk with a hole in the middle which was issued in strings of 1000. In the year 1000 the state issued 1.3 million such strings. Such coinage is cumbersome. Printed credit notes were used by merchants as a convenient alternative during the Tang period. But proper paper money – promissory notes backed by the equivalent metal reserves – began in the Song period. To guard against forgery the notes were issued in three and, later, six colours. They were valid for three years. At the end of their term they could be replaced, but only at a discount. Thus they had a built-in inflationary effect. By 1126 the Song had issued notes to the value of 70 million strings of coins.

This was a period when China was harassed by pastoralist tribes. Song emperors were always on the defensive. Competition always drives innovation but, despite all their innovations, the Song emperors were never able to master the nomads. Harassed by a succession of pastoralist tribes, Tanguts, Khitans, Uighurs and Tibetans, north China eventually fell to the Jurchen. In 1126 the Jurchen sacked the capital of the Song empire, Kaifeng. The Song leadership moved south and set up a new southern kingdom with a capital at Hangzhou at the southern end of the Grand Canal and continued the brilliant Song civilisation. However from 1200 A.D. onwards, the Mongol invasions started. At first the northern pastoralist kingdoms were overwhelmed. Then Genghis' successor, Kublai, took up the fight against the Song and in 1276 Hangzhou fell. A great civilisation was stopped dead in its tracks by a brutal, ruthless enemy.

The Mongol empire and the transfer of technology

Expansion of Mongol power after Genghis Khan's death in 1227 continued under his third son, Ogidei. By the time of Ogidei's death in 1241 the Mongol empire included most of China, all modern day Russia and most the Middle East, excluding only Egypt and part of Anatolia. It was over twice as large as any empire seen up to that time. For a period of more than 100 years Mongols held the pivotal position between the great civilisations of the east, west and south. The Mongols produced few of their own goods and were reliant on others for many essentials. As a result Mongols had always supported merchants and trade. Merchants provided Genghis Khan with information about neighbouring states and served as diplomats. As the empire grew merchants received protection as they travelled through Mongol realms, and intercontinental trade prospered. The Silk Road from the Middle East to China was revived. Marco Polo travelled along the Silk Road and brought back fabulous stories of the East, which were met with disbelief in the West. However, it was rare to travel the complete journey from the Middle East to China. Merchants moved products like a bucket brigade, transferring them from one middleman to another, each adding their margin until the goods commanded exorbitant prices at their destination.

Trade also allowed technologies to be transferred from East to West. The transfer of paper technology predated the Mongol era. According to legend the Arabs first learnt the secret of papermaking from two Chinese prisoners from the Battle of Talas in 751. This led to the first paper mill in the Islamic world being founded in Samarkand. Certainly the Arabs were using paper by the ninth century. The first instance of paper-making in Europe was in the mid-twelve century in Muslim Spain. However, paper-making spread slowly through Europe and it did not really take off until the first printers started using woodblocks in the fourteenth century.

Knowledge of the use of gunpowder passed to the Mongols and from them spread independently to Europe, the Middle East and India. Thereafter canon and musket technology developed rapidly in all three centres of civilisation. Its first serious practical use in Europe was as cannon to demolish walls during a siege and to fire on infantry during battle. It was used at the battle of Crécy in 1346.

The use of gunpowder was instrumental in the expansion of the Turkish Ottoman Empire in the Middle East. The famous Janissary corps of the Ottoman army used matchlock muskets as early as the 1440s. The Ottomans were probably the first to use muskets aboard ships.

The first reported use of the magnetic compass in Europe was 1190. Improvements were made around 1300 when a dry compass was invented in which the needle swung freely in a box with a glass window. Compass directions were written into the box, allowing a relatively accurate indication of a ship's bearing. The use of the compass made for more accurate charts. In Italy we see the first of the pilots' charts or rutters, which were navigational notes based on compass directions and estimated distances observed by the pilots at sea. The use of these charts by the Portuguese led to the first stage of European exploration down the West Coast of Africa.

In the fourteenth century disunity among competing Mongol khans led to the gradual disintegration of central authority and by 1400 the empire was a mere rump of its former self. The fall of the Mongol empire led to the collapse of political, cultural and economic unity along the Silk Road. Trade and information no longer passed freely and the land-based route from East to West fell into decline. However before the East–West gate closed, it brought one of the most devastating pandemics in human history, the Black Death. The disease is thought to have travelled along the Silk Road from China or Central Asia. From there, it was probably carried by Oriental rat fleas living on the black rats that were regular passengers on merchant ships. Spreading throughout the Mediterranean and Europe, the Black Death is estimated to have killed 30 to 60% of Europe's population. Whole villages were wiped out. Towns were emptied through people fleeing to perceived safe havens. All in all, the plague reduced the world population from an estimated 450 million to between 350 and 375 million in the fourteenth century. In the end, for all the devastation caused by the Mongols in warfare, this paled into insignificance compared to the effect of facilitating the transfer of the Black Death.

Recovery and political cultures

During the fifteenth century, improved military technology using gunpowder with cannon and hand-held firearms spelt the end of the military advantage of highly mobile cavalry. Mongol power declined. They were ejected from China by the native Chinese Ming dynasty in 1368 and expelled from the Middle East by resilient Turkish emirs by 1400. This left the Mongols controlling none of the centres of civilisation, just modern day Russia and Siberia.

The age of triumph of the pastoralists was over. It was clear that of the great empires of the past, only the Chinese Empire had survived intact. However the old empires' religious cultures and philosophies lived on. The Catholic Christianity of the Roman Empire thrived in Europe, Hindu culture had developed in southern India, Neo-Confucianism had returned to China. Only Zoroastrianism had disappeared. It was replaced by Islam, which had established a predominant position in the Middle East, North Africa and North India. Even though the great empires of the past had disappeared, their religious culture had persisted and even grown in importance. The pastoralists may have destroyed cities, despised scholarship and generally murdered and pillaged, but the faith of the people had remained resilient, loyal to a limited number of core religious cultures.

Civilisation began to re-establish itself, regional and city-states began to grow and thrive again. The past experience of empires had been absorbed and one type of mediaeval society emerged which was adopted across the civilised world (see Figure 6.1).

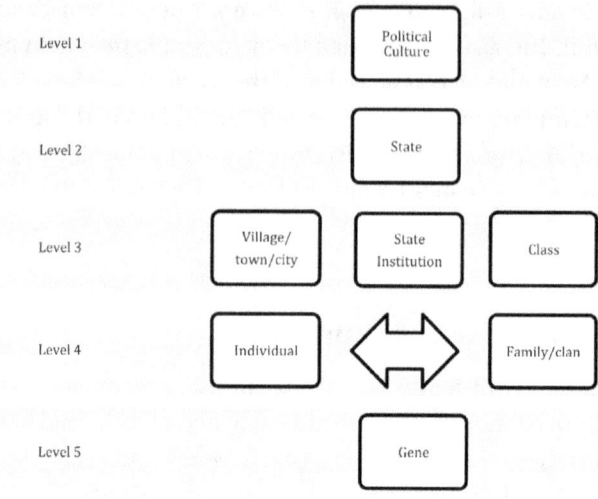

Figure 6.1 Evolutionary structure of mediaeval society

At the top level of the evolutionary structure were the communities of states sharing the same political culture. By a political culture I mean the underlying conventions and assumptions that define how individuals, classes and state organisations relate to each other and how power was exercised. It is a set of relational and cultural memes, a combination of the shared values of a community, often expressed in a religion or a philosophy of life, and the conventions of state administration. It covers shared beliefs, individual rights and responsibilities, social conventions, the way commerce is managed, the administration of the law and the organisation of government. The old empires had fractured into a number of states. However, having the same religious background, these states shared a common political culture. As a result four political cultures now dominated civilisation: Catholic monarchies, Muslim sultanates, Confucian empires and Hindu kingdoms.

Consider as example the political culture of a mediaeval Catholic monarchy. Their belief system was founded on the Catholic religion, whose priests were considered God's representatives on Earth. The words of the Bible were taken to be literally true. If you led a good life, you would be rewarded in heaven; otherwise you would be condemned to hell. Their responsibilities had developed from the feudal system. The peasants were bound to the land and owed a duty of service to their barons. The barons owed a duty to the king. Freemen were liable to tax by the state. A system of individual rights had built up over the centuries. Law was based on Roman law or common law. In both, there was an established principle that before being punished for a crime, there must be a legal process which examines the facts of the case. Civil cases, however, could be resolved locally according to precedent. Both the aristocrats and the church had special exemptions. Aristocrats could only be tried by their peers and the church had its own independent legal process to try and sentence the priesthood. Social conventions like eating fish on Fridays, attending church on Sundays and fasting during Lent were commonly derived from the church calendar. Commerce was strictly controlled. Trade fairs had to be approved by government statute and all movements of goods were subject to tax. In the towns, craft guilds operated a closed shop with specific rules for apprentices, journeymen and masters. The state was governed by the king and his appointed council. All laws had to be sanctioned by the king. If the king had sons the eldest son would automatically inherit the throne on his death. Church matters were separately administered by the Church and large cities would be governed by a city corporation established by royal statute. Otherwise local administration was in the hands of the barons.

Even though there were many Catholic monarchies in Europe at this time, it is my contention that they shared this same political culture. Similarly, there was a common political culture for the many Islamic sultanates or emirates. Memes developed within states were quickly copied by others of the same political culture. From this point onwards, even as states battled with other states, the ultimate deciding factor in the direction of human evolution would be determined by which of the developing political cultures would be successful.

Only one other major change has been made from the structure shown for empires. Clans were becoming less significant as a competitive organisation. They were becoming more of a network of contacts useful for the young and ambitious to obtain training and support in developing their careers. I have therefore shown the individual and family or clan on the same level as mutually supportive entities.

Most states were monarchies. The few city–states had the same mediaeval evolutionary structure as monarchies but were ruled by an oligarchy of leading families. Artisan guilds formed played an important role in city administration.

Outwardly the four different political cultures that emerged, Catholic monarchies (including city–states), Muslim sultanates, Confucian empires and Hindu kingdoms were similar. However, the relational memes between classes had evolved differently, which had important implications for the future. In Catholic monarchies, the Catholic Church was politically powerful. The church was a substantial property owner through its network of monasteries. In addition, some states were directly run by bishops and the pope himself was the temporal sovereign of a territory around Rome. The church was an independent centre of power within the state. In addition, the aristocracy was anxious to maintain local independence. Cities also jealously guarded their hard-earned privileges. As a result, class competition in Catholic monarchies was very strong. Agreements on rights and privileges, such as the Magna Carta signed between King John and his barons, formed the basis of a modus operandi between the classes. There began to emerge independent legal systems to make judgements of rights in the case of disputes. This degree of independence of legal systems was unique to Catholic monarchies; in all other political cultures the power of the state was not controlled by any independent legal process. In Catholic monarchies the monarch did not have a totally free hand to exercise his power. In many countries the monarchs even set up parliaments to consult with the classes before imposing new laws and taxes.

The monarch was also bound by religious laws set by the pope. Kings were allowed only one wife, not a harem, as in other parts of the world. Kings were not allowed to marry cousins or divorce their wives without approval from the pope. The church's ultimate threat was excommunication, being banned from the Church, and hence risking eternal damnation. In those times superstitious religious belief was profound and the fear of the supernatural intense. The threat of excommunication was a real deterrent.

Succession was governed by the rule of primogeniture; the first born male inherited the throne. As a result contested succession to the crown occurred less often than in other political cultures.

Although Latin was the lingua franca of Europe, each state had its own vernacular language. In the case of France, England and Castile, just one local language gained common usage. This led to a greater bonding between rulers and subjects and, as a result made some states of Europe more robust entities with more significance than a collection of noble land-holdings.

In contrast Muslim sultanates were governed by Turkish or Mongal (Mughal) rulers and aristocrats. The local peasants were of a different stock. As a result, the bonds between subject and ruler did not exist in the same way in Muslim states. Neither the religious scholars nor the cities had the same access to power as in Europe; the class war had been won by the aristocracy during the pastoralist invasions.

As in Christianity, superstition and religion dominated village life. However through Shari'a law, Islam came not only to dictate peasants' religious life but also to provide a legal framework for resolving disputes. Sunni Islam is a community-based religion. People were trained to read

Arabic in mosques. Some went on to become imams or religious scholars. To a great extent the communities in Muslim countries were self-governing and the imams acted as judges for individuals breaching the law. Ibn Battuta, the famous fourteenth century Berber scholar and traveller from Tangier, was able to pass freely throughout the Islamic world and be recognised, received and consulted by the authorities for his scholarship. He was able to act as a magistrate in a place as remote as the Maldives, due to the common acceptance of Shari'a law. Although the sultanates did not adopt Shari'a law in its entirety, they had to observe its tenets, and in most cases the sultans were content to collect their taxes and leave the communities to run themselves.

The lack of a central independent religious authority, however, meant that the sultans did not have to share overall power with the religious authorities. This left them free to pursue their own whims. There was no possible legal redress to any decision they made. Islam allowed men to have more than one wife. Sultans had their harems and their eldest born did not necessarily inherit the throne. The resultant machinations of palace intrigues, eunuchs and contested successions seriously weakened the state.

At the same time, Islam did not display the same abuses of power as Christianity and therefore was not as vulnerable to the same backlash as would occur in the Reformation in Europe.

China was the one remaining empire to survive the invasion of the pastoralists. There was none of the competitive jousting between states as in other political cultures. The Chinese state was totally centralised; all power resided in the emperor. The emperor in China had the mandate from heaven; he was both the temporal ruler and the high priest. No independent state religion was recognised and no priest class held power. Most nobles did not control their estates and were dependent on state tax collection for their income. The class struggle had been won by the scholars who administered the empire. They were selected by a sophisticated exam system based on their ability to absorb and express neo-Confucian logic. The law was also administered by the scholar class, who were known as mandarins in the West. Local mandarins had considerable autonomy to manage their own affairs. The only opportunity for redress against a mandarin was to appeal to the emperor, which was a long, costly and uncertain process.

Chinese emperors had many consorts who lived in harems with the attendant eunuchs. Eunuchs became very important in Chinese society. They had direct access to the emperor and were the only effective political counterweight to the scholar class. The success of the empire depended totally on the ability of the emperor. A strong emperor could harness the power of the state and achieve great success. A weak emperor could lose control of his administration and local corruption would increase.

In Hindu kingdoms, the lack of any strong central administration meant that local communities were largely self-governing. The fundamental unit of Hindu society was the village. Taxes were generally paid collectively in a manner organised by the village headman. The beneficiaries were the nobles (zamindars) who had a hereditary right to village revenues.

The Hindu religion had institutionalised a class structure. The Brahmin class maintained their place on top of the social scale, acting as high priests. Below this level, the organisation of the other three Varnas was more complex, as they were further subdivided into thousands of Jatis (castes). These Jatis could be considered as professional groups who exercised a monopoly on their craft, similar to the guilds in mediaeval towns in Europe. The distinct difference between them was that only those of the same caste or kin were allowed to take up the craft. The chandellas (untouchables) remained at the bottom of the social structure.

All this made for a very rigid society with restricted opportunity for social advancement and change.

Summary: situation in 1450 A.D.

Civilisation had survived the depredations of the pastoralists. One evolutionary structure and four political cultures now dominated the world. Neo-Confucianism was the political culture of the East. The Ming dynasty had re-established central native Chinese rule and, with the powers of the pastoralists in decline, China was set for a period of stability with increasing wealth and population growth. Beijing was the largest city in the world with a population of about 650,000. Local variants of neo-Confucianism also formed the political culture of Vietnam, Korea and Japan.

In south-east Asia both Buddhist and Hindu culture had been spread by traders and the first native states with writing skills appeared there in the seventh to ninth centuries A.D.: Pagan in Myanmar, Khmer in Cambodia, Cham in Vietnam and Srivijaya in Indonesia. One Hindu state, Vijayanagara, dominated south India, Its capital, of the same name, had a population of 500,000. Today Vijayanagara is deserted and is one of the great ruined cities in the world. Buddhism had lost out in India to Hinduism.

Muslim sultanates dominated the Middle East and north India. Led by Turkish elites imbued with Persian culture, the region was recovering from the turmoil of the Mongol invasions. They had entered a transitional stage of regional state formation as a prelude to the great empires in the sixteenth century. Cairo in Egypt with 400,000 people was the largest city and the capital of the Turkish Mamluk sultanate. From the fourteenth century onwards Arabic traders spread Islam to Malaysia and Indonesia replacing Buddhist and Hindu kingdoms.

Catholic Christian monarchies dominated West Europe. Parts of East Europe and Russia still recognised the spiritual authority of Orthodox Christianity, with its patriarch in Constantinople. Orthodox Christianity had declined in importance as the Byzantine Empire had shrunk to a rump state around its capital.

China, Europe, Middle East and India had access to the same technologies, with the key exception of printing, which was still only present in China. In Europe and the Middle East there is no doubt that there was less wealth and scholarship than in the great days of the Roman Empire. However, ancient Greek and Roman knowledge had still managed to cling on in books translated by the Muslims.

Despite the vicissitudes of the pastoralist invasions and the Black Death, world population had increased over the past thousand years. By 1450 A.D. the population of the world had reached 400 million, about the same as the population of South America today. Life expectancy had increased marginally to around 25 years.

In one final flourish of success, tribal societies reached the last large uninhabited islands on the globe: Iceland in the ninth century and New Zealand around 1250 A.D. However, with the spread of civilisation to south-east Asia, most of Europe, Asia and North Africa were now controlled by mediaeval state societies. From this point on it is these states that are the dominant political organisation in Asia and Europe.

CHAPTER 7:
SCIENCE AND THE SECULARISATION OF THE WEST (1450–1780)

The next stage in human development was led by the Catholic monarchies of Europe. After the Black Death a shortage of labour caused the old feudal system to begin to break down. More agricultural workers paid rent rather than give days of labour to their lord; some became independent yeoman farmers. This resulted in much greater agricultural efficiency. Farmers were now helped by horse power. Selective breeding, the improvements in harnesses and the development of horseshoes allowed carthorses to pull heavier ploughs. More effective than oxen, horses began to be widely used by all sections of society, not just aristocrats. The level of manufacturing increased to supply the new technologies imported from the East. Iron was required for making the new canons and muskets; paper mills were needed to meet the explosion in demand for printed materials. As the level of manufacturing and trade began to increase an independent middle class of merchants, manufacturers, bankers and shopkeepers began to develop. These were people who owned property and could afford to buy food produced by others. This new independent urban class was commercially innovative and were the prime drivers for the development of the memes that would eventually allow Western culture to dominate the world.

For the first time since the first millennium B.C., economic competition between states became important. It led to a huge increase in wealth and power of Western nations and a gradual urbanisation of society. Table 7.1 shows the estimated population of West Europe and its major cities from 1500 to 1800. Overall, the percentage living in the major cities increased by 80 percent from 2.9 to 5.3% over this period. As noted in Table 6.1 in the previous chapter, an increasing proportion of the population living in the major cities is a measure of expanding trade and wealth. There were few official censuses in those days and countries have changed geographical areas, so inevitably the figures have to be viewed with caution. However, the data show a picture that I shall be developing over this chapter. Different countries at different times developed new memes that allowed them to become wealthier and hence more urbanised. Firstly, Italy had the advantage with trading and banking skills. Then the Iberian countries advanced with the development of seaborne trade and the discovery of America. Then France, with its land-based power and the Netherlands, with its commercial strength, became more urbanised. Finally, the British and their American colonies with their more democratic processes start to develop the infrastructure that presaged the Industrial Revolution. Rather than introduce all the new memes at the start of the chapter, new memes will be highlighted in bold as they are reviewed in the text.

Italy	1500	1600	1700	1800		Benelux	1500	1600	1700	1800
Naples	150	281	216	427	Antwerp	40	47	70	60
Venice	100	139	138	138	Ghent	40	31	51	51
Milan	100	120	124	135	Brussels	35	50	80	74
Florence	70	70	72	81	Bruges	30	27	38	32
Genoa	60	71	80	91	Amsterdam	14	65	200	217
Rome	55	105	138	163					
Bologna	55	63	63	71					
Palermo	55	105	100	139					
Total 8 cities	645	954	931	1245	Total 5 cities	159	220	439	434
Total Italy	10500	13100	13300	19000	Total Benelux	2350	3100	3900	5400
% cities	6.1	7.3	7.0	6.6	% cities	6.8	7.1	11.3	8.0

Iberia	1500	1600	1700	1800		United Kingdom	1500	1600	1700	1800
Granada	70	69	62	55	London	40	200	575	865
Valencia	40	65	50	80	Birmingham				71
Lisbon	30	100	165	180	Manchester				75
Barcelona	29	43	43	115	Liverpool				82
Cordoba	27	45	28	40	Glasgow	3	6	12	70
Seville	25	90	96	96	Edinborough	12	30	60	80
Madrid	0	49	110	167					
Total 7 cities	221	461	554	733	Total 6 cities	55	236	647	1243
Total Iberia	7800	9340	9770	14500	Total Iberia	3942	6170	8565	15500
% cities	2.8	4.9	5.7	5.1	% cities	1.4	3.8	7.6	8.0

France	1500	1600	1700	1800		Germany/Austria	1500	1600	1700	1800
Paris	100	220	510	581	Nurembourg	36	40	40	27
Lyon	50	40	97	100	Cologne	30	40	42	42
Rouen	40	60	64	81	Lubeck	24	23	30	40
Marseille	20	40	80	110	Danzig	20	50	50	40
Lille	10	20	35	53	Augsburg	20	48	21	28
Bordeaux	20	40	50	88	Vienna	20	50	114	231
					Munich	10	15	24	50
					Berlin	10	9	50	172
Total 6 cities	240	420	836	1013	Total 8 cities	170	275	371	630
Total France	12000	16000	15000	23000	Total	14000	18500	17500	22500
% cities	2.0	2.6	5.6	4.4	% cities	1.2	1.5	2.1	2.8

	1500	1600	1700	1800
Total 40 cities	1490	2566	3778	5298
Total countries	50592	66210	68035	99900
% cities	2.9	3.9	5.6	5.3

Table 7.1 Urbanisation in Western Europe – population in thousands

Italians and banking

Table 7.1 shows that by 1500 the top eight cities in Italy accounted for more than 6% of the population. The city-states of Venice, Genoa and Florence were the most important commercial centres in Europe. Venice and Genoa had established themselves as the predominant maritime powers in the Mediterranean; they were the principal European intermediaries in the trade of spices with India and silks and other materials with China. Florence was the centre for banking.

It was in the field of **banking** that the Italians provided the longest lasting and most valuable memes. Firstly Fibonacci (1170–1250) introduced Europeans to simpler ways of financial accounting, using Arabic rather than Roman numerals. Fibonacci was the son of a Pisan merchant based in a port in what is now Algeria. As a boy working with his father, he recognised the advantages of using Arabic numerals in arithmetic. Fibonacci travelled throughout the Mediterranean and studied under the leading Arab mathematicians of the time. On his return in 1202, at the age of 32, he published the *Liber Abaci* (Book of Calculation) which is the basis of modern arithmetic. It showed how Arabic numerals were so much easier to use in fractions and long multiplication than Roman numerals. Being the son of a merchant he demonstrated its use with practical examples from the world of trade, including bookkeeping, the conversion of weights and measures, the calculation of interest and money-changing.

Arabic numerals were soon widely used in banking. The original European banks were merchant banks managed by Jews who lent money as an advance to farmers in the Lombard plains. The Jews had one great advantage over the locals. Christians were strictly forbidden to practice usury (charging interest on loans). Jews were also forbidden to commit usury but there was a get-out clause; it states in the book of *Deuteronomy* that 'unto a stranger thou mayest lend upon usury'. In other words, a Jew could lend to Christians but not to another Jew. The native Lombard banking trade grew when it was able to avoid the practice of usury by advancing payment on the security of a future delivery of grain. The bank made its profit by discounting the sale price rather than charging interest on the loan.

Long-distance trade was made much easier by the development of **bills of exchange**. Bills of exchange are documents guaranteeing the payment of a sum of money once a trade transaction has been successfully completed. These bills made it possible for merchants to transfer large sums of money without the complications of hauling around large chests of gold and hiring armed guards to protect the gold from thieves. The system extended out of Lombardy to cover long-distance trading at the mediaeval trade fairs. The most famous fair was at Troyes in the Champagne area of France, which acted as the trade crossroads of Europe. Troyes gave its name to a standard system of weights and measures that was in widespread use in Europe up to the early nineteenth century.

In the middle of the thirteenth century, groups of Italian Christians invented legal ways of getting around the ban on usury, for example, by offering a loan with a condition that insurance was also taken out at a cost equivalent to the interest. The Christians practising these legal fictions became known as the pope's usurers, and reduced the importance of Jews as the money lenders to European monarchs. In 1403 charging interest on loans was ruled legal in Florence, despite the traditional Christian prohibition of usury.

The basis of good banking is the meticulous keeping of accounts and the control of risk. The Medici Bank in Florence was one of the first to use double-entry bookkeeping to track debts and credits and ensure they matched. Risk was partly allayed by the spread of their operation. With branches in the major centres in Italy, London, Bruges and Lyons, they were truly international bankers. From the founding of the bank in 1397 to about 1460, the bank managed risk well. Banking was clearly a very profitable business and in that short time the Medici became one of the wealthiest families in Europe. They joined the ranks of Renaissance princes as the leaders of Florence. However, the good times did not last; a number of poor loans and consequent defaults, in addition to Lorenzo de Medici's expenditure on art and self-aggrandisement, meant that by 1494 the bank was effectively insolvent. The Medici Bank's remaining assets and records were seized and distributed among its creditors; all the branches were dissolved. However, the innovations of the Italian banking system became a model for later banks in north Europe and would support the trading empires of the Netherlands and Britain in future generations.

Portugal, ships and navigation

From 1300 onwards instead of supplying north Europe overland from Italy, Italian merchants sent their goods to north Europe by sea, linking with already well-established sea routes in the North Sea and the Baltic. This started the Atlantic sea trade, which was later transformed by the Portuguese, a tiny nation on the Atlantic fringe of Europe with a population of no more than a million people. Beginning around 1415, Portuguese ships explored and traded further and further down the coast of Africa until in 1488 Bartolomeu Dias rounded Cape Horn and entered the Indian Ocean. Thereafter, they established fortified trading posts in Iran, India, Sri Lanka and Malaya and by the mid-1500s were trading with China and Japan. It is an astonishing story. What gave them the confidence to attempt such feats of arms and seamanship?

It appears to be partly a matter of religious motivation and partly one of greed. The religious motivation came from the continuation of the crusading zeal against Islam that led to the re-conquest of the Iberian Peninsula. In 1415 they crossed the Mediterranean and took the town of Cueta in North Africa,. In the event the local states were too strong militarily and the conquest could make no further progress on land. However the expertise acquired in the development of ships suitable for carrying the army to Morocco allowed further exploration of the African coast and the Atlantic Islands. The economic opportunities that this opened up were seized upon by the Henry the Navigator, the third son of King Joao I of Portugal. He gained a number of monopoly trading rights that allowed him to set up a fort in Mauretania, trading in slaves and gold, and to settle Madeira, initially for logging (the Portuguese word for wood is madeira) but later to produce sugar cane. He showed that there was profit to be had by trading outside Europe. On his death many of the monopoly rights he had won were taken back by the crown, which proceeded to sponsor voyages further down the coast of Africa, culminating in Vasco da Gama's arrival in Calicut in India in 1498 after rounding the Cape of Good Hope. This made it possible for Portugal to take over the Asian spice trade to Europe by supplying it directly using the newly discovered oceanic route, cutting out all the profits taken by middlemen.

Bizarrely though, it appears the chief motivation in the mind of King Joao II and his son Manual in sponsoring these ocean voyages was some sort of pincer movement on Muslim territory by coordinating anti-Islamic operations with the mythical king, Prester John. The legend of Prester John as a Christian king in the East had been told in Europe since the twelfth century. In the early days of exploration Portuguese sailors were told to ask about the whereabouts of the kingdom of Prester John wherever they landed. They managed to reach Ethiopia, the assumed home of Prester John, in 1521, only to find the state of the kingdom in no condition to fulfil their crusading hopes.

Thus far, after the death of Henry the Navigator, the Portuguese crown had financed the voyages of exploration around Africa and in the Indian Ocean. A small country like Portugal could not possibly provide the investment necessary to exploit the Asian trading opportunity. Over time Portugal concentrated on setting up a number of trading posts and selling the right to trade to individual merchants. This allowed the private sector to finance the required increase in shipping at the same time as meeting the crown's need to raise revenue. There was a dramatic increase in private Portuguese merchants trading on their own behalf. Perhaps the acme of their feats was to set up a triangular trade route between India, China and Japan. Starting in Goa, European and Indian goods were taken to Macau to be traded for Chinese gold and silk, which were taken to Nagasaki and exchanged for Japanese silver; then the ships returned to Macau again to trade the silver for more Chinese goods for India and the West. At one stage this involved 200 or more private traders, all from the tiny nation of Portugal, trading with the world's major empires on the other side of the world.

This Portuguese success was based on **technical developments in ships and navigation.** In mediaeval Europe ocean-going ships were simple and robust,with a single mast and a square-rigged sail. They were too slow and cumbersome for long-distance travel. Prince Henry the Navigator sponsored the development of a new design of ship known as a caravel. This design was based on that of local fishing boats and used lateen sails (triangular sales set at right angles to the mast) to make it easier for the ship to sail into the wind. Caravels were used by the Portuguese for oceanic exploration during the fifteenth and sixteenth centuries. However these were small, light ships and did not have much cargo-carrying capacity. A larger ship, the carrack, was developed with more masts and sails; it was usually square-rigged on the foremast and mainmast and lateen-rigged on the mizzenmast.

Carracks were ocean-going ships, large enough to be stable in heavy seas and roomy enough to carry the provisions needed on long voyages. This was the design of ship in which the Spanish explored the world in the fifteenth and sixteenth centuries. The carrack was one of the most influential ship designs in history and though other larger, more specialised ships were developed, the basic three or four-masted design remained constant throughout the age of sail.

The Portuguese also made major progress in navigation, developing new instruments and improved charts. We have already seen how the invention of compasses and the use of rutters helped navigation. However, mariners still needed to calculate their position in terms of latitude and longitude. In the northern hemisphere the latitude could be estimated from the altitude of the pole star, but this was of little use south of the equator. The Portuguese developed the mariner's astrolabe for measuring the angle of the sun to the horizon at midday. Using tables constructed by the astronomer Zacuto in the 1470s, they were able to calculate the ships' latitude anywhere in the world. The mariner's astrolabe was used by Vasco da Gama when he sailed to India in 1497. However, the measurement of longitude remained imprecise until the eighteenth century.

As European knowledge of world geography grew, atlases and charts of the newly discovered lands were developed. However, there was a problem in how to represent a spherical surface on a flat piece of paper. Mariners needed to plot a compass bearing that was a straight line. This was solved by the Flemish mapmaker, Gerard Mercator, in 1569. In a Mercator projection, parallels of latitude and meridians of longitude cut each other at right angles. Bearings can be calculated directly from the map.

Spain and America

In the late fifteenth century trade with Asia was not only a goal for the Portuguese. The gains to be won by opening up the oceanic route to Asia were very obvious to the Italians; they had been the principal European beneficiaries of overland trade in the fifteenth century. One man, Cristoforo Colombo, from Genoa in Italy, had a novel idea for getting to Asia. He envisaged avoiding the long journey around Africa by sailing west and reaching Asia from the other direction.

It seems that it had long been known that planet Earth was a sphere; the Greeks in the third century B.C. had even worked out the length of its circumference. However, up to now no-one had dared to sail into the western ocean towards Asia with only enough food and water for one direction of the journey. Though he was not a learned man, Cristoforo had studied relevant books and atlases and believed it could be done. This was because he had falsely deduced that the distance from the Canaries to Japan was about 3700 kilometres, whereas the true distance was 19,600 kilometres. Cristoforo had begun by working for a Genoese family involved in the oceanic trade from Italy to North Europe. He later traded down the coast of Africa among the Portuguese. He had lined up half the capital required from Italian investors but now needed royal sponsorship. The Portuguese turned him down, but the Spanish provided the rest of the funding. In 1492 he set off for India and duly discovered West India (or, as we know it, the West Indies) and its native population of Indians. It is unclear when he realised that India was still many thousands of kilometres away and that he had actually discovered America.

The initial result of subsequent trading and settling in the West Indian islands was modest. However, all that changed when two small bands of rapacious adventurers lead by Hernán Cortés and Francisco Pizaro conquered respectively the Aztec empire of Mexico in the 1520s and the Inca empire of Peru in the 1530s. It was a case of the Iron Age meeting the Stone Age and the Native Americans simply could not compete in battle against the iron weaponry, body armour and horses of the Spanish. Nevertheless one has to admire the sheer chutzpah of these conquistadors; the story of the intrigue and duplicity of how they both won through is truly amazing. What gave them the confidence to act as they did in a foreign country far from any support from their own countrymen? Greed and the lure of gold, certainly played a part, but there was also arrogance, a belief that whatever they were doing was right. They had little understanding of or interest in the cultures they met. The Native Americans were barbarians; they did not know the true religion and therefore deserved and received brutal treatment from the 'civilised' Westerners.

The reward for the Spanish was gold and silver. Soon the lure of gold brought other Europeans, principally the Portuguese, British and French. The results for Native Americans were catastrophic. Amerindians did not have immune resistance to diseases like measles, smallpox, influenza and typhus. These were Eurasian diseases that had developed from domesticated animals. The Caribs, the inhabitants of the West Indies at the time of Columbus, were wiped out. Incipient civilisations in south-east USA and the Amazon disappeared. Altogether, it has been estimated that the Amerindian population fell over 90% in the two centuries following the arrival of Columbus. This had significant implications for the colonisers of America. As there was very limited native resistance to the new settlers, the newly arrived Westerners were mostly competing among themselves for territory. Unlike the Portuguese in Asia, who were limited to setting up a number of trading centres, the Spanish, British and French were able to set up farming communities throughout America. The 'discovery' of the New World vastly increased the West European cultural sphere of influence. Two continents, North and South America, had been depopulated by Western diseases and had essentially become virgin lands, ripe for development. Australia and, to a lesser extent, New Zealand would later suffer the same fate.

Depopulation also meant that there was little acclimatised local labour available who were accustomed to work in strong sunlight. African labour was suitable and African coastal kingdoms had the ability to capture slaves. Soon the Portuguese, Dutch and British were shipping huge numbers of slaves into America to act as farm labourers.

Protestantism: scholarship and the Reformation

Just as the Catholic monarchies reached the height of their powers, the Catholic Church lost its monopoly on religious practice in Europe. The immediate cause was the development of printing. The resulting spread of information and knowledge created a new level of European intellectual activity which challenged the authority of the Catholic Church. Eventually this was to lead to a new type of political culture in which rationality ruled over superstition. The origin of this more rational approach had begun earlier with the creation of universities and the education of a new lay breed of scholars, administrators and lawyers. Before universities were formed, the only higher education available was in cathedral and monastic schools. The University of Bologna broke the mould in 1088. It began as a law school teaching Roman law. From the start Bologna was an autonomous academic institution awarding its own degrees and compiling its own courses, making it the oldest continuously operating institution independent of kings, emperors or direct religious authority. Paris (c. 1150) and Oxford (1167) copied the idea and as the Middle Ages advanced, universities were founded all across Europe as rulers and city governments vied to attract the best scholars.

In Europe, rich young men proceeded to university where they completed their study of the trivium, grammar, rhetoric and dialectic (logic) and the quadrivium, arithmetic, geometry, music, and astronomy. The universal language was Latin and the direction of learning was strongly influenced by the rediscovery of Greek texts, particularly the works of Aristotle. The general teaching style was known as scholasticism. This focused on preparing men to be doctors, lawyers or professional theologians. Based on the disputations of ancient Greek philosophers, scholasticism placed a strong emphasis on dialectical reasoning. It often took the form of a question, counterproposal and then rebuttal. Its subjects were often traditional and the reasoning was pedantic. By modern standards, the study was dry and sterile, producing little of benefit to humanity; Aristotle's view of science was a dead end and the medical knowledge was not far removed from quackery.

Tentatively, scholasticism began to attempt to reconcile Christian theology with classical philosophy. This work accelerated towards the end of the fifteenth century when a new style of learning, Renaissance humanism, began to evolve. This sought to teach students to be able to speak, write and reason with eloquence and clarity and thus to engage successfully in civic life. This was to be accomplished by teaching the humanities: grammar, rhetoric, history, poetry and moral philosophy. Renaissance humanism emphasised logic and reason, drawing inspiration and strength from the Greek philosophers. Initially the Catholic Church was supportive of the new style of learning. Humanist scholars went back to original texts of the Bible to discover the true meaning of obscure Latin translations. This combination of logic and study of biblical texts would eventually set the scene for the Reformation.

The arrival of printing dramatically increased the pace of humanist enquiry. **Printing with movable typefaces** arrived in Europe at the end of the fifteenth century. Johannes Gutenberg produced the first major work – a Latin Bible in 1453 in Mainz. It was immediately obvious that this was an epoch-changing event. Books, that formerly were the reserve of nobles and bishops, could now be made available to anyone who could read. From a single print shop in Mainz, printing had spread to no less than 270 cities in Europe by the end of the fifteenth century and the printing presses had already produced more than 20 million documents. Individual authors could now produce bestsellers; Erasmus (1466–1536), the most well-known Renaissance humanist, sold 750,000 copies of his published works during his lifetime. The church immediately lost its monopoly on the dissemination and interpretation of knowledge. For the first time ordinary people were free to read and judge for themselves.

Translations of the Latin Bible had been made into the vernacular before printing was available; the translations had been declared heretical by the Church, anxious to maintain its monopoly on the propagation of religious knowledge. In the early days of the Reformation, the impact of newfound availability of books and pamphlets took princes and the papacy by surprise. In the period from 1518 to 1530 original printed translations of the Bible were made available in German, Dutch, French and English. All were banned by the Catholic Church. The difficulty for the Catholic Church was that not only did different translations give a different insight to the text, but the translators also had access to the original Greek texts which exposed errors in the existing Latin version.

The Catholic Church at this time was powerful and wealthy but, it seems, was not content with its enormous income from tithes and monasteries. Its most profitable additional source of income in the Middle Ages was the exploitation of the idea of purgatory. Purgatory, as a sort of assessment centre before a person's spirit would go to heaven or hell, has a long history in the human imagination. However, this idea was refined in the later Middle Ages. It was envisaged that some souls were not wicked enough to be destined for hell and had not shown enough remorse for their sins to be allowed into heaven. Such souls, ultimately destined to be united with God in Heaven, must first endure purgatory, or purification. Purgatory could be remitted only by suffering penance during life, buying indulgences from the Church or paying for masses and prayers to be said on the sinners' behalf after death.

The first to exploit this idea were the monasteries. Monks offered to say masses for the souls of their financial patrons after their death. Later, rich people were encouraged to make provision in a will for priests to pray for them in a specific building (chantry). Not content with income from prayers for the dead, Pope Leo X authorised the sale of indulgences (remissions from sins) to help finance the renovation of St Peter's Basilica in Rome. As a result, in Germany, when parishioners came to confession, they presented their indulgences and claimed they no longer had to repent of their sins to avoid purgatory. The fact they had paid money for what was theirs by right as a free gift from God outraged an academic priest from Wittenberg, Martin Luther. He felt compelled to expose the fraud and announced a public debate on the issue by posting a memorandum on the 'Ninety-Five Theses on the Power and Efficacy of Indulgences' on the door of the castle church of Wittenberg.

This act lit the touch paper of a powder keg of grievances against the venal practices of the Catholic Church. Within a short time a number of new breakaway religious sects had formed, none of them recognising the authority of the pope. All these sects were able to access the Bible in their vernacular language and could take their own view on the most appropriate way of worship. Their religion was collectively known as **Protestantism**.

The Catholic Church reacted violently to this challenge, demanding that the Protestants be arrested and charged with heresy. Heresy is a particular issue of the Christian Church. In religions such as Judaism and Islam, doing the right thing (orthopraxy) is more important than espousing the right beliefs (orthodoxy). The issue of heresy goes right back to the early days of Christianity, when divisions arose as to whether Jesus was a god or merely a prophet. This is when the three in one formula of the Holy Trinity (God the Father, God the Son and God the Holy Ghost) was devised and determined to be the true belief. Many groups in Syria and Egypt were unable to accept this compromise, and according to Gibbon in his *'The History of the Rise and Fall of the Roman Empire'* the divisions that resulted were a cause of the failure of Byzantium.

In the thirteenth century the Catholic Church set up the Inquisition, a special institution to investigate whether those holding unorthodox Christian beliefs were guilty of heresy. This was a brutal institution, comparable to the secret police in the communist or fascist states in the twentieth century. Suspects were tortured until they confessed, then forced to recant. One famous example of false wrong thinking was Galileo's arrest for suggesting that the earth circulated the sun. Galileo chose to recant rather than face torture. He was lucky that he was not guilty of more serious heresies. Protestant believers, in countries where the Inquisition was allowed to operate, faced death by burning at the stake.

Eventually individual states chose whether to be Catholic or follow a particular Protestant belief. At the time Germany was a collection of many independent states under the nominal control of the Holy Roman Emperor. The Emperor, Charles V, was an ally of the Catholic Church and also happened to be the King of Spain. He tried to reimpose Catholic beliefs in Germany through war and failed. At the peace of Augsburg in 1555 the rulers of the German states were allowed to nominate whether to have the Catholic or the Lutheran Protestant faith as their state religion. The monopoly of Catholic Christian power in Europe had been broken, but the age of personal religious freedom was still far away. Citizens had to either accept their leader's choice of religion or emigrate. Religion and power were so tightly knit that no state could yet tolerate a sizable religious minority in their countries.

This was just the start of a number of religiously inspired wars. The Protestant Dutch rebelled against the Spanish. The French suffered civil wars between Catholics and the Protestant Huguenots, and war in Germany started again. The Thirty Years War (1618–1648) developed into a broad regional conflict and became one of the most brutal wars in European history. Huge changes in military technology had taken place in the last hundred years. The **musket** had become the main infantry weapon. Battles between columns of infantry now involved shooting at a distance as well as hand-to-hand fighting. With a bayonet attached to it, the musket also replaced the pike in defence against cavalry. Specialised archers disappeared, to be replaced by **mobile cannon**; a tactic first used with destructive success by the Swedes at this time. The combination of lethal military might and religious zeal in the Thirty Years War led to the extensive destruction of entire regions by foraging armies and subsequent episodes of famine and disease. The German population declined by a third. At the end most of the combatant powers were bankrupt.

At the following Peace of Westphalia, Calvinism (a more fundamental form of Protestantism) was recognised as a permitted religion and the right for Germans to practice the religion of their choice was established. Christians living in German states where their religious affiliation was not to the established church were guaranteed the right to practice their own faith. This was the first step on the way to the modern secular state. Needless to say the Catholic Church was not pleased; Pope Innocent X called it 'null, void, invalid, iniquitous, unjust, damnable, reprobate, inane, empty of meaning and effect for all time'.

Science

The wars of religion were a negative result of the arrival of printed books in the West. On the positive side it enabled the start of modern science, the systematic study of the natural world through observation and experiment. The Greeks had initiated scientific discovery but it had stalled. Without paper and Arabic numerals, the ability of a scientist to record and calculate results from large numbers of observations had been limited. Without printing, the dissemination of scientific ideas had been slow and expensive. Printing gave individuals the ability to record facts, express complicated ideas and circulate them to a wide audience.

The first scientific discipline to develop was astronomy: in 1515 Nicholas Copernicus observed the earth had an eccentric orbit around the sun; Tycho Brahe (1546–1601) made the first accurate observations of the planets; Johannes Kepler (1571–1630) showed that the planetary orbits were elliptical; Galileo Galilei (1564–1642) improved the **telescope** for astronomical observations, and was the first to observe the satellites of Saturn. Then came Isaac Newton (1642–1727). His three laws of motion laid the foundation for classical mechanics. He first identified a force called gravity that acts between any two bodies and this force varies according to their mass and the inverse square of the distance between them. He then went on to prove that this force explains the elliptical orbits of planets around the sun. In doing so he laid the foundations for mathematical astronomy.

It is hard to understate the significance of this event. Man the animal had, using the language of mathematics, managed to describe the physical forces of the universe. This language could be used to predict what would happen when anything was subject to a force: the distance a shell would travel when fired, the frequency of oscillation of a pendulum, an eclipse of the sun. There was no reference to god in this, no possibility for god, for instance, to send in a comet to show his displeasure of the actions of mankind, as had been believed in the past. This was an animal created by nature that was beginning to understand the world around him without reference to a super-being. At the time this did not create doubts about religion; god was just viewed as a wonderful creator. Newton himself was a profoundly religious man. It would take several generations before the nature of scientific discoveries would start to appear to conflict with the fundamental beliefs of religion. However, this newfound process of scientific enquiry finally allowed centuries of superstitious belief in the natural order of the world to be questioned and where wrong jettisoned. The Reformation had finally broken the monopoly of the Church on explaining the workings of nature.

Later, in the age of Enlightenment in the eighteenth century, people in general gained the confidence to use reason to determine truths. A widespread interest in natural philosophy developed: people acquired fossils, observed nature in action and collected species of plants and animals from the newly 'discovered' lands. Those of an inquiring mind developed a broad range of interests and many inspired enthusiastic amateurs made significant discoveries.

By way of an example, the life of Joseph Priestly (1733–1804) covered an extraordinary range of academic interests. Born to a Calvinist family in Yorkshire, Priestley had exceptional intellectual gifts. He attended local schools where, in the educational tradition of his time, he learned Greek, Latin and Hebrew. He was a minister by training, but his broad-ranging mind soon led him into teaching, philosophy, theology and scientific studies. He is credited with a seminal work on English grammar, an aid to the teaching of history in publishing original historical timelines and a number of influential political and religious texts. His metaphysical texts are credited with inspiring the foundation of a new Christian denomination, Unitarianism. In addition to all this, he made major contributions to science. After producing an influential work on electricity he turned his attention to chemistry and is credited with isolating a number of gases including nitrous oxide, nitric oxide, ammonia and, most importantly, he was among the first to discover oxygen. He tested oxygen on mice and found it was 'five or six times better than common air for the purpose of respiration'.

Priestley was an example of a self-made man who came to prominence by his own genius. However, there are also examples of wealthy men with time of their hands, who indulged their own interests to make significant discoveries. One such was Henry Cavendish (1731–1810), related to the Duke of Devonshire and one of the wealthiest men in Britain. He was also painfully shy and his only social outlet was the Royal Society, whose members dined every week. Otherwise he preferred to spend his time conducting scientific experiments in the freedom of his own home. He is credited with the discovery of hydrogen and observed that oxygen and hydrogen combined to make water. He also calculated that 'common air' was made up of four parts nitrogen and one part oxygen, together with a residual gas which accounted for no more than 1/120th of the volume. He was also the first to measure the force of gravity and to calculate the Earth's density. This was extraordinary enough but, being a secretive man, he did not publish all his work and did not disclose many of his findings to his fellow scientists. In the late nineteenth century, long after Cavendish's death, James Clerk Maxwell looked through Cavendish's papers and found seminal works and discoveries later credited to others, such as Ohm's law (electric resistance), Dalton's law of partial pressures, and Charles's law of gases. A manuscript 'Heat' describes a 'mechanical theory of heat,' anticipating the science of thermodynamics.

A few extraordinary amateurs were even clever enough to found a science. One such individual was another aristocrat, the Frenchman Antoine Lavoisier (1743–1794). Although Priestly and Cavendish had discovered oxygen and hydrogen, they both believed in the old theory that a substance known as phlogiston was released on burning (they called oxygen dephlogisticated air and nitrogen pholgisticated air). Lavoisier developed the theory of the conservation of mass during chemical reactions and conclusively disproved the phlogiston theory. With others he assigned the modern names of oxygen and hydrogen and put the nomenclature of chemistry on a modern basis by assigning rational names to elements and compounds. He also was involved in developing the metric system of measurement. Lavoisier's' *Traité élémentaire de chimie* is considered to be the first modern chemistry textbook. It presented a unified view of new theories of chemistry. In addition to the law of conservation of mass, he postulated that all chemical substances are combinations of a limited number of elements. This synthesis of chemistry theory led to him be called the father of modern chemistry. This honorific, however, would not help him during the French Revolution. As a former tax farmer for the ancien régime, Lavoisier was branded a traitor during the Reign of Terror and in 1794 tried, convicted, and guillotined in Paris, at the age of 50. By then he had created the science of chemistry. The dubious practices of alchemy were forgotten and the commonly held belief in four elements earth, water, wind and fire of the sixteenth century was now perceived as nonsense.

The Netherlands and companies

Up to the start of the sixteenth century oceanic trade had been monopolised by the Catholic countries of Spain and Portugal. This had been enshrined by a papal bull in 1493, which allowed Spain to claim all new territories west of a line drawn 100 leagues to the west of the Azores, and Portugal to claim those to the east. This somewhat vague division of the spoils was formalised in the Treaty of Tordesilhas, in which a more precise definition of the Spanish and Portuguese spheres of interest was agreed. By this the Portuguese could lay claim to Brazil and, on the other side of the world, the Spanish could claim the Philippines.

Needless to say, this was not agreed by the Protestant states to the north who now took up the competitive challenge to gain their share of world trade. They successfully competed both militarily, in seaborne warfare, and economically, with the invention of the joint stock company.

In the Middle Ages naval tactics involved close combat, ramming and boarding. With the advent of the cannon, the aim became to sink the opposing vessel. At first **naval guns** were fired from the superstructure and the size of ships was enlarged to maximise firepower. However, these ships proved very unstable. The Spanish developed a more successful design around 1550, the galleon. This was a fast and manoeuvrable ship that fired cannon through ports cut in the hull. The Dutch found them effective against the large carracks used by the Portuguese in their Asian trade. Over time the number of guns carried on board increased; and by the time of the Napoleonic wars the largest ships were capable of firing 100 cannons simultaneously.

The newly formed country of the Netherlands was one of the first to make a major indent into the Iberian monopoly. In the late sixteenth century Dutch ships were the dominant force in the north European Baltic trade and this gave them opportunities to extend their trade to India in direct competition with Portugal. However, they needed investment in ships and trading posts to compete with the Portuguese in Asia and individual merchants were limited in the amount of money they could raise. The solution was to create the Dutch East India Company, known by its Dutch abbreviations as the VOC (Vereenigde Nederlandsche Geoctroyeerde Oostindische Compagnie). This was a monopoly federation of Dutch traders who could raise sufficient money, not only to finance trade but also to use force to support their merchants.

It proved a spectacular and brutal success. The Portuguese were completely unable to compete with the dynamism and financial muscle of this new ultra-aggressive entity. Within a short time the Dutch had taken over the spice trade. The VOC was open to investment by all financiers, and very soon it had a large number of subscribers, over 1000 in Amsterdam alone. Over the course of time many of the characteristics of a modern company emerged: trading shares on a stock exchange, publishing annual reports and accounts, declaring regular dividends, electing directors for fixed periods and auditing accounts by third parties. The VOC, however, was untypical in that it could also use force and thus was able to compete directly and successfully with states. In 1619 an employee of the VOC, Jan Pieterszoon Coen seized control of the small Javanese port of Jakarta, renamed it Batavia and started a company-run Dutch empire in the East Indies.

Although, eventually the VOC would disband and the political ownership would come back under Dutch government control, the idea of a **joint stock company** lived on. This new concept would prove the ideal vehicle to facilitate the future Industrial Revolution. It allowed investors to make a long-term investment in a business with the prospect of a regular dividend. At the same time, their risk exposure was limited to their invested capital; they could not be held responsible for any consequential losses of the business. In addition, they could withdraw their capital at any time by trading shares. The joint stock companies themselves had sufficient secure starting capital to employ and train people and buy the necessary tools and plant. It is probably true to say the Industrial Revolution would not have had such a dynamic impact without the creation of this type of organisation.

The English were early adopters of joint stock companies. The famous East India Company was granted a Royal Charter by Elizabeth I in 1600. However, progress towards the modern limited company in the UK was set back as a result of the South Sea Bubble in 1720. The South Sea company was founded in Britain in 1711 and gained a monopoly on the trade to South America. There was never any realistic chance that it was going to make significant profits. However unrestrained speculation in their shares and their inevitable collapse ruined many investors. As a result, the law in Britain was changed so that companies could be incorporated only by royal charter or a private act of parliament. Because of this, in the rapid expansion of capital-intensive enterprises during the early phase of the Industrial Revolution in Britain, many businesses came to be operated as unincorporated associations or extended partnerships. This was resolved in the Joint Stock Companies acts of 1844 and 1856, which established a limited liability company with a distinct legal personality, separate from that of its individual shareholders. The basic shape of a vital new form of human organisation had acquired all the chief characteristics which would allow it to accelerate technological development in the next centuries.

Technical developments in Britain

In the sixteenth century the Portuguese and Spanish were the principal world traders. In the seventeenth century, the Dutch dominated the spice trade to the Far East. In the eighteenth century the British and French competed for domination of world trade. Britain triumphed and became the foremost commercial power in the world. In doing so the British solved the last two remaining issues in long-distance ocean travel, the calculation of longitude and the health risks of scurvy. In the early eighteenth century longitude could still be estimated only from the ships' speed and bearing. What was needed was a clock that could keep accurate time over several years in a variety of climatic conditions. By noting the time when the sun was at its apex and comparing it with Greenwich Mean Time an accurate **measure of longitude** could be made. In 1714 the British Navy offered a prize of £20,000 for the development of such a clock or chronometer. John Harrison (1693–1776), a carpenter by trade and a clock repairer in his spare time, took up a challenge. It took 25 years of effort, and in 1760 he made a chronometer (about twice the size of a pocket watch), which was unaffected by the movement of a ship and changes in the weather. Captain Cook, the famous British explorer and cartographer, used a copy of Harrison's watch on his 1772 trip round the world. When he returned to Plymouth three years later, the cumulative error in longitude was less than 8 miles. Harrison, however, was not a gentlemen. The Navy demurred and put his results down to luck. He never got his full £20,000 reward and was awarded only £8500, three years before his death.

Scurvy is a potentially fatal disease caused by a lack of vitamin C in the diet. During the eighteenth century it killed more British sailors than enemy action. Captain Cook became the first captain not to lose crew members infected by scurvy on a long-distance voyage. He achieved this by enforcing strict shipboard cleanliness and the frequent replenishment of fresh food. However, it was not until the early nineteenth century that the use of fresh lemons in sailors' diet finally eliminated scurvy from the navy.

This was the final hurdle to safe ocean voyaging. With the improved precision of navigational instruments, maps had become more accurate and detailed coastal surveys were completed. Dietary issues involved in long voyages had been overcome. The incidence of shipwreck had fallen significantly. The stage had been set for the large-scale development of trade and colonisation by Europeans in the next century.

The British also made crucial improvements in financial systems and transportation infrastructure which enhanced the country's competitive position. Over history, British kings in financial difficulty had been eventually driven to either debase the currency or default on their debts. This made state borrowing very expensive. At the end of the seventeenth century British government finances were brought under scrutiny and control by Parliament. Parliament copied an idea invented by the Dutch and created a market in **bonds**, which are bundles of government debt paying a guaranteed interest rate. Existing debts were consolidated into non-redeemable bonds, known as consols. These consols finally put state finances on a solid basis. After 1672 the British government never defaulted on its debts and investors had a strong degree of security. Consols were arguably the most successful bonds ever issued. One of the causes of the

French Revolution was the French king's difficulties in controlling France's debts. In contrast, during the subsequent Napoleonic wars, British finances remained sound despite Britain being the main financial supporter of the anti-French coalitions. Even though the price of existing consols fell by as much as 30% the British government was still able to issue new bonds. From then on British bonds were always regarded as a secure investment and as a result government debt could be financed at lower interest rates than in most other countries.

In the eighteenth century, whereas the convenience of shipping by sea had improved dramatically, land transport remained slow and difficult. Since Roman times Britain's roads had suffered 1300 years of neglect and were in a shocking state of repair; as a result in 1700 it took about four days to reach York from London by carriage. The solution was to set up Turnpike trusts, a private enterprise system whereby a board of trustees was granted, in a private Act of Parliament, the right to construct and maintain a particular length of road. The trustees raised loans to pay for the construction of the road and were repaid by charging tolls on those who used it. A trust would normally run for 21 years. Between 1750 and 1790, some 1600 acts of Parliament were passed setting up turnpikes covering some 24,500 miles of road.

Road engineering also gradually improved as a result of the pioneering work of the great road builders: John Metcalf, John McAdam and Thomas Telford. As a result, by 1830, just before the arrival of the railways, the journey time to York had been reduced to 20 hours.

For bulky industrial goods such as coal, water transport was much more effective than roads, yet Britain lagged far behind France and the Netherlands in building canals. This changed in 1760 when a coal owner, the Duke of Bridgwater, commissioned the engineer Richard Brindley to build a canal from his mines at Worsley to Manchester. Brindley's canal included an aqueduct over the River Irwell, which was considered an eighteenth century engineering marvel. The canal was immediately successful, reducing the costs of shipping coal on horseback. The Duke of Bridgewater followed this up by building a canal from Liverpool to Manchester, which reduced the cost of transport from 12 to six shillings a ton. Having shown the way, others followed and by 1820 a network of canals was established linking the major cities of the UK. The effect was dramatic, the price of coal fell 75% in Manchester after the Bridgwater Canal had been built. Without these new canals and roads the British Industrial Revolution would not have been possible.

British political culture

By the end of the eighteenth century, Britain had become the wealthiest country per capita in the world. As a consequence its people were healthier and had a higher life expectancy than others in Europe. This commercial success was ultimately due to the fact that Britain was better governed than other European countries and its citizens were better able to exercise individual initiative.

The British system of government was a **constitutional monarchy**. In other European monarchies all power was ultimately held by the king; in Britain power was shared between the king and parliament. All laws, taxes and policies had to be approved by an elected group of representatives meeting in the House of Commons and ratified by peers in the House of Lords. The king was still formally in charge of the executive; he selected ministers and had the power to veto legislation. Increasingly, however, the king delegated his powers to a chief or prime minister who could prove he could assemble a working majority in the House of Commons. In these early days of democracy there were no formal organised parties as now, but a number of factions centred on influential individuals. The chief minister had to negotiate with key individuals to gain approval for the budget and pass necessary legislation.

The British Parliament first met in 1264. Parliament's initial purpose, as seen by monarchs, was to gain public approval for extra taxation. Many similar bodies, such as the Cortes in Spain or the Etats Généraux in France emerged in Europe in the same period. Although strong monarchs could often govern directly, over time the convention grew in Britain that significant changes in taxation needed the approval of Parliament. The House of Commons separated from the House of Lords in 1341 and consisted of non-aristocratic representatives from the shires and boroughs of England and Wales. In the seventeenth century members of the House of Commons became more assertive and started initiating legislation that made demands of the king. This came to a head in the reign of Charles I, when the disagreement between king and parliament led to the English Civil War (1642–1651). This resulted in the execution of Charles I and Oliver Cromwell becoming the country's leader. On three separate occasions thereafter, Parliament determined who would rule the country: in 1660 they arranged the restoration of the monarchy with Charles II; in 1668 they approved the coup d'état of William and Mary; and in 1714 they appointed George I, a Protestant, to the crown. At each successive phase the English Parliament increased its power. By the eighteenth century the king's ministers and parliament were the most important force in running the country and the king was increasingly peripheral to the critical decisions of government.

This British parliamentary system of government proved more successful than the absolute monarchies in Europe. Its prime ministers had to prove their ability in the job or risk a vote of no confidence in Parliament. The problems caused by ineffective monarchs had been overcome; even though none of the four king Georges from 1714 to 1830 was particularly competent, the country managed to thrive. Further, the necessity for members of parliament to be re-elected ensured that Parliament did not develop into a self-serving oligarchy, like in the Venetian and Dutch republics. This was not democracy as we would recognise it. The voting franchise was limited to less than 10% of the population; the election of members of parliament in the so-called rotten or pocket boroughs was decided by the local landowner. Bribery, corruption and nepotism still hindered efficient government. Nevertheless it was clear across all Europe that better policymaking in Britain gave the country a substantial competitive advantage. Whereas other European countries became embroiled in land-based wars for a small territorial gain, the British military strategy was based on improving its commercial advantage. British territorial expansion concentrated on exploiting its sea power to monopolise trade in the rapidly developing markets outside Europe.

Alongside the development of constitutional monarchy, there had developed new protections against capricious acts of the state, giving greater rights for an individual to be treated fairly in law. The English judicial system had developed in a rambling way over centuries from Anglo-Saxon times. The system of common law was established by precedent; judgements in the past formed the basis for judgements in the future. Since the time of Henry II in the twelfth century the law was nationally applied by judges, who were figures of great power and importance. From the start they were professional appointees, not directly beholden to the king or any other crown official. By the Act of Settlement 1701, judges, once appointed, could only be removed by Parliament. However, they were still likely to be conservative upholders of the status quo, putting the commoner at a disadvantage. A tradition gradually developed that felonies (serious crimes) should be tried by a jury of 12 ordinary men of good character. This provided some protection to commoners, but juries could be cajoled by a judge into handing down his preferred verdict. The turning point came in 1670 when a certain Edward Bushell, a member of a jury that acquitted two Quakers indicted before a judge in London, was fined for recording a verdict against the judge's direction. He refused to pay and the judge committed him to prison. Bushell applied for habeas corpus and was freed by the Chief Justice. Since that time no-one serving on a jury has been called to account for giving a verdict according to their own judgement.

From this point onwards the British legal system developed into one in which the judiciary was effectively independent of the government. Judgements were based on the facts presented at the hearing according to a precedential interpretation of the law. New laws could only be set by Parliament. Anyone charged with a felony had the right to be tried by a jury. By these means the English government had to act within the law and no-one could be detained without due legal process. Commoners of few other nations had such protection from the arbitrary abuse of state power at this time.

This became clear in the greater freedom available to the British press to criticise its rulers. It was particularly effective in cartoons. Both royalty and politicians were ridiculed in grotesque caricatures. For example, a popular cartoon of the 1730s depicted the naked royal backside of George II descending towards the face of Prime Minister Walpole. One of the most effective satirists was James Gillray (1756–1815). He was able to caricaturise the royal family with relative impunity, depicting the future George IV as a fat, self-indulgent lover of buxom women. Such specific and wounding criticism would never have been possible in an authoritarian regime with no protection for freedom of expression.

The development of individual legal rights in Britain had encouraged individual initiative and limited the potential of the government and the aristocracy to abuse their position of power. These rights did not yet include total religious freedom in Britain. Catholic civil rights were severely curtailed and Catholic priests were liable to imprisonment. Neither Catholics nor non-Anglican Protestants could become civil servants. Nevertheless, it was clear that individual initiative was encouraged in Britain more than in the rest of Europe. Not only was Britain commercially successful but the British were foremost in technical and scientific developments as well. The beliefs of the Enlightenment had become part of the British political culture. It was accepted that the natural world was governed by rational forces; religion did not have a monopoly on knowledge and the power of the church in state affairs was marginalised.

All this rationality did not mean that religion was any less important, as religion continued to structure most people's lives. Religions actively competed with each other for congregations. In the eighteenth century popular evangelists like George Whitfield and John Wesley addressed huge open-air meetings, beginning the Protestant evangelist tradition that continues in the USA today. English towns and large villages now had a church and a chapel, one for the traditionalists and one for the dissenters.

USA and republicanism

In North America the political changes begun by the British were developed further to create a fully democratic system of government. The British colonisation of North America had always been different from the Portuguese and Spanish approach. The native Indians were Neolithic farmers with few domesticated animals, no access to gold or silver and little except food and furs to trade. From the start the British government made little direct investment. Colonisation was all financed privately, sometimes by joint stock companies, sometimes directly by private investors. The British set up farming and trading settlements; there were no quick and easy fortunes to be made. Native Indians with their corn, squash and beans and slash and burn agriculture had made a limited impact on the environment and could provide only limited resistance to Western advance. This was effectively virgin farming country. Using their knowledge of Western technology there was a huge opportunity for ordinary people to exploit the land. Some new foodstuffs from America like the potato and tomato could be produced in the temperate climates of Europe, but tobacco and sugar could only be supplied from the New World. Much of the land was potentially arable and rewarded all those who were willing to work hard. In North America colonists were largely families; both men and women came from Britain. This was unlike the Spanish settlements where there was large-scale inter-marriage between Spanish men and Indian women. In North America, from the start, the Indians were separate from the Westerners and always on the losing side. A few missionaries tried to convert the Indians to Christianity, but there were no mass conversions of natives, as under Spanish rule. The British were creating another Western society in the USA, not a mixed society like in Latin America.

Life in British America particularly appealed to minority religious groups such as Calvinists, Quakers, Anabaptists and Roman Catholics. They saw it as an opportunity to escape religious discrimination in Britain. Initially in some colonies such as Massachusetts, there was a state religion that was intolerant of all others. However, the independent nature of the settlers made a strict theocratic style of government impossible to sustain. In 1649 Maryland, which was created as a Catholic colony, became the first colony to officially support religious toleration for all, making Baltimore for a time the fastest growing town in America. Others soon followed, as they saw its success in attracting dissenting minorities.

From the start the colonies were largely self-governing, all with charters from the crown. They all established law-making institutions similar to the House of Commons (often called the House of Burgesses). In general, all male freeholders elected local representatives who met regularly to decide critical issues. They were all abreast with the developments in the UK and the steadily increasing power of Parliament. However there was one major difference; they had all developed constitutions defining their rights and procedures. The Fundamental Orders of Connecticut (1639) was the first written constitution, not only in America, but in the entire world.

At the start the British Crown was content to let the colonies govern themselves, but towards the end of the seventeenth century the initial charters were revoked and governors were appointed. However, the governors proved to be in a very weak position, with little money at their direct disposal. The British government was reluctant to spend any money and the local parliaments were reluctant to impose taxes. Following the lead of the House of Commons in the UK, by the end of the eighteenth century all the colonial 'houses of commons' had established effective control over the governor and his executive. As result British North America was the least taxed area on Earth. There was no aristocracy to maintain, no state church to support, no local king to maintain in his pomp. These were states run by the local farmers who owned the land and practised a self-sufficient independence. There was no standing army or navy; local volunteer militias fought any necessary battles against the Indians and the British army and navy fought the wars against neighbouring French, Dutch and Spanish colonies. This was a whole new egalitarian democratic society. Nothing like it had been seen in the world before.

The British government never fully engaged with its North American colonies. No senior member of the UK government ever visited America. Britain persisted in seeing the colonies as subordinate to the UK and tried to regulate their trade to the benefit of Britain. Manufacturing was discouraged; the British idea was that the colonies should supply the raw materials for UK manufacture and import finished goods in return. When, eventually, Britain tried to tax its new wealthy colonies, rebellion broke out, and through general ineptitude, poor local knowledge and lack of will, and vision the British lost their prize possession.

In 1787 the founding fathers of the United States agreed on a new constitution. This was a blueprint for a new type of state, a **republican democracy**, in which all male citizens bar slaves had the vote. Careful attention was given to ensure that individual freedoms were maintained and that there was little possibility of the exercise of autocratic power. There were regular elections at specified times. Executive power was in the hands of an elected president. Legislative power was held by two bodies – the Senate, which had an equal number of representatives from each state, and the House of Representatives, composed of members whose constituencies were equal in terms of population. The executive and the legislature were mutually interdependent. The nation's budget was proposed by the executive but had to be approved by the legislature. Legislation approved by the two houses had also to be signed by the President. The judiciary was fully independent. As an amendment to the agreed constitution, there was a legally enforceable Bill of Rights to ensure certain basic freedoms. The important First Amendment prohibited the violation of the citizens' basic rights to the freedom of religion, assembly and speech, freedom of the press and the right to petition.

In the Declaration of Independence the wording of the second paragraph was inspirational:

> We hold these truths to be self-evident, that all men are created equal, that they are endowed with certain inalienable rights, that among these are life, liberty and the pursuit of happiness.

This was a clarion call to all the nascent middle-classes of Europe to demand similar liberties of their autocratic masters. However, is not easy for a state to adopt a democratic form of government. It relies on those in power to hold fair elections and to be willing to abide by their result. It requires disciplined debate, party organisation and the willingness to respect your opponents, even if you do not agree with them, and to compromise with them if necessary. Crucially, the British North American colonies had more than a century of experience of local democracy and thus were able to make the new state-wide democratic system work well from the start. Other countries, as we shall see, would find this much more difficult.

Political cultures by 1780

Following the Reformation, the political culture of the Catholic monarchies had fractured. There were now four types of Christian political cultures: those of Britain and the USA, the Protestant monarchies of north Europe and the remaining Catholic monarchies. The three remaining political cultures mentioned in the last chapter, Islamic sultanates, neo-Confucian empires and Hindu kingdoms, declined in importance. Another new culture developed in Russia as the Mongol empire declined in power. Developments in each of these are summarised below.

Protestant monarchies

There were Protestant kingdoms in Germany, Scandinavia and the

Netherlands. Initially, the Protestant states were just as intolerant of other religions as the Catholic states. However, while they were always very suspicious of Catholicism, they learned to live with the existence of many shades of Protestant opinion. This was important. Up until now the presence of different religious groups in a country had led to internal conflict. There began to be a separation of church and state and a respect for other people's religious opinion. In contrast, the Catholic Church continued to be intolerant of other faiths; the Inquisition, whose purpose was to identify and punish heretics, was active in Spain, Portugal and Rome until the nineteenth century. Spain and Portugal expelled the Jews and the Moors, France expelled the French Protestants (the Huguenots) and there was a very significant transfer of the Flemish population to the Netherlands following the suppression of Protestants in Belgium by the Spanish authorities. All these groups had been significant contributors to the local economy and the Dutch and the British, in particular, benefitted from the technical knowledge and ability of Protestant immigrants.

Protestant culture emphasised the belief that all individuals had it in their own power to succeed. This was manifested in the so-called Protestant work ethic, which identifies hard work and thrifty behaviour as an integral part of the Protestant faith. It became an important reason for the economic advance of protestant communities. In the power struggle between states, Britain and the Netherlands had proven the competitive importance of trade and capitalism, alongside more traditional armed warfare. A new organisational entity, the limited company, had been developed, which allowed risk capital to be used for investment. Partly as a result, manufacturing and trade in north Europe grew to such an extent that a new middle class acquired a significant proportion of wealth.

With the Protestant religions had come a new enquiring mind, free from superstition and the dead hand of Catholic orthodoxy. The eighteenth century was the age of Enlightenment, when the intellectual elite first consistently challenged ideas grounded in tradition and faith through the scientific method. Its practitioners promoted scientific analysis, scepticism and intellectual interchange and were completely opposed to superstition and intolerance. Medical and other traditional practices not based on a logical approach were exposed to ridicule. It was an age when people dared to think objectively about God. Few people were atheists but several intellectuals espoused deism, the belief in a supreme being, the creator of the world, but one who does not intervene in the workings of the universe. They rejected the tenets of organised religion, with its belief in miracles and supernatural events and the use of religious revelation as its authority. The movement had no significant following, however, among the masses. Religion, in ordinary life, was still as important as ever.

Catholic monarchies

Catholic countries in Europe were now starting to fall behind the Protestant states in terms of economic development. In France, the lack of a political change, in part due to the intense conservatism of the Catholic Church, led to a sizable anti-clerical movement among the middle classes. This was to have a significant impact in the French Revolution.

After the trauma of the Reformation, the religious wars and the breakaway of the north European protestant religions, the Catholic Church had reinvigorated itself. It developed a strong proselytising drive, both to recover 'souls' lost to the Protestant movement and to convert non-Christians, particularly in Latin America, but also in India, China and Japan. At the forefront of this drive were the Jesuits. The Jesuit order was founded in 1534 and Jesuits were notable for their work in education, their intellectual integrity and their missionary efforts. Jesuits were rigorously trained in both classical studies and theology and founded schools throughout Europe. They were instrumental in the Counter-Reformation in Europe, overcoming the drift towards Protestantism in Poland and southern Germany. Jesuits were also sent out as missionaries to the French, Spanish and Portuguese colonies across the globe with the aim of converting the indigenous population to Christianity. In both Japan and China they had some initial success, impressing the emperors with their academic abilities and Western knowledge. They had even greater success in Latin America and Goa, converting large numbers of the local population. However their continual challenging of temporal authorities and several disputes with the pope eventually led to them being suppressed in 1773. By then the missionary credentials of the Catholic Church had been established. The recovery of their position in Europe and their dominance in South America meant that, despite the loss of adherents to the Protestant cause, the total numbers of its congregations had increased dramatically during this period.

Islamic sultanates

The sixteenth century saw the height of Islamic expansion. Islam spread to Malaya and Indonesia with Arabic traders. Three great dynasties, the Ottoman, Safavid and Mughal, were established in Turkey, Persia and India, respectively. Thereafter the Islamic sultanates failed to absorb the new memes developed in the West, except those involved in warfare. Hence, towards the end of the period all Muslim states were beginning to lose out to the West, both economically and technically.

In India, the Mughal empire was founded by Islamic invaders from Uzbekistan in 1526. Its leaders were initially tolerant of the Hindu religion and Mughal culture became a brilliant cultural fusion of Islamic, Persian and Indian influences. At its height, it controlled all but the southern tip of India. However, at its roots nothing had changed; nobles, priests and emperors still lived off the labour of the Indian peasant. Administration and organisation were still weak, and once again, when the line of competent emperors stopped after Aurangzeb in 1707, another Indian empire fell apart. None of the successor states were a match for the military might of the British East India Company and by 1850 the company had replaced the Mughal empire and established control of the whole Indian subcontinent.

The Ottomans were a clan of Turkic peoples who, from the fourteenth century on, established an Islamic state in Anatolia and the Balkans, profiting from the weakening of the Byzantine empire. Eventually in 1453, the Ottomans took Constantinople and put an end, after one and a half millennia, to the final remnants of the Roman empire. The Ottoman Turks went on to control a large area of the eastern Mediterranean, North Africa and south-east Europe.

Key to their military success was the establishment of a standing army of motivated slaves, known as Janissaries. Non-Turkish children, usually Christians from the Balkans, were seized, subject to strict discipline and indoctrinated in a system known as devşirme (changing). These boys (usually aged between 6 and 14) were taken from their parents and given to Turkish families in the provinces to learn Turkish language and customs. The recruits were then taught Islam, and supervised 24 hours a day. They were subjected to severe discipline, and prohibited from growing a beard, taking up a skill other than soldiering, and marrying. As a result, the Janissaries were extremely well-disciplined troops. Known as the slaves of the Porte, Janissaries formed their own distinctive class and owed allegiance only to the Sultan. As such, they were powerful political rivals to the aristocracy.

Like the Mughals in India, the Ottomans were traditional Islamic dynastic empires with all the intrigues of the harem, the political machinations of the eunuchs and the brutal fratricidal fighting to acquire the succession. As an example, Murad III (1574–1595) sired more than a hundred children and was survived by 20 sons. His successor, Mehmet III, began his reign by strangling his 19 brothers and murdering seven pregnant women in his father's harem. The chaotic nature of Islamic succession and the debauchery of court life was one reason for the eventual decline of the Ottoman empire. However, we also have to ask why the revolution caused by the printed press in Europe did not happen in the Middle East. Movable typefaces were just as practical for the Arabic alphabet as the Roman. It appeared that the religious turmoil caused by printing in Europe scared the Ottoman sultans. In 1483 Sultan Bayezid II prohibited printing in the Arabic script on penalty of death. Sultan Achmed III (1703–1726) gave his permission for the establishment of the first printing press for Arabic secular works only. However, printing activities in the Ottoman empire did not really lift off until the nineteenth century a full 400 years after printing arrived in Europe.

The Shi'ite sect of Islam had so far been restricted to separatist communities on the fringes of the Islamic world. Their leaders or imams had been systematically eliminated by the Sunni powers until the twelfth Imam disappeared altogether. Thereafter Shi'ites believed a Hidden Imam would return as leader. Shah Isma'il (1501–1577), founder of the Safavid dynasty, claimed that he was the hidden Imam and launched a successful invasion of modern-day Iran. The subsequent Safavid rulers and their successors the Qajars retreated from the claim to be the heirs of the hidden Imam and allowed the Shi'ite religious community to develop a considerable degree of autonomy. As a result, Persia became a state with a strong independent clergy, similar to the situation in Catholic countries.

Neo-Confucian empires

After the Mongols were defeated, the newly established Ming dynasty (1368–1644) was modelled on its illustrious Tang and Song ancestors with neo-Confucian philosophy, an educated administrative class and a centralised government system.

China during the Ming dynasty experienced a period of political stability and population growth. Beijing became the first city with a population of over a million since Rome at the peak of the Roman empire. However, it was also a period of institutional atrophy. China totally failed to make any of the organisational advances that enabled capitalism to develop in the West: banking, the rule of law, the protection of property, trading companies and democratic government. It also failed to make any of the scientific and technical advances of the West.

Around 1400, both in Europe and in China, scholars were studying the wisdom of old texts and emphasising the need for rational study. Yet scientific reasoning developed rapidly in Europe, while scholarship remained backward-looking in China. The ease of printing alphabet scripts compared to those with Chinese characters, the spread of literacy and the development of arithmetic based on Arabic numerals had a major role to play in the advance of Europe. However there was something beyond this. The independent middle class failed to grow in China or have any significant influence on power; the promise of economic progress of the Song dynasty had been unfulfilled. This must have had an economic and political cause.

One of the reasons may have been the lack of competitive threats within and without Chinese society. For memes to develop there have to be winners and losers. Those with the good ideas must be able to thrive and those without ideas must fail. There were simply no competing organisations outside the administration that could benefit from change; there was no independent aristocracy, no independent cities and no threatening foreign states. Even the competing philosophies of Confucianism, Daoism and Buddhism had been merged into the new philosophy of neo-Confucianism. There were simply no organisations outside the government bureaucracy that could gain from new ideas.

The government, itself also failed to initiate change. The first Ming emperor, Taizu, made a disastrous decision which condemned the dynasty to strategic inaction. Taizu abolished the office of chancellor and forbade any of his successors to re-establish the post. This left the overall coordinating role of the administration to the emperor who had to deal directly with tens of state ministries and agencies. The emperor turned to his eunuchs for advice and they started to review the performance of the administrative hierarchy. This developed in a sinister way into a form of secret police, known as the eastern depot and eventually grew out of control. By the end of the Ming empire an estimated hundred thousand eunuchs were associated with the palace. One emperor tried to establish a Eunuch Rectification Office to bring the eunuchs back into line, but this was piling one centralised system on top of another.

With no clear strategic direction and little motivation to change, the process of government drifted on. The classic example of the atrophy of Chinese government at the time is the Wanli Emperor (1573–1620). This is from the *Rough Guide Chronicle for China*, p. 255:

> Wanli is China's one long-reigning emperor about who nothing good may be said. He had as little to do with his responsibilities as was humanly possible. In 30 years between 1590 and 1620, he met with his Grand Secretaries a mere five times. At a court hidebound by rigid protocol, officials and foreign envoys regularly kowtowed before an empty throne.... In ministry after ministry positions fell vacant.... Wanli dissipated ... vast amounts of revenue on a life of private luxury and indolence. Wedding clothes for his many children became a main

item in the national budget. Ministers who complained were flogged. Ministers who protested their punishment were flogged harder. Obese and bloody-minded, he ringed himself with 10,000 eunuchs, whose idle hours were passed sequestering imperial revenues. Yet perhaps the most alarming aspect of Wanli's reign was that nothing was done to get rid [of] him.

The Chinese government was effectively a centralised bureaucracy, run by clever officials with little strategic direction and little prospect of change. The absence of any mechanism of local political accountability made the Chinese central system of government unalterable without a strong powerful emperor.

In 1644 the incompetent Ming dynasty was overthrown by another foreign people, the Manchu, who rapidly re-established the emperor's authority. With a series of competent emperors, including one of the world's greatest rulers the Kangxi Emperor (1654–1722), they expanded the boundaries of China to their greatest extent, incorporating Manchuria and Inner Mongolia from the area originally controlled by the Manchus and adding Taiwan, Tibet, Mongolia and the Moslem Turkish territories to the north-west in what would become Xinjiang Province. However, eventually the Qing (Manchu) emperors came to rely on the same administrative system as their predecessors with the same lack of local accountability. In time the Qing dynasty also displayed the same lack of initiative, curiosity and economic development that had beset its predecessors. From 1700 onwards it became clear that European military technology and organisation was superior. The Chinese empire became again vulnerable to military threat. The Kangxi Emperor, recognised this. In a secret memorandum to his family and ministers written shortly before his death he wrote:

> Like other Europeans the Russians, Dutch and Spanish succeed in whatever they determine to do. They are fearless, clever and opportunistic. As long as I am alive China need not fear them. I respect them, they respect me and they do their best to serve me. The rulers of France and Portugal have sent me excellent subjects, skilled in the sciences and arts, and they benefitted the empire. But should civil war break out, or should we be invaded by Mongols what then would become of the empire? The Europeans would do to China as they please.

This was prophetic, as we shall see in the next chapter. However he need not have worried about the Mongols. The time that the pastoralists could threaten farmers had passed. The development of guns and cannon had destroyed their military advantage. The power of the steppe nomads vanished. Their empire in the east was taken by the Chinese and in the west by the Russians.

Japan has not been mentioned thus far in the story of the advance of civilisations because, by and large, developments there were peripheral to the main events in Eurasia. Because of Japan's later importance it is necessary to set the scene by describing Japanese society at this time. Japan had a form of neo-Confucian political culture with significant differences to that of the Chinese. It was a stable, relatively urbanised and rigidly controlled society. Following more than a century of disunity, from the end of the sixteenth century, military rulers (shoguns) established complete control over the kingdom. From 1603 onwards the Tokugawa family of shoguns established a system of internal controls so complete that there were no wars in Japan for 250 years.

Japanese society was divided into five broad classes: the daimyo (aristocrats), the samurai (warriors), the town dwellers (merchants and artisans), priests and peasants. The Japanese emperor, who lived in Kyoto, had a religious and ceremonial role and the shoguns, who governed the country, resided in Edo (Tokyo). Shoguns controlled the daimyo by requiring them to attend court in Edo on alternate years and when they were not in attendance, the daimyo had to leave their wives and children behind. This effective hostage-taking maintained the Shogun's power over the aristocracy, preventing rebellions and coups. The samurai, with little fighting to do, were dependent on stipends from their daimyo and gradually changed into bureaucrats, dependent on academic ability for promotion. Villagers and townspeople had to register at their local Buddhist temple and were not allowed to move house or even travel without permission.

Foreign influence was rigidly restricted. Christian missionaries from Spain and Portugal had been expelled. The use of guns was prohibited, and apart from limited trade with the Netherlands, links to the West were effectively cut off. In this splendid isolation, Japan, like China, initially thrived economically and became urbanised, with three of the largest cities in the world: Osaka, Kyoto and Tokyo. However, also like China, Japan stagnated intellectually, failing to keep up with the ideas being developed in the West.

Hindu kingdoms

The last major Hindu state, Vijayanagara, was defeated in 1565. Although the Hindu religion continued to thrive, it appeared that as a political culture,

Hindu kingdoms could not survive in military competition with Islamic sultanates. Why was this so? Fighting on their own soil, in a land in which the majority were Hindus, against a frequently religiously intolerant foe, one would have thought that ultimately the Hindus would triumph. It could be that the caste system, which prohibited the non-kshatriya varna from becoming warriors, limited the ability of the Hindu states to exercise their full power. A glimpse that this might be part of the cause can be seen in the success of the Sikhs in the nineteenth century.

Sikhism began with Guru Nanak in the early sixteenth century in the Punjab region of India. At that time Punjabis were predominantly Hindu peasant farmers who had been subservient to Muslim overlords since the eleventh century. Under the first five of its gurus Sikhism was really just another sampradaya – a Hindu institution centred round a guru in which a set of theological and ritual traditions are maintained. Sikhs believe that, by meditation and through interior devotion to one indivisible and formless god, one can achieve liberation from the cycle of death and rebirth. As such, it was a gentle, inward-looking faith. The new faith was tolerated in the relatively liberal outlook towards other faiths in the reign of Emperor Akbar at the height of the Mughal empire. However, his successors did not share the same religious tolerance and the sect was transformed after two of its gurus were executed or, as the Sikhs would say martyred, for their beliefs. The sixth guru, Hargobind, was a very different character to his predecessors. Rather than composing hymns like the first five gurus, he enjoyed hunting and encouraged his followers to bear arms. He established political as well as religious power and wore two swords symbolising spiritual and temporal authority. The tenth and last guru Gobind Singh instituted the khalsa in which men and women dedicated themselves to the Sikh cause. As part of the khalsa tradition the Sikhs gained a uniform, which included a turban covering uncut hair and a curved sword. The religion now had a more military bias and martyrdom to the cause was venerated. The Sikh leaders gradually became increasingly successful in establishing a state in north-west India incorporating all the Punjabi peoples. Their cause was helped by Sikh opposition to the caste system; they allowed all followers, not just the warrior caste, to fight for the cause. The culmination of their success was the formation of a Punjabi empire led by Ranjit Singh in the early nineteenth century. At long last, a native Indian religion and culture had turned against the debilitating caste system, united Indians and thrown off centuries of foreign Muslim rule.

Unfortunately for them, the British arrived soon after with their superior military hardware. By 1849, after two wars the Sikhs had been subjugated. However the British always had a respect for the martial qualities of the

Sikhs and allowed them to adopt the outward symbols of their religion, including the turban, while incorporating them into the British army.

Orthodox Russia

The Orthodox Christianity of Russia was quite different from other branches of Christianity. It had developed in close alliance with the autocratic power of the (Byzantine) emperor and was a religion characterised by elaborate rituals and ceremony. When the Byzantine empire fell, Russia took on the mantle of the leading Orthodox Christian state. Russian emperors saw themselves as the new protectors of the Christian faith and their leaders as Caesars, shortened to csars. The Church, however, remained largely unchanged from mediaeval days, relying on the inspirational effect of its chanting and liturgy to inspire the peasantry.

The new Russian state had been the creation of Ivan III 'the Great' (1462–1505) the leader of the Grand Duchy of Moscow. He established the primacy of Moscow among the other northern Russian principalities and in 1480 he ceased paying tribute to the Mongols. His grandson Ivan IV 'the Terrible' (1533–1584) established the strong autocratic style of csarist leadership and began the process of conquering and colonising former Mongol lands. However, Russia might have remained an eastern autocratic state with little development potential if it had not been for Peter the Great (1682–1725). He saw the importance of Western culture and military technology. Using decrees, forced reforms and Western advisors, he transformed the military capability of the Russian army and navy. During the Great Northern War (1700–1721) he secured Russia's place among the great European powers by establishing a Russian presence on the Baltic Sea. In doing so he protected Russia's western flank and allowed the Russian empire to expand across the steppes into the homelands of the former pastoralist invaders without any European competition.

While Peter also attempted to westernise Russian culture, his successes were only superficial. Shaving off Russian beards and adopting Western style dress did not change the deeply conservative orthodox Russian culture. Russia remained a strongly hierarchical society with a poorly developed trade and industrial base.

Summary: situation in 1780

In a period of transition before the advent of the Industrial Revolution, the most important development was the advance of science in the West.

Humans made real progress in understanding the workings of the natural world. In the period of the Enlightenment ordinary people began to look with open eyes at the world around them and question ancient beliefs. The power of superstition on the human mind had been reduced. People began to believe they could control their own destiny without asking for divine help at every step of the way. The religious monopoly of the Catholic Church had been broken and a new type of secular state began to emerge.

Outwardly, however, the overall structure of advanced human society remained the same. The land was still the principle source of income and aristocrats and kings still lived off the labour of peasants. Mediaeval state societies still dominated the human world. The pastoralist tribes of the Russian steppe had been defeated. America had been invaded and colonised. Only in Africa and Oceania were tribal and hunter–gatherer societies still in the ascendant.

The exceptions to this unchanging order of society were the USA and the UK. In the newly created USA there were no kings and aristocrats and ordinary people had a say in a new democratic style of government. In Britain the power of the king had been limited by parliament and a more effective system of government had evolved. By 1780, world population had reached 900,000 and was doubling every 300 years. Life expectancy was showing its first significant increase and was estimated to have risen to 28 years, as shown in Figure 7.1.

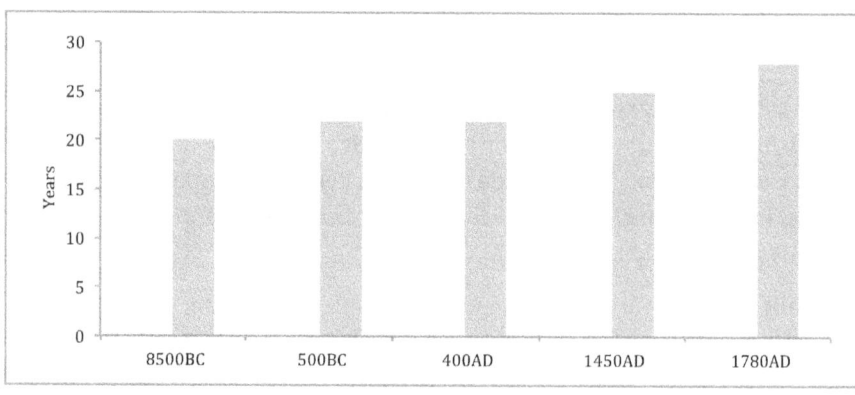

Figure7.1 World life expectancy up to 1780 A.D.

Wealthier countries with a developing middle class such as Britain, USA and Germany had average life expectancies approaching 40. For the first time a significant number of middle-class people were able to live a life that was less physically demanding than the normal subsistence lifestyle. They lived in more spacious, healthier dwellings than agricultural labourers and were therefore not as vulnerable to the spread of infectious diseases. Stable countries without a strong middle class, such as China and France, had life expectancies in the low thirties and unstable regions, such as India, still had life expectancies in the twenties.

The increase in life expectancy was due to several factors in addition to the developing wealth of the West. The devastating effect of pastoralist raids on Asian populations had diminished. States were better organised to provide support to survive natural disasters such as droughts, floods and storms. However, the most important factor was the increasing human resistance to infectious diseases in Eurasia. Since the opening up of trade between East and West, the devastating effect of the diseases, spread from one region to another (such as the Black Death), had lessened. It was still not eliminated, as the plague of 1665 in London showed, but it seemed that fewer people were vulnerable.

The exception to this picture of growing population and wealth was the Americas, where the native population, reduced by 90% following the discovery of the New World, had only recovered to 20% of its former population levels after Western colonisation.

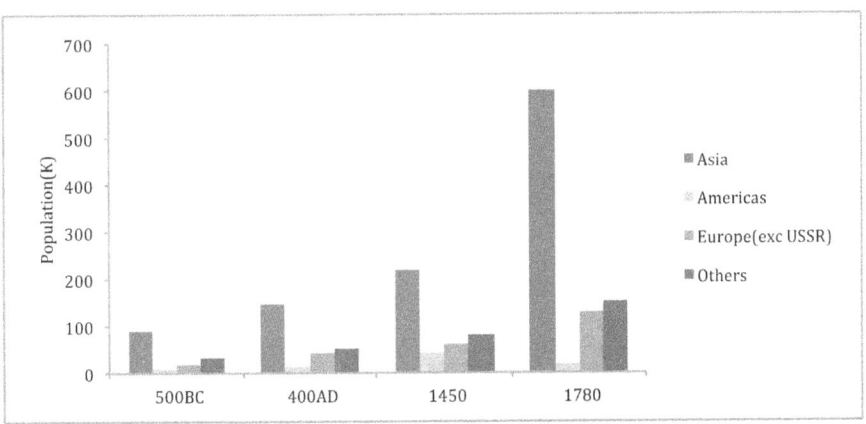

Figure 7.2 Population by region to 1780 A.D. (thousand inhabitants)

Figure 7.2 shows that, in contrast, Asian populations had grown significantly. Asia now accounted for 67% of world population and 60% of the world's wealth. China's population had risen dramatically from about 100,000 to about 300,000, recovering from the Mongol invasions and benefitting from the political stability from the Ming, then the Qing dynasties.

However, the apparent dominance of Asia in the world's economy was about to be challenged. In the UK improved techniques of production were being developed; in future manufacturing and trade rather than land would become the most important source of wealth. The days of the landed aristocracy living off the fruits of peasant labour would be numbered. New structures of society would be created in which aristocrats would lose their privileged status in society.

CHAPTER 8:
THE DEVELOPING INDUSTRIAL REVOLUTION
(1780–1870)

Since the start of the Neolithic revolution agricultural productivity had been gradually increasing. The selective evolution of domesticated plants and animals was improving yields. Whereas once ground had to lie fallow every few years to maintain its fertility, now improved crop rotation allowed its continuous use. Increasingly powerful horses and oxen allowed bigger fields to be maintained. Wind and water mills not only ground grain but powered iron-making furnaces. These in turn made stronger ploughs and new, more effective, agricultural machinery.

Improved methods of agriculture allowed the Earth to support higher population levels. Figure 8.1 shows how population growth rates had speeded up during the human story. The growth rates in hunter–gather times were very slow; on average it took 3500 years for the population to grow 10%. After the start of the Neolithic revolution around 8500 B.C. it took only 200 years to add the same percentage. The peak growth period was from 3500 B.C. up to 500 B.C. Thereafter, population growth rates declined due to the depredations of the pastoralists and the spread of infectious diseases, culminating in the calamity of the Black Death. From 1450 onwards life expectancy crept up as the effects of war and disease lessened. Particularly in Europe, an increasingly efficient agriculture sector could support a rising urban population. In the period from 1450 to 1780 human population increased 10% every 40 years.

In the eighteenth century Britain started to dramatically increase the rate of technical innovation. This was the start of the Industrial Revolution, the third and most dramatic step change in human life. It would transform human existence, permitting a huge surge in population levels and life expectancy. From now on the population would increase 10% every 5–10 years. Britain was at the forefront of technical developments with new materials, new processes and new equipment:
Portland cement for construction, **improved processes for iron manufacture, steam engines** to power factories and automated **textile machinery** to produce cheap cotton fabrics. British inventions also transformed transportation with **railways** and **steam-powered iron ships** driven by **propellers**. Communications were revolutionised with the invention of the **telegraph** and **improved postal systems**. The British began to understand the principles of **capitalism**, espoused free trade and became the economic powerhouse of the world. By 1870 this small country would dominate the world stage in a way that had never been seen before.

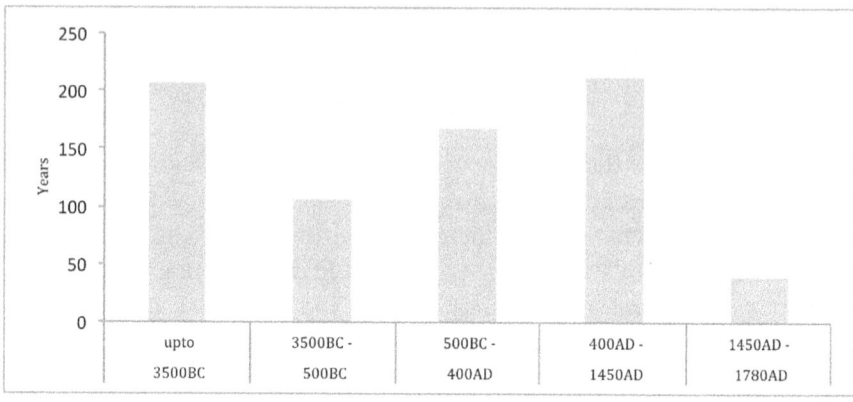

Figure 8.1 Number of years required for the population to increase by 10%

The rest of Europe struggled to adapt to the new democratic ideas coming from Britain and America. The French Revolution was a profound shock to the established order. Monarchies in Europe struggled to hold back the increasing political power of the middle classes. Ideas of freedom and democracy gradually gained ground; **slavery was abolished** in the Western world. Feelings of **nationalism** became stronger and the new nation-states of Germany and Italy were formed. Outside Europe and North America the other political cultures failed to keep pace with Western developments. They were now even losing out militarily; Western armies were becoming increasingly technically sophisticated with the widespread adoption of the **rifle** and **machine gun.**

Britain and the Industrial Revolution

In early eighteenth century Britain, iron production began to decline because there was a shortage of suitable wood for making charcoal. The Industrial Revolution began when it was discovered how to make iron using coke. Coke could be made from coal by driving off its volatile components in an airless furnace, in a process parallel to making charcoal from wood. It was first used in the brewing industry in order to roast malt without ruining its flavour. Many individuals attempted to use coke in iron-making but it was not until the 1750s that the first successful use of coke in a blast furnace was achieved by Abraham Darby in Coalbrookdale in Staffordshire. There, great quantities of iron were produced for the manufacture of various artefacts, such as cast pots, fire-backs, boilers and water pipes and it was also there that the first cast-iron bridge was erected over the River Severn in 1779. However, cast iron is brittle and there was still a need to convert cast iron to wrought iron which takes greater levels of stress. The ultimate solution would have to wait for steel making in the next century. However, in 1784 a process known as puddling was devised, to convert cast iron to wrought iron. Working in an iron mill at this time must have been like a vision of Dante's inferno. The pig iron was heated in a furnace through which holes had been made in order to allow bars in to stir the metal. This is an extract from the *Annals of Agriculture* describing what happened during the puddling process:

> When melted [the iron] spits out in blue sparks the sulphur which is mixed with it. The workman keeps constantly stirring it about, which helps disengage the sulphurous particles; and when thus disengaged, they burn away in blue sparks. In about an hour after melting, the spitting of these blue sparks begins to abate (the workmen stirring all the time), and the melted metal begins to curdle, and lose its fusibility, just like solder when it begins to set.

One can only imagine the men working in intense heat, deafened by the noise of combustion, breathing sulphurous fumes and straining to stir the heavy molten iron. If there is an image of the birth of industrialisation this is it; human ingenuity, by brainpower and intense labour, was able to extract the primary material of the Industrial Revolution from molten rock. Coalbrookdale, where this happened, now exists as part of the Ironbridge Industrial Museum. It is a vastly more important site in the history of human achievement than any of the traditional sites of visitor pilgrimage like Stonehenge or the Pyramids.

Steam engines (known as Newcomen engines) were first used in the early eighteenth century to pump water from mines. They were very inefficient and the cost of coal transportation made them uneconomic to use outside coal-mining areas. James Watt (1736–1819) made the technical breakthrough necessary to create the first commercially successful steam engine. Watt was a natural engineer. When his father died at the age of 18 he went to London to study instrument-making. On returning to Glasgow he was lucky enough to establish a small business using contacts at Glasgow University. This was despite opposition from the Guild of Hammermen, as he had not served a 7-year apprenticeship. To supplement his income he also repaired instruments; in the winter of 1763–1764 he was sent a small Newcomen engine for repair. He soon made an improved working model of an engine, with a separate condenser and a piston stroke driven by steam. This new design of the steam engine was to become the workhorse of the first stage of the Industrial Revolution. However, it took until 1776 to make the first practical scaled-up version, and it required another Watt invention to convert steam power to rotary power which could be used in textile mills. It was not until 1785 that orders started coming in large numbers. Before long the steam engine had largely replaced water and windmills to become the principal source of power for the first industrial nation.

With the Industrial Revolution came a decisive change in human social organisation. Unlike the rest of the animal kingdom, most British would not now be directly involved in food production. Former farmers moved away from the land to work in factories making textiles, pottery and iron goods. They worked for cash and bought their food in the towns. A new class evolved; former peasants became part of a newly emerging working class.

The transport of industrial goods by cart and canal was slow, and it took the invention of the railways to take industrialisation to its next phase. As the engineering of steam engines developed, with boilers capable of working at high steam pressures and thus generating more power, mobile steam power became possible for the first time. Although moving engines were demonstrated in the early 1800s, it took until 1825 before the world's first working public railway was started from Stockton to Darlington and it was not until 1830 that the first practical steam engine, the Rocket, worked on the Liverpool and Manchester railway.

Constructing railways was expensive. By nineteenth century standards the investments were huge. Joint stock companies raised the capital, bought the land, hired the navies to build the railway and invested in the locomotives and rolling stock. The railway companies were the biggest commercial organisations in the early industrial period. The railways revolutionised transport. By 1850 there were 6000 miles of railway track in Britain, transporting many more passengers at much greater speed than road transport by coach. We saw that in 1700 it took 4 days to travel by coach from London to York, and this was reduced to 1 day by road improvements. The same journey could now be covered in 8 hours. More than that, people were travelling in large numbers; 50 million passenger journeys were being made each year in Britain by 1850.

By 1870 there were about 14,000 miles of railway in Britain and the movement of freight raised more revenue than passengers. The competition between railway companies and canal owners for freight traffic had been won by the railways. The most important freight was coal, which could now be transported cheaply to industrial plants throughout the UK. Lower transport costs also made coal the most important source of power for domestic heating. As a result coal production mushroomed from 25 million tons in 1830 to over 100 million by 1870.

It was in the North of England that industrialisation took off. The most important industry was textile manufacture, both spinning and weaving. New mills were constructed, powered by steam. Labour flocked to newly burgeoning textile towns such as Bradford and Manchester. Other towns specialised in other goods, such as Sheffield for cutlery and Stoke for ceramics. The population of Britain grew rapidly from about 7 million in 1780 to 21 million in 1870. This was a far higher rate than had ever been seen before in human history. Liverpool, Birmingham, Manchester and Glasgow, previously minor towns, now became the most important towns in the British Isles after London. New towns required new housing construction methods. The material chosen was brick; poor people's houses could now be made of cheaply transported brick, instead of the traditional mixture of wood and wattle and daub. In new cities like Manchester, labourers now lived in back-to back brick dwellings in a grim rectangular pattern of streets.

Brick construction became more practical following the invention of mortar made of Portland cement. The ability to make cement in Britain had been lost since Roman times. Whereas simple traditional lime-based mortars degrade with exposure to water, those made of Portland cement harden. The inventor of Portland cement was another unheralded hero, Joseph Aspdin (1788–1855), from Leeds in Yorkshire, who ran a small building business and, like his father, was a bricklayer by trade. Portland cement is manufactured by heating limestone with clay in a kiln. From its discovery in 1824 Portland cement has become the most commonly used building material in the world.

In addition to the railways, ships were developed that could be powered by steam. Isambard Kingdom Brunel (1806–1859), who built the Great Western railway from London to Bristol, was the great pioneer of modern ship design. Brunel worked out that the larger the ship, the more efficient it would be. To test this theory, Brunel offered his services for free to the Great Western Steamship Company. His first ship, the Great Western, was the longest ship in the world at 72 m. It sailed regularly from Britain to North America and held the record crossing time of 13 days westbound and 12 days 6 hours eastbound. It was powered by paddle wheels, an ancient technology adapted from existing water wheels. The service was sufficiently commercially successful for Brunel to be asked to design another vessel. The resulting Great Britain is considered the first modern ship, being built of metal and driven by propeller. When launched in 1843, Great Britain was by far the largest vessel afloat. The Great Britain carried thousands of immigrants to Australia until 1881. The ship was scuttled in 1937, but later recovered and refurbished. In 1970 it returned to Bristol, where it was first built, and became an award-winning visitor attraction.

Britain's economic growth was helped by the widespread adoption of principles advocated in the *Wealth of Nations* by the Scottish political economist Adam Smith (1723–1790). This is one of the few philosophical works that has had an immediate positive practical effect. Its application has advanced the wealth of mankind. Smith's belief was that economic progress depended on three major factors: the pursuit of self-interest, the division of labour, and freedom of trade. In advance of Darwin's idea of 'natural selection' Smith identified the 'invisible hand to promote an end which was no part of his intention'. By this he meant that each individual pursuing self-interested gain for their own profit benefits society as a whole. It is probably more succinctly put in another famous phrase:

> It is not from the benevolence of the butcher, the brewer, or the baker, that we expect our dinner, but from their regard to their own interest.

Economists have shown mathematically that, in a perfect competitive environment in which consumer choice is strictly made on grounds of price and utility, the overall wealth of society is optimised. Smith, however, was of the real world and was not blind to industry's faults and he repeatedly warned of a 'conspiracy against the public or in some other contrivance to raise prices'. Smith states that the interest of manufacturers and merchants 'in any particular branch of trade or manufactures, is always in some respects different from, and even opposite to, that of the public'. In other words, the invisible hand needs regulation and policing to ensure that competition works effectively.

In the second key factor for economic progress, Smith saw that mankind benefitted from increasing specialisation in human productive activities, which he called 'the division of labour.' One telling example he gives is the pin-maker:

> ... the trade of the pin-maker; a workman not educated to this business ... could scarce, perhaps, with his utmost industry, make one pin in a day, and certainly could not make twenty. But in the way in which this business is now carried on, not only the whole work is a peculiar trade, but it is divided into a number of branches, of which the greater part are likewise peculiar trades. One man draws out the wire, another straights it, a third cuts it, a fourth points it, a fifth grinds it at the top for receiving the head; ... and the important business of making a pin is, in this manner, divided into about eighteen distinct operations ... they could, when they exerted themselves, make among them about twelve pounds of pins in a day.... Each person, therefore, making a tenth part of 48 thousand pins,

might be considered as making four thousand eight hundred pins in a day. But if they had all wrought separately and independently, ... they certainly could not each of them have made twenty, perhaps not one pin in a day; that is, certainly, ... not the four thousand eight hundredth part of what they are at present capable of performing, in consequence of a proper division and combination of their different operations.

The advantages of specialisation did not affect only industry, they applied to agriculture as well. Increased speed and the decreased cost of travel allowed more specialisation in farming and therefore more efficiency. For, example, fresh milk could be brought from the West Country by train and sold in London. This allowed West Country farmers with their wetter climate to specialise in dairy farming, while allowing local London farms to concentrate on arable crops.

Free trade was the third of Adam Smith's principles and it was in this area he was to have the greatest effect on state policy. States throughout the world had, over centuries, seen trade and manufacture as an opportunity to raise taxes. The passage below is from the *Oxford History of the French Revolution*, p.4 and describes the system of internal custom barriers in France in the eighteenth century:

> The salt tax, the notorious gabelle, was levied at six different rates according to area, while six other specially privileged districts, including Brittany, were exempt. And the whole country was criss-crossed with innumerable internal customs barriers, whether at gates of towns, along rivers or between provinces, where excises, tolls and tariffs could be collected – again at a bewildering series of rates, on a limitless range of items. Goods shipped down the Saône and Rhône from Franche Comté to the Mediterranean, for example, paid duty at 36 separate customs barriers, some public and some private, on the way.

The situation was even worse in Germany, which by the eighteenth century, had fractured into a myriad of states and there were an estimated 1800 customs barriers. Even in the largest German state, Prussia, there were 67 customs barriers. Adam Smith captured the advantage of free trade thus:

> It is the maxim of every prudent master of a family, never to attempt to make at home what it will cost him more to make than to buy.... If a foreign country can supply us with a commodity cheaper than we ourselves can make it, better buy it of them with some part of the produce of our own industry, employed in a way in which we have

some advantage.

Britain's last internal custom's barrier was abolished in 1707 with the union of England and Scotland, which gave British manufacturers an immediate advantage in the import and export of goods. Initially, however, Britain had a restrictive view on trade within the empire. It was perceived as the supply of raw materials from the colonies to support the manufacture of finished goods in Britain, which were then freely traded in the colonies. Predictably, while this benefitted British manufacture, it had a negative effect on the cottage industries overseas. For instance, cotton manufacture in India was badly hit as the English textile industry grew; imports of cotton fabrics into India increased from 1 million yards in 1814 to 995 million yards by 1870.

However, during the nineteenth century Adam Smith's ideas became increasingly accepted and absolute free trade became an underlying principle used to promote the world dominance of British manufacturing. In 1846, the Corn Laws restricting imports of cereals were repealed. The increasing size and speed of ships made it possible for commodity foods to be transported across continents. Bulk grain could be imported to Britain from the USA, reducing the price of bread for British families and at the same time reducing the risk of famine. The result was a further transfer of the British economy from the agricultural sector to manufacturing.

The Industrial Revolution, the advantages of the British political culture and the application of Adam Smith's capitalist principles were three reasons for the UK's commercial success. Sea power and its colonial strategy enabled Britain to ensure this commercial advantage could be exploited globally. Britain's influence was felt from Latin America to China, together with its ability to impose its will on any country outside Europe and North America.

Having lost part of its empire in North America, Britain had been able to expand in other territories. The most important was India where, after taking control of the East India Company in 1858, the UK established an empire including the whole of modern-day India, Pakistan and Bangladesh. This was the largest empire ever established in Indian history, uniting the whole of north and south India for the first time. India was the second most populous area in the world and Britain managed to control it with relatively few administrative staff from a distance of several thousand miles away.

During the American wars of independence, the Canadians to the north had remained loyal to the British Crown. Now Canadian territory was consolidated and extended across the continent to British Columbia. In addition, the British conquered the former Dutch colonies in South Africa,

Malaysia and Sri Lanka, when the Dutch republican government sided with the French in the early days of the Napoleonic wars.

Although the Dutch had first discovered Australia in the seventeenth century and called it New Holland, it was the British who first established a colony there in 1788. At first it was a penal colony. However the discovery of gold and the development of agriculture soon led to the establishment of a number of self-governing states. The native population were Stone Age hunter–gatherers and could not offer any effective military resistance. They were rapidly defeated and their lands seized. As with America, contact with Europeans introduced new diseases which soon dramatically reduced the native population.

Finally, a British colony was established in New Zealand from 1840 onwards. However, the local Maori, being tribal farmers from Asiatic ethnic groups, were less crippled by foreign diseases and able to offer more organised resistance. As a result, the local population were able to fare better under British rule than the Australian Aborigines.

Not only did Britain's own population increase significantly as it enjoyed the benefits of increased industrialisation, its citizens emigrated to the USA, Canada and Australia to exploit the territories taken from the natives. This left the British, in the mid-nineteenth century, as the only global Western power with interests in all five continents. No empire of its like had ever been seen before.

With this global empire, good communications were vital. The British developed three major improvements: the reform of the post, the electric telegraph and newspapers. The post was the brain-child of another great Englishman Roland Hill (1795–1879). Hill noticed that most of the costs of the existing postal system were spent in the administration of pricing, charging and sorting the letters rather than in their transport. He reasoned that the costs could be reduced dramatically if postage were prepaid by the sender in the form of stamps. On becoming Postmaster General, in 1840 Hill introduced a new uniform post rate of one penny for all deliveries within the UK. Communication by letter was transformed. The system provided a model for quick and efficient personal communication that was soon copied by every country of the world.

During the Napoleonic wars the ability to transmit messages over distance was enhanced by a system of semaphore signals transmitted by line of sight from one semaphore station to another. This allowed short messages to be sent reasonably quickly over considerable distances. Napoleonic France, for example, had a network of 556 stations stretching a total distance of 4800 kilometres. This optical telegraph was replaced in the middle of the nineteenth century by the first significant practical use of electricity: the electric telegraph. Powered by the newly invented electric battery, small surges of current could be sent down a copper wire. These surges were coded to represent letters, the most famous coding system being the American-invented Morse code. The first commercial telegraph in 1837 connected Paddington to West Drayton on the Great Western Railway. As telegraph wires could be conveniently constructed alongside railway lines, the development of railways went hand in hand with the development of telegraphy. The technology improved rapidly and it was soon possible to transmit messages across the sea by laying cable on the ocean floor. By 1870 the British had completed a telegraphic link to India. Messages that would have taken weeks to transmit could now be sent in minutes: the world had suddenly become dramatically smaller.

Newspapers and magazines, first started in the seventeenth and eighteenth century, now added to the speed of communication of events and ideas. Using the new electric telegraph, the Crimean War (1853–1856) was one of the first wars to be documented extensively in written reports by newspapers, notably by William Russell (for *The Times* newspaper). This was the first war in which the public were kept informed of the day-to-day realities of conflict. Russell, dressed in semi-military clothes, mixed with junior officers and other ranks to gain his information. He jingoistically praised British heroism but equally condemned the logistic problems and medical bungling that caused great suffering to the troops.

Russell's reports led to attacks on the government, claiming that out of a British contingent of 32,000, two-thirds were unfit for duty due to their ill health. Russell reported that a large number of soldiers were dying of cholera. Prejudice against women's involvement in medicine was overcome by the public outcry resulting from Russell's articles, and Florence Nightingale was eventually given permission to take a group of 38 nurses to the military hospitals located in Turkey. The results of her work led to the establishment of a trained nursing profession, to the great benefit of the health of the nation overall. When a proposal was passed to create an enquiry into the condition of the British Army, the prime minister was forced to resign. For the first time the fast dissemination of information was affecting events in a very significant way.

The period of up to 1870 was one of astounding British industrial predominance. This is from *The Rise and Fall of Great Powers*, p. 193 by Paul Kennedy:

> Around 1860, which is probably when the country reached its zenith in relative terms, the United Kingdom produced 53% of the

world's iron and 50% of its coal and lignite, and consumed just under half of the raw cotton output of the globe. With 2% of the world's population.... It alone was responsible for one-fifth of the world's commerce, but for two-fifths of the trade in manufactured goods.... It was no surprise that mid-Victorians exulted at their unique state being now ... the trading centre of the universe.

The UK, however, was still not a true democracy. Only upper and middle-class men were enfranchised and the British aristocracy continued to exercise undue influence and control over the country.

France and the growth of the nation-state

The growth in British power was viewed with a mixture of respect and envy by the other kingdoms of Europe, subjecting them to mounting political pressure to match British performance. The French government was weak and cracked under the strain, in the most momentous political event of the age: the French Revolution.

Ironically, it all started with the French government's default on the loans raised to fight the British during the American War of Independence. King Louis XVI was warned that French support for a rebellion against a fellow monarch might be counter-productive, and so it proved. After a number of unsuccessful attempts to impose a new tax regime by traditional means, Louis XVI was forced in 1789 to convene the Etats Généraux. This was a traditional advisory body and consisted of three separately elected chambers representing the clergy, nobility and commoners. The inspiration of the American War of Independence, which had spread the heady ideas of liberty, democracy and equality, emboldened the third estate, the commoners, to declare itself a National Assembly and to set about changing the power structure in France. When the Paris mob stormed the Bastille on 14 July 1789 and the King was advised that he could no longer rely on the army to put down riotous behaviour, all France knew the French monarchy could not survive in its current form. Revolutionary committees were set up in the towns. The peasantry ransacked the muniment rooms where the records of feudal obligations were held, and refused to pay tithes and dues. In August 1789 the Assembly legalised the peasants' actions and swept aside centuries of feudal tradition. All the privileges of the nobility and clergy were annulled in the 'Declaration of Rights of Man and the Citizen' and in the new constitution. It is a tribute to the power of the new American democratic ideas that all this was achieved with an amazing amount of consensus across the country. The election of representatives to the new

Legislative Assembly in 1791 was conducted in an orderly fashion. The aristocracy were unable to prevent the changes to their status and were forced to choose between exile and compliance.

The role model for the new constitution was the British constitutional monarchy. The king would select his ministers and the National Assembly would enact any legislation. However, unlike the British system, the king was forbidden to select ministers from the National Assembly. The Assembly and the king's ministers were thus independent entities and prone to conflict. Given Louis XVI's proven incompetence, the new constitution was unlikely to work. When the King tried to escape by making a dash for the Belgian border, the cause of constitutional monarchy was lost.

The end came after the new government declared war on Austria. The war initially went badly for the French and panic and recrimination swept Paris. A mob of sans-culottes, Parisians claiming to be ordinary patriots without fine clothes, invaded the Tuileries, the King's residence in Paris. The Assembly had no power or will to support the government, the King was deposed and a Republic was declared. The Assembly effectively took over the government of the country, but the institutions of power had not been fully developed. Leadership and control was difficult to exercise. Worse, the Paris mob had the Assembly in its power and was able to influence events by mass provocation. It was from this point that the French Revolution entered a deeply disturbing period, later to be repeated in fascist and communist states in the twentieth century. In the Terror some 16,000 people were killed, labelled as enemies of the state. To have spoken a different view from that espoused by the government of the moment risked a visit to Madame Guillotine. Paranoid leaders, urged on by a bloodthirsty mob, conducted a cold-blooded, quasi-judicial culling of their opponents.

The attempt by the French to copy British and American democratic ideas had failed. Other governments were severely frightened by the events in Paris. The cause of democracy had received a setback. It would be another 40 years before democratic ideals would be taken up again in Europe.

The French Revolution launched another evil – the patriotic war. After the initial military reverses, the French revolutionary government proclaimed a levée en masse in 1793, conscripting close to a million Frenchmen to arms. The French army, imbued with patriotic revolutionary fervour, was able to overwhelm the armies of other nations by sheer force of numbers. This degree of fighting intensity had not been seen since the religious wars of the seventeenth century. The revolution had been able to harness national patriotic energies, martial resources and mobilise mass citizen armies on a scale never seen before. The scope of conflict in Europe had been enlarged to be not just a fight between princes who employed professional armies, many of whom were mercenaries, but between nations that were able to call up their citizens to fight with fervour in the national cause.

Eventually the French Revolution developed a more traditional course and a military strongman, Napoleon, seized power. In 1804 he dissolved the Republic and re-established royal rule. The French Revolution developed into a fight to establish a French empire across Europe. Ultimately Napoleon was defeated, but the reaction to the French Revolution would overshadow political developments in Europe for decades to come.

After the Napoleonic wars, the Christian monarchies of Europe followed the British lead and invested in railways, roads, coal and iron. The resulting improvement in communications transformed the working conditions of the peasantry. The peasant lifestyle had been the normal existence of most of mankind from the Neolithic period onwards. It was based on a local village community. Isolated by poor roads, each village had its own customs, culture and even language. In France in the early nineteenth century the French language was only the language of government, educated people and most of northern France. At the edges of France, peasants still spoke in unrelated foreign tongues: Flemish, Breton, Basque, Catalan, German and Italian. In the South of France peasants spoke Occitan, a romance language somewhere between French and Catalan. As a result about a quarter of French people did not speak French at all. French officials had to employ interpreters to make themselves understood in their own country.

The peasants were aware they were French, but the concept of France was of something foreign and outside village life. It meant dealing with officials from the towns. Being often illiterate they were in constant fear of being cheated by magistrates, lawyers, tax collectors and the police. The diet was poor. There was sometimes bread and pancakes but the stable meal was soup. Soup was cereal, chestnuts, cabbage, turnips or potatoes boiled in water. Meat was rarely eaten. Their bread was miserable stuff, baked in large batches every two or three weeks to save fuel.

Peasants were unconcerned with hygiene. They bathed when and where they could, sometimes in a stream but frequently in a stagnant pond which was often contaminated with sewage. Washing clothes was a complicated logistical exercise done as infrequently as possible. A poor diet, hygiene and cleanliness inevitably led to poor health. This was compounded by diseases due to inbreeding. In 1830 the minimum height for army recruits was lowered to 1.54 metres. Even then, a large proportion of peasants were not fit for military service.

Critically dependent on suitable weather for the growth of their crops, peasants were always liable to experience periods of famine. Limited in the amount of food they could store or sell in a good year; peasant farmers were always seeking to minimise their risks and reluctant to change proven practices. Ancient ploughs continued to be used.

This lifestyle was typical of peasants all over the world, living a largely self-sufficient existence in a cash-poor economy. They were always prey to famine and disease and were at the mercy of richer members of society. Their loyalties were to their own villages and all strangers were threatening. Change only came when roads, canals and railways linked the villages to the rest of the world. When transportation was sufficiently cheap, farmers were able to sell their produce to a wide market; this changed the whole economy of the peasant society. The economic objective of peasant farms became to maximise an income from their crops or animals, not just to provide sufficient food for their family. Farmers could specialise in crops that were suitable for their lands. Once the farmers' objective was to make money, efficiencies naturally flowed. They no longer needed to make cloth and clothes. Fresh bread could be brought from the local bakery rather than stale bread made infrequently. The land could be used for its most suitable purpose. They could invest in new equipment, buy fertiliser and improve their land.

Traditionally, French peasants had been deeply superstitious. Their religion, though formally Catholic, harked back to an earlier era with festivals and practices evoking local saints and traditional lore. This ancient religious culture was lost in the change; the local patois disappeared along with traditional dress, festivals, stories, songs and dance. Culture became regional and national. To new peasant farmers being French no longer meant the imposition of a foreign state. France provided education and opportunities for advancement as well as demanding tax payment. As peasants joined in with the rest of the country, their allegiance changed from the village to the country as a whole. France had progressed from an area in which the population owed allegiance to a king; it had become a nation-state, with the whole population bound by strong communal feelings.

The development of roads and railways had the same effect in other nations in West Europe. Hitherto, the land boundaries of states had been defined by the power exercised by a king or prince over his lands. In the case of a few countries such as England, the Netherlands and Castile, the king's realm contained mostly one people speaking one language, so that the boundaries of the king's lands corresponded with the boundaries of the nation. Both Castile and England were able to increase their power by dynastic union with an adjoining country (Aragon and Scotland, respectively) to create the new nations of Great Britain and Spain, with a combined population that retained a relatively undiminished level of patriotic support. This was not true in the rest of Europe. The further east one went in Europe, the less were the empires, kingdoms and principalities tied in with the concept of nationality. German speakers could be found in Austria, Prussia, Switzerland and 38 other independent states. Italy consisted of a patchwork of nine states, of which the largest was the Kingdom of Sardinia. The Austrian, Prussian and Russian empire contained communities of Slav language-speakers to the east (such as Czechs, Poles and Croats) who had no state of their own.

During the Napoleonic wars conquered states were initially enthused by French ideas of liberté, egalité and fraternité but soon learned that they were expected to fully support the French empire. In turn this provoked nationalist feelings among the Germans and Italians. Popular liberal and national opinion in both countries was enthusiastic about the political unification of those who spoke the same language. This was exploited by ambitious politicians of Prussia and Sardinia in a way that allowed their states to expand and encompass the rest of the country. Prussia became the dominant state in the new German empire. The Kingdom of Sardinia became the Kingdom of Italy.

The political cultures of Europe had changed again. The Protestant monarchies of north Europe and France and Italy had all become nation-states. These new nation-states not only copied successful British technologies; they started to copy British political culture as well. The change was sometimes painful. The middle classes had trouble wrestling political power from the monarchies, who were torn between being frightened of the effects of reform, as seen in the French Revolution, and seeing the benefits of more democratic government, as demonstrated in Britain and the USA. The French symbolised this struggle to find the right leadership system. Over the period to 1870 they tried absolute monarchy, constitutional monarchy, democratic republic and empire; none achieved long-term stability.

Europe now consisted of three major nation-states (Britain, France, and Italy), two empires with a dominant national group (Germany and Russia) and the multinational Austrian empire. All six states were intensely competitive and keen to make their mark on the world stage. With developing nationalism and huge patriotic armies, the stage had been set for the competition between these six to grow until they erupted in that great disaster for mankind, the First World War.

USA and the elimination of slavery

Britain's daughter state, the USA, had thrived in its first 100 years of existence. Except in the developing north-east, the USA was still a predominantly rural economy. The nation had concentrated on exploiting the opportunities of expansion in a largely depopulated continent. As a result of the Louisiana Purchase from Napoleon, wars against Mexico and agreement with the British on their northern border, the USA was able to expand from the original 13 colonies right across the continent to California and Oregon. Thus Western culture made a huge territorial expansion which was equal in size to Europe itself. The Native Indians, their population shrunk by disease and war, would soon be confined to a few miserable reservations.

However, one major issue had developed with the new Constitution. Although all men had been created equal, according to the Declaration of Independence, this manifestly did not include slaves. The legality and morality of slavery divided the new nation. The issue raised great passions between northerners and southerners and prevented the nation from uniting behind a common political culture. Slavery had been a condition of mankind from Neolithic days and was commonplace in most of the ancient empires. It had withered away as an institution in Western Europe in the Middle Ages but expanded greatly with the discovery of the Americas. In all of central and south America (including the south of the USA) local native labour was scarce due to death from disease. This led to the importation of slaves from Africa to work in the mines and on the farms. By the nineteenth century substantial proportions of the population of Brazil, the West Indies and the USA were slaves of African origin.

Slavery could not be easily reconciled with the conscience of liberal or religious opinion in Europe and the northern USA. As the British poet William Cowper wrote in 1785:

> We have no slaves at home – Then why abroad? Slaves cannot breathe in England; if their lungs receive our air, that moment they are free, they touch our country, and their shackles fall. That is noble, and bespeaks a nation proud. And jealous of the blessing. Spread it then, And let it circulate through every vein.

The first to abandon slavery in America were the French. In the height of the revolution, inspired by egalitarian ideals they abolished slavery in the French West Indies. This was later reversed when Napoleon came to power, but the slaves in Haiti rebelled and managed to overthrow the French and establish their independence, becoming the first Black-led state in the Americas.

Slave trade was made illegal by the British Parliament in 1807, and later in 1827 Britain reinforced this by declaring that the slave trade was piracy. Between 1808 and 1860 the Royal Navy seized approximately 1600 slave ships and freed the 150,000 Africans who were aboard. Slavery was abolished in the British empire in 1838. £20 million compensation was paid to slave owners but nothing to the slaves. France followed with abolition in 1848.

This left America as the major slave-owning country in the world. Slave labour was considered crucial to the cotton trade which was vitally important to the economy of the southern USA. Until the end of the eighteenth century, Western people had worn linen and wool garments, which were difficult to wash and dry and therefore were often filthy. Cloth made of cotton offered a solution. However, until industrialisation, cotton cloth was expensive because it took 12 to 14 man-days to make a pound of cotton compared to 1–2 days for wool. By 1812, as automated spinning developed in England, the cost of cotton yarn had fallen by 90% and by the 1860s the price of cotton cloth was less than 1% of what it had been at the start of the Industrial Revolution. Cotton products, chiefly made in Britain, clothed the world.

The American Eli Whitney (1765–1825) invented the cotton gin which automated the otherwise tedious process of separating the cotton seed from its fibre. The southern states of America now had the formula for profitable cotton manufacture: cheap land, slaves and the cotton gin. The first cotton bale from America arrived in Liverpool in 1784 and by 1860 over 90% of the cotton imported into England came from Southern plantations. To overcome the issue of the lack of new slaves coming from Africa, the southern states resorted to the obscenity of slave-breeding farms, in which mates were provided for nubile women slaves. At this period the value of slaves themselves accounted for 35% of the working capital of the South. Culturally, the South was committed to slavery and it was not long before religious arguments were invented to justify this stance. In 1822 the Southern Baptist Association produced a biblical defence of slavery and the Bishop of Charleston found theological arguments to ease the conscience of southern Catholics.

The difference between the north and the south reflected in many ways the difference between a more modern industrial democratic state and the older, aristocratic-style of social organisation. The gentlemen-farmers of the South with their slave plantations had much in common with the landed gentry in Russia and their estates worked by peasants. The Southerners felt their way of life threatened by industrialisation and the egalitarian liberal philosophy of the North. Eventually, they made a futile attempt to break away from the North and brought the USA to civil war. Religious sects divided on questions of principle between south and north; both sides sought God's support. The major religions: Presbyterians, Episcopalians, Lutherans and Baptists tried to avoid public discussion but Revivalists and Evangelists naturally went towards extreme positions from both sides. This is from *The History of the American People* p. 479 by Paul Johnson:

> To judge by the hundreds of sermons and especially composed church prayers which have survived on both sides, ministers were among the most fanatical of the combatants from beginning to end. The churches played a major role in dividing the nation and it may be that splits in the churches made a final split in the nation possible. ... Southern clergymen were particularly responsible for prolonging the increasingly futile struggle.

This is yet another illustration of the way religion combines with state policy to create a political culture. The political culture of the South was that of a privileged elite in a rural society. Across the world this culture was being challenged by the changes resulting from industrialisation. When the North won, slavery was abolished in the USA and a major break had been made with the past. From this point onwards, slavery as an institution began to disappear across the world.

Discrimination on the grounds of race, however, would continue. Hatred between different racial groups would remain endemic in the US South and other parts of the world. Other non-Western ethnic groups were perceived as inherently inferior. Unfortunately, racists found intellectual support in the newly developed theory of evolution. They incorrectly deduced that it was genetic differences, rather than the combination of memetic development and serendipity, which had determined which group of humans had won out in the battle of survival of the fittest.

Evolution

Christians at this time believed, as it says in the Bible, that the world was created in six days and man was created in God's image on the last of these days. They also recognised just one catastrophic geological event, the Flood, which was survived by Noah in his Arc. Archbishop Ussher had used the source material in the Bible to calculate the creation of the Earth as 23 October 4004 B.C. Gradually, this simple view of the world began to unravel as humans began to look with Enlightenment eyes at the world and the animals around them.

It was in the field of geology that the first chinks began to be found in the biblical creation story. Fossils found in rocks were recognised as the petrified remains of once living things. Puzzlingly, some of the remains were from species that appeared to be extinct – how could it be that God created animals, only for them to die out?

The first recognisably modern geologist was James Hutton (1726–1797), who came from a middle-class background in Edinburgh, had a medical degree and took a broadly Enlightenment view of natural phenomena, including geology. His interest was awakened when examining the strata exposed during excavation of the Forth and Clyde canal. He published his seminal *Theory of the Earth* in 1785. In this he made the deduction that if fossils of sea animals are found in rock, then at one point in geological time this rock must have existed below the sea. He deduced that 'the land on which we rest is not simple and original' and 'that before the present land was made, there had subsisted a world composed of sea and land'. He reasoned that there must have been a period of time where loose material gathered and was compressed at the bottom of the sea and at some stage in the past the land below the sea must have been elevated. This was the first coherent explanation that the world had not been created in one event during the six days, but must be continually changing. Further, the process by which mountains rise and erode away and ocean bottoms are elevated to become land must have involved many more years than the 6000 or so allowed for by Archbishop Ussher.

The first geological map was made by a very practical man, a true working-class hero, William Smith (1769–1839), a canal surveyor and son of a blacksmith. Working in the canals and mines of Somerset, he observed that rock layers were arranged in a predictable pattern and that the various strata could always be found in the same relative positions. Additionally, each particular stratum could be identified by the fossils it contained and the same succession of fossil groups could be found in many parts of England. By 1801 he had sufficient knowledge of the geological composition of Britain to draw his first sketch of the 'The Map that Changed the World' – the first geological map of Britain. Unemployed, he travelled the length and breadth of the country to complete his work and in 1815 published his magnum opus. Unfortunately his maps were plagiarised and he ended up in a debtor's prison. His relatively humble education and lack of connections delayed the official recognition of his achievements. However very late in his life he was awarded a medal by the Geological Society of London, and acknowledged as the father of English geology.

While the science of geology was developing, the study of plants, botany, also took a step forward. In Europe plant compendiums had been produced describing plants and their medicinal use. Now plant collections were being brought back from all parts of the world by the British and Dutch, in particular. Many tried to classify the plants but none was successful until Carl Linnaeus (1707–1778). He assumed that if nature was a divine creation it must have a rational order and that he had been privileged to see the outline of the creator's plan. His technique was outlined in a slim pamphlet called the *System of Nature* in 1735; this later grew to a multi-volume classic, covering many hundreds of families of plants and animals.

His structure was simple. He grouped similar species into genera and grouped genera into broader families. His grouping method was based largely on the reproductive organs of flowering plants. He freely admitted that eventually better ways of grouping plants would be found and that his classification would be refined and superseded.

In 1758 he extended his system to animals. A simple example of his hierarchical structure of family, genus and species is the cat family (Felidae); this includes the genus *Felis*, which contains the species *catus* (cat) and *lynx* (lynx), and the genus *Panthera* which the species *leo* (lion), *tigris* (tiger) and *pardus* (leopard). The overall Linnaean classification divided the world into plants, animals and minerals. Animals were further divided into mammals, birds, amphibians, fish, insects and vermes, which was everything not covered by the first five classifications.

Most botanists at this time believed that God created species and once created, the species were fixed for all time. Linnaeus recognised that hybridisation is common in the plant world and that it is possible that some new species could be created by hybridisation. He therefore suggested that God might have created the first of the family and the remaining members were created later by a hybridising process. This was the first glimpse of the idea of evolution.

The next major step forward was made by Georges Cuvier (1769–1832), who developed a great skill in comparing the anatomies of animals from their internal structure, as revealed during dissection. He began to understand that animals could be grouped in a more profound way according to their internal structures. He envisaged four types of animal

Vertebrates	Creatures with a backbone (including, mammals, birds, fish and reptiles)
Molluscs	Creatures with no backbone , but sometimes an external shell (such as oysters, clams etc)
Articulates	Creatures with segmented bodies (insects, spiders, worms etc)
Radiates	Creatures with a circular organisation (starfish, sea urchins etc)

Table 8.1 Cuvier's types of animal

Although there have been refinements to this structure and the types (or phyla) have been redefined and subdivided, the basic idea has remained valid to this day. Cuvier was also able to use his anatomical skills to review and classify fossils. He was the first to prove that distinct species have become extinct with his work on mastodons and mammoths. Like William Smith he realised that the fossils contained within each geological stratum were distinct. He also realised that the older the strata, the stranger the fossils became. Strata (and by extension, geological time) could in fact be ordered according to the type of fossils they contained. He thus launched the science of palaeontology, the study of fossils to determine their evolution on earth. By 1850 the palaeontologists had identified three broad eras: cenozoic (the age of mammals), mesozoic (the age of reptiles) and the paleozoic (the age of fish).

This already gave an indication of the direction of evolution, but the scientific community was not yet ready to admit anything other than the existence of fixed species created by God. God was perceived to have created a natural order in which man and animals pursued their divinely determined destiny.

One man who punctured this vision of a static God-created environment was Thomas Malthus (1766–1834), a clergyman from a village near London. In his 1798 essay on the principle of population, he postulated that population growth is exponential, whereas the food supply growth is linear. This means that natural population growth is always checked by the ability of the environment to support it. This could be by positive checks: famine, war or disease or preventative checks: delayed marriage or sexual abstinence. As a true Protestant Malthus subscribed to the work ethic and believed that the constant threat of poverty and starvation served to teach the virtues of hard work and moral behaviour. He saw the increase in the population at the time of the Industrial Revolution with horror, predicting that if the working and agricultural classes did not cease to procreate they would suffer famine, death and disease. In this of course he was entirely wrong; the benefits of industrialisation had dramatically increased the human ability to support a higher population density from natural resources.

Nevertheless, his theory that population growth is naturally checked by the environment was one of the critical inputs in Charles Darwin's theory of evolution. Darwin's book *The Origin of the Species*, published in 1859, is probably one of the most influential books ever written. It had a long gestation. Charles Darwin (1809–1882) was the fifth son of a wealthy, middle-class family. He trained as a clergyman and read works on the divine design of nature. The defining moment in his life came when in 1831 he gained the post of naturalist accompanying the voyage of HMS Beagle. This 5-year cruise around the world, including a prolonged spell in the Galapagos Islands off Ecuador, provided the background examples from which he could speculate on the processes by which species developed. By 1844 it is clear that Darwin had developed the outline of his theory that one species could evolve into another over a long period of time. The mechanism of evolution was perceived as natural selection whereby those variants of the species that were best adapted to thrive in an environment would be naturally selected over time. Thus, for example, those that could run faster, or had longer necks or sharper teeth will procreate in a specific niche at the expense of those less endowed. Over hundreds of thousands of years the cumulative effect of small changes would produce very different animals adapted to new environments in novel ways. Eventually they would no longer breed and would become new species. Evolution was thus perceived as a branching process in which several species evolved from a common ancestor.

Darwin was, however, a very cautious man and spent several years in correspondence testing his theory and doing his own field work, particularly on racing pigeons and barnacles, to provide more concrete examples. In the end, it was only when a rival, Alfred Wallace, reached similar conclusions that he was persuaded to publish. The book was a sensation; his arguments were sufficiently persuasive for the scientific community to accept evolution of species over time as the way life developed.

Natural selection was not, however, generally accepted as the mechanism of evolution. It seemed too haphazard and did not seem to allow much scope for God to influence events. Most scientists preferred the idea that the physical changes acquired during the lifetime of an animal could be passed onto its progeny. Thus animals could, helped by God, initiate their own self-improvement process. It was well into the twentieth century with the development of genetics before natural selection became seen as an uncontested means of evolution.

In the meantime, evolution theory left several questions unanswered concerning the evolution of humans and racial groups, which were of potent political importance. The pseudo-science of eugenics advocated improving the human species by the selective breeding of positive traits. Practitioners recommended that those with 'less desirable characteristics' such as the disabled, promiscuous women, homosexuals, or even racial groups, should be sterilised. In the twentieth century the Nazi party in Germany even used eugenic arguments to justify the mass murder of Jews and Gypsies. It was not until the second part of the twentieth century that these perversions of the evolutionary idea were finally proved to be false.

Health care in Europe and North America

Until the nineteenth century in Europe it was hard to distinguish between medicine and quackery. Doctors were trained in the entirely false Greek theories of the four humours: blood, yellow bile, black bile and phlegm. In an ill patient these humours were thought to be out of balance and the patient was therefore subject to bloodletting or purging to bring the humours back into line. Doctors were fully trained in this useless theory and considered themselves a cut above surgeons, who dealt with the practical issues of broken bones and abscesses resulting from physical injury.

Concerned with the need to treat the large numbers of soldiers injured in the wars of the French Revolution and prepared to break with long-established traditions, the French Assembly put its trust in a scientist, Antoine François, Compte de Fourcroy (1755–1809). Fourcroy, who had made his name as a chemist, set the blueprint for a new style of medical school that emphasised the scientific method of practice and observation over trumped-up theories of no proven validity. Students were encouraged, in his words 'to read little, see much, do much'. He combined the professions of doctor and surgeon and thus allowed surgical intervention to be a possible medical treatment. He also insisted that doctors should practise in hospitals where they could see, treat and observe a large number of patients. Up to that point doctors always practised in their own surgery or at the house of their rich patients; hospitals were places for the poor sick only. Three of these new medical schools were set up in Paris, Strasbourg and Montpelier, and were the forerunners of modern hospitals.

French hospital medicine started a new discipline using the three pillars of medicine: diagnosing an illness, relating the clinical symptoms of patients to their pathology on death, and evaluating the effectiveness of medical interventions. The diagnostic technologies available to them were limited but were nevertheless scientifically applied. Medical students were taught the four methods of physical diagnosis: inspection, palpation (touching), percussion (tapping) and auscultation (listening to breathing). The latter was considerably helped by the use of the stethoscope invented by René-Théophile-Hyacinthe Laennec (1781–1826). He was particularly interested is diagnosing tuberculosis or, as it was then known phthisis, a disease that later claimed his life.

The newly trained French doctors were their own pathologists, trying to correlate their observations of a disease's progress during a patient's life with its physical manifestation in the morgue. They were intent on observing lesions – the physical damage connected with the disease. A new classification of diseases came to be based on the major organs affected. Tiny lesions called tubercles were identified as the hallmark of tuberculosis and swollen lymph nodes in the membrane of the large intestine were associated with typhoid fever. Limited cures were available but valuable groundwork was being made to identify diseases.

The first attested clinical trial was done by Pierre Louis (1787–1872). In his *Researches of the Effects of Bloodletting in Some Inflammatory Diseases*, Louis attempted to evaluate the effectiveness of different therapies by dividing similar patients into groups and comparing the results. This and other studies soon led to the abandonment of bloodletting as a treatment. The old quack doctors could no longer practise with impunity.

The British made two major advances in public health, by developing vaccines for smallpox and by bringing cholera under control. Smallpox was common throughout the world in the eighteenth century, and carried a significant risk of mortality. It was recognised that those who recovered from the disease never caught it again. Edward Jenner (1749–1823), a general practitioner from Gloucestershire, noticed that a similar disease in cattle, cowpox, sometimes induced a pock blemish on the hands of milkmaids and they, too, seemed immune. Jenner conducted the ethically dubious experiment of injecting some of the cowpox lesion from a milkmaid into a boy and then deliberately infecting him with smallpox. Fortunately, the boy survived in good health and Jenner published his results. The first scientifically based vaccination had taken place and the practice was quickly taken up in the Western world. It was not until 1979, however, that the disease was finally eradicated.

Cholera had been present in the Ganges area of India from early times, but in the nineteenth century the disease spread, through a series of pandemics, to affect most of the world. Millions of people died. However for the first time in human history, the cause and mechanism of the transmission of the plague were identified. John Snow (1813–1858), a doctor from London, proposed that the disease was transmitted by human faeces in water. He was able to demonstrate this in two studies. In the first, he correlated the incidence of the disease with people drinking water in central London at a particular pump with a contaminated water source. In the second he compared the incidence of the disease between those people drinking filtered water upstream of London with those downstream drinking unfiltered water from London sewers. He was able to show that there was a 13-fold difference in the incidence of the disease and thus demonstrated that it was waterborne.

The obvious solution was to build a proper sewage system for London, separating drinking and waste water. However, Parliament did not feel the problem was sufficiently urgent until the 'Great Stink' of 1858. With more people using flushing toilets, the existing cesspits could no longer cope and raw sewage spilled into the drains and into the Thames. During the hot summer of 1858 the bacteria multiplied and the smell became unbearable. Eventually business at the House of Commons on the Thames was no longer possible, and Parliament was relocated upstream to Hampton Court. The next year work commenced on the first modern sewage system for carrying wastewater away from London. Over one hundred miles of intercepting sewers were constructed to take wastewater away from the river Thames at London. These were fed by 450 miles of main sewers and 13,000 miles of local sewers. Drinking water soon improved and once another source of contamination from the river Lea had been eliminated in 1867, cholera ceased to be a problem disease in London.

Developments in other political cultures

Political cultures outside Europe and North America struggled to keep pace with the new advances in technical innovation and military prowess developed in the West. A brief summary of developments in the other political cultures is given below.

Latin Catholic cultures in Iberia and Latin America

Both Spain and Portugal were effectively ruined by the French invasions during the Napoleonic wars and thereafter struggled to find a stable system of government. During these wars they also lost control of their Latin American empires. When Spain tried to reassert control after the war, local revolutionary groups rebelled. At first they met with little success, but gradually the forces of rebellion gained the upper hand and by 1821 all the former Spanish Latin American territories had achieved independence. It was in this period that important new states such as Chile, Colombia, Mexico, Peru and eventually Argentina were established.

Brazil also emerged from Portuguese colonial dominance at the same time. During the Napoleonic wars the King of Portugal Joao VI was evacuated to Brazil and on his return to Portugal in 1822 he left his son Dom Pedro as the governor of Brazil. Later that year Dom Pedro declared independence with himself as Brazil's new emperor. The Portuguese were powerless to change this fait accompli.

From the start these new Latin American states showed none of the political stability and economic dynamism of the USA. The independence movement had been led by the landed classes. There was none of the same egalitarianism of the American colonies. Society was divided into five distinct classes: those of only Spanish or Portuguese parentage, land-owning creoles, other mixed race locals, Indians and slaves. There was no question of a democratic form of government pulling the new countries together. Further, the newly formed states did not have the same experience of self-government before independence as the ex-British colonies in the USA. In consequence, their early years of independence were chaotic, with different interests struggling for power. In Mexico, for example, there were 56 governments in 40 years after 1822.

The Islamic sultanates of the Middle East and North Africa

The long period of decline in Islamic power was now well underway. Islamic states never took on the innovation brought in by Western powers. The Ottoman Empire and Qajar Persia were forced to yield land to Russia. The Islamic states in India were effectively controlled by the British and in the Far East the Dutch and the British eliminated the remaining Islamic states.

The neo-Confucian culture of China, Japan, Korea and Vietnam

This was the period when the military weakness of China compared to Europe was first brutally exposed. In the first of the opium wars with Britain (1839–1842), China had a population of 410 million people, compared to 27 million in the UK, but China was unable to beat the British on its own soil. The British warships were totally dominant at sea and the Western artillery and muskets (which were now rifled and shooting conical bullets) were markedly superior to the Chinese ordinance. In the second opium war (1856–1860), a British and French force of only 18,000 was able to advance overland to Beijing, defeat superior numbers of Chinese on the way, burn the emperor's summer palace and prepare to lay siege to Beijing.

With their obvious deficiencies in military technology, this should have been a wake-up call for the leadership to modernise its military forces. But, imbued in neo-Confucian philosophy, the Qing dynasty had no idea how to respond to a technically superior foe.

The British and French insisted in setting up concessionary trading ports like Hong Kong, with free trade rights into China and privileged legal status to protect their nationals from the capricious nature of Chinese law. Westerners insisted on the legal right to try their own people in their own courts; Chinese people living in the concession were also to be subject to Western Laws. This was to be the first of many losses of territorial sovereignty that the Chinese were forced to endure in the nineteenth and twentieth century.

The Orthodox Christian empire of Russia

Only the Orthodox Christian Russians managed to match the Western powers militarily. They expanded their influence in the West by participating in the dismemberment of Poland, and to the south by gaining the Crimean territory from the Ottoman Empire. In the east they made a huge territorial gain, filling the vacuum left by the declining influence of Mongols and other nomadic peoples.

For all its military might and large population, Russia made only slow progress in organisational development. It was predominantly an agrarian society; up to 1861 peasants were still bound to the land in a state of serfdom. The isolated, brutal and superstitious peasant lifestyle survived in Russia and 80% of the population remained peasants in 1913. The csar was still an autocrat and the landed gentry still lived on income from the peasants on their estates. Russia was over a century behind political developments in the West of Europe.

Summary: situation in 1870

The Industrial Revolution had started in Britain and industrialisation was spreading to Europe and North America. An increasing proportion of wealth was coming from manufacturing and trade. As a result the social order of the Western World was in a state of flux. The mediaeval society of king, aristocrat and peasant was struggling to adapt to the new industrial society of managers and workers. In the UK more people worked in the factories than in the countryside. Even in the countryside, the role of the subsistence farmer was disappearing. With improved transport the farmer sold his output for cash and bought, rather than made, his consumables. All western states were becoming more integrated and engaged. The days when states were defined by the area of land owned by a king or prince were numbered. The purpose of states was changing from just promoting the interests of a few towards looking after the interests of all its citizens. These citizens were now demanding a voice in how the government should be run.

By 1870 the world population had reached 1.2 billion which is about the current population of India. Population growth since 1780 had averaged 0.4% p.a. (doubling every 180 years) which is faster than that in any previous period. This accelerated growth occurred only in the Americas and Europe. The American population grew at an unheard of 1.6% p.a. and Europe grew at 0.7% p.a. As Figure 8.2 shows, other regions continued to grow at a slower rate.

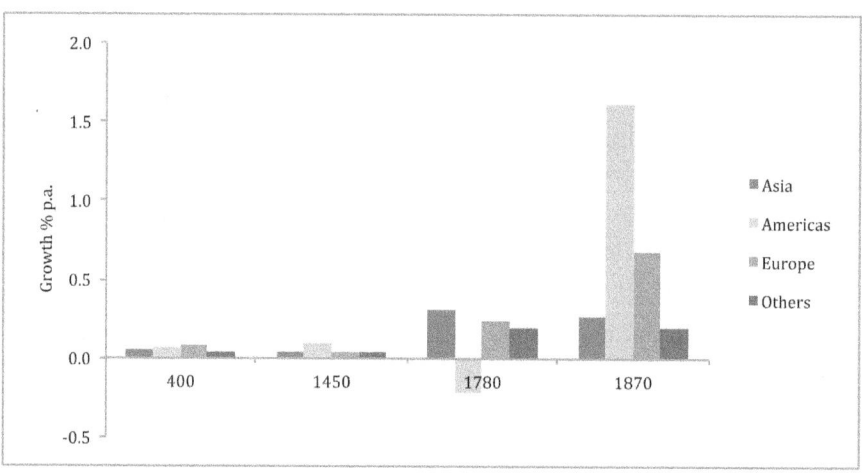

Figure 8.2 Population growth rates by region to 1870 (% p.a.)

The size of the cities had changed remarkably. Whereas only Beijing had a population of over a million in 1800, now London had reached a population of around four million. New York City and Paris were around two million each and Berlin and St Petersburg had also joined the one million club.

CHAPTER 9:
WESTERN STATES COMPETITION (1870–1920)

This period saw the spread of industrialisation throughout Western Europe and North America. **Electricity** was the symbol of this modern age: **light bulbs** illuminated houses, **electric motors** powered new appliances (**refrigerators** in the home and **power tools** in the factory), **telephones** were used for direct person-to-person communication and **trams** provided cheap urban transport. **Gas** and **oil** were developed as an alternative fuel to coal. The **internal combustion engine** provided a new, more versatile source of power than the steam engine.

The political culture of Western nations evolved and converged to a new form which I call **secular capitalism**. By the end of the period all **men had the vote** and **free primary school education** was provided by the state. The capitalist approach developed further; huge companies based on new principles of **industrial management** emerged. The working classes became more influential, **trade unions** pressed their case for better terms and conditions, and ideas of **socialism** fuelled the political debate.

As industrialisation spread, countries became increasingly urban; cities grew substantially with skyscrapers, bridges and tunnels all made of **reinforced concrete**. Now that the peasant lifestyle was disappearing life expectancy improved dramatically. Surgery could now be undertaken with less pain due to the use of **anaesthetics** and there was a much better chance of survival due to the implementation of **aseptic surgery.**

Western political cultures were totally dominant across the world. All breakthrough meme development took place in Europe and North America. Fuelled by nationalism, competition between Western states was intense. They competed economically, politically and ultimately militarily in the disaster of the First World War.

Industrial society

The Industrial Revolution started the third great change in human lifestyle after the Neolithic and Upper Paleolitithic revolutions. The period is usually defined quite tightly by historians as between around 1760 to 1840. In reality the momentous changes resulting from these technical developments are still being built on and absorbed today. I am going to use the term technological revolution to cover the period from 1760 onwards when humans began to embrace technology, improve their lifestyle and live longer. The great advances in technological memes led to an explosion of change in other types of memes. Warfare memes became so potent that the human race became capable of destroying itself. Communicative memes became so fast that individuals could talk to each other instantaneously anywhere in the world. Relational memes were transformed as the aristocracy could not maintain its privileged status. Styles of dress, art and music were revolutionised. However, attitudes and beliefs changed much more slowly. Old ideas of class and status took a long time to diminish and religions clung to their old traditions.

The technological revolution began in the UK and spread to Europe, North America and Japan before the First World War. The revolution brought with it a new political culture which expressed optimism and hope for future advancement. Based on the system of constitutional monarchy in the UK and the democratic republic in the USA, there began to emerge a synthesis of their political cultures that I am calling secular capitalism. This culture developed its own beliefs, rights, commercial rules and methods of governance that, unlike political cultures of the past, were not tied to any particular religion. The beliefs of secular capitalism embraced the scientific advances of the Enlightenment. It took as read that the natural world is governed by forces obeying natural laws, not gods, demons or spirits. It had an optimistic view of the future, that the development of science and technology would improve the lot of mankind.

Individual rights were of prime importance. The middle classes wanted freedom from arbitrary government. They wanted to be treated fairly and to have the opportunity to compete on equal terms with the aristocracy, even if this meant extending these rights to the working classes. In secular capitalist societies it was expected that all individuals, no matter what their race, class or religion should be treated as equals in the eyes of the law and everyone should be free from arbitrary arrest. Individuals should also have the right to express their own opinions, subject to libel laws.

It was essential that, for individual rights to be maintained, the legal system should be administered as independently as possible from the government. The extension of this principle from the USA and the UK into Europe was relatively straightforward as, since the days of the Catholic monarchies in the Middle Ages, all countries in Europe had developed extensive legal systems with a degree of independence.

The freedom to practice the religion of one's choice had been a harder fought battle. The USA had led the way, where there was a developing religious plurality. In Britain, the 1828 test acts swept away the requirement that only those professing to the faith of the established church could hold public office, and in 1829 restrictions against Roman Catholics were eased.

Jews also now began to be accepted in Europe as ordinary citizens without all the traditional residential and legal restrictions on their movements.

The Protestant religious culture of the middle classes had had a vital role to play in the development of this consensus. Unlike the Catholics, Protestant belief had more confidence that individuals control their own destiny. The Protestant middle classes believed that hard work brings its rewards and that individuals are responsible for their future. They trusted in the power of the capitalist system to improve their lot.

Secular capitalism was a political culture that aimed to harness the dynamics of capitalism to improve human life; it both enabled capitalism's positive effects and prohibited its excesses. In secular capitalist countries the government supported the development of commerce. They enacted laws for property rights, patents and the formation of limited companies, which enabled **commerce** to function freely. At the same time, governments controlled the excesses of capitalism both by setting legal requirements for safe operating practices and by allowing the operation of trades unions to support workers' rights.

One of the most important manifestations of this hard work and fair-play attitude was the reduction of the corruption of state officials. Since states were created, gaining a government post had been an invitation to make money by bribery and corruption. In the nineteenth century this practice abated for the first time. A new standard of behaviour in public service came to be accepted, enabling those that still used venal practices to be exposed by the press and prosecuted.

The system of **government** was an elected democracy in which all adult men had the right to vote for their government every few years. Democracy was a key part of secular capitalism because it was an expression of equality of opportunity for all. However, democracy in its own right did not ensure a government that supports a secular capitalist political culture. The political culture that developed in the nineteenth century was not the only possible form of culture that could emerge in an industrial society. Other cultures would develop in the next century and secular capitalism itself would change and adapt to the times. In an industrial society these political cultures competed on the first level of the evolutionary structure (Figure 9.1).

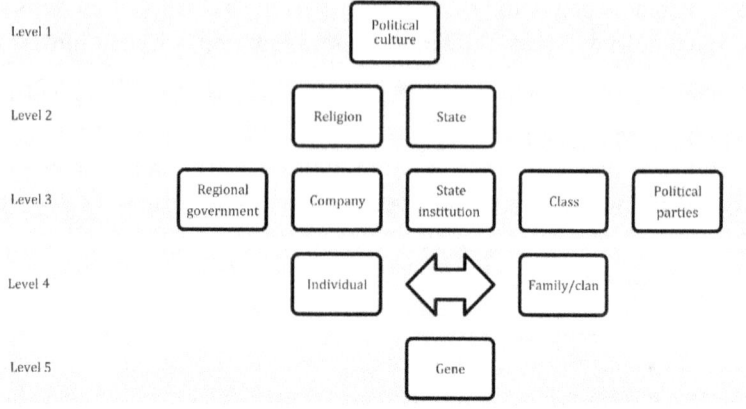

Figure 9.1 Evolutionary structure of industrial society

The spread of technology went hand-in-hand with three important developments in relational memes, secularisation, capitalism and democracy. Three more communities were added to the evolutionary structure of mediaeval states: religions, companies and political parties. In mediaeval times, religion was part of the political culture of the state and religious affiliation was closely controlled. Now there were many religions operating in a state and they actively competed with each other. Despite the secular nature of the political culture, religion was still extremely important in peoples' lives and defined many of their customs. Competition between religions was strong. In the UK it could be characterised as Church against Chapel. In the USA there was much more plurality. New religious sects were still being invented. Mormons trace their origins to the visions of Joseph Smith in the early 1820s. In a strange echo of the divine initiation of the prophet Mohamed who claimed that the Quran had been dictated by God, Smith said an angel directed him to a buried book written on golden plates containing the religious history of an ancient people. This became the book of Mormon, their supposed prophet. This new variant on Christianity was hounded by other religious groups forcing the Mormons to move west across the USA until eventually they set up their base in the state of Utah.

Christian Science also started during this period. It was developed in nineteenth-century New England by Mary Baker Eddy (1821–1910), who argued in her book *Science and Health* (1875) that sickness is an illusion that can be corrected by prayer alone. This religion became one of the fastest growing American faiths in the early twentieth century.

Companies also featured as a community in the evolutionary structure of industrial society. They had their own memes for business organisation and culture and trained their workforce in technologies and skills. They competed economically with each other in a battle for survival of the fittest. The best companies developed new technologies faster, made products cheaper and more reliably, honed the skills of their workforce and competed to attract the best labour. Thus, they quickened the material advance of humanity. As companies operated within the boundaries of a single state, it was in the state's interest to encourage their growth.

The class structure of industrial societies changed in a profound way. The same class organisation of aristocrats, priests, scribes, artisans, merchants and peasants that evolved after the formation of the first states was no longer sustainable. More wealth now came from business activity than land ownership. The new elite, the wealthy businessmen, bought land, built sizable country houses and mimicked the lifestyle of the aristocracy. Professional people of middle income, including the clergy, became a new middle class. These now played an important role in society and gained some of the reins of power in the more successful countries.

It was the artisans such as the weavers, shipwrights and wheelwrights who were most adversely affected. Historically, these trades had been organised in guilds, which operated as a closed shop protecting their working conditions and livelihoods. In Britain these guilds emerged from village associations stretching back before the Norman Conquest. Most were what would be later known as friendly societies, providing support for members in periods of poverty, sickness, old age and death, as well as regular social activities. These associations were organised on democratic lines, with considerable power to impose their will on their members. In urban areas, associations of craftsmen were formed on the same basis but, by virtue of establishing a local monopoly for their craft, they were able to regulate product standards and conditions of work. These craft guilds, as they were called, were the harbingers of local democracy, establishing procedures for meetings, minute taking, and election of leaders. In time members of guilds became part of the local town government. All across Europe it was common to have three levels of craftsman; the master who owned a workshop and was registered by the guild as sufficiently skilled in his craft to train others; the apprentice, who lived, worked and was trained in a masters workshop and journeymen, who had qualified through an apprenticeship and were able to charge a master for their work on a daily basis. One of the major functions of guilds was to set journeymen's wage rates. Because of the obvious conflicts of interest in setting these rates, journeymen were permitted to establish yeomanry sections within the guild. These bodies increasingly took direct responsibility for dealing with new members arriving in their area, collecting dues and distributing welfare. As the industrial age approached and workshops became larger, the yeomanry section of the guilds were increasingly transformed into independent associations of wage-earners. These were effectively the first trade unions with the ultimate power to strike. The woollen workers were among the earliest to act in this way, with a strike recorded in Trowbridge, Wiltshire in 1677.

Weavers were the most numerous trade and initially thrived in the Industrial Revolution as the development of industrial scale spinning led to the availability of cheap yarn. However, weaving began to be organised on an industrial scale. Before weaving machinery had been developed, weavers were organised as outworkers for a merchant; the merchant supplied the materials and ordered the work done. The master craftsman found himself competing with journeyman, prices were driven down and his income and status reduced. In time his entire job was mechanised. The weaver's artisan status disappeared and he became a factory worker. For some crafts, such as wheelwrights, the level of skill remained high. However for the majority, the skill base was permanently lost. Many of the jobs in weaving could now be undertaken by women and children who were paid even lower wages.

Journeymen, smallholders, labourers and others at the poorer end of society flocked to the newly burgeoning towns to become industrial workers. The work was hard, the hours were long and the wages were low. Their accommodation was in cheaply constructed high-density housing with poor sanitation. In 1833 P. Gaskell described the effects of factory life on workers:

> Any man who has stood at twelve o'clock at the single narrow doorway which serves as the place of exit for the hands employed in the great cotton- mills must acknowledge that an uglier set of men and women, of boys and girls, take them in mass, it would be impossible to congregate in smaller compass. Their complexion is sallow and pallid – with peculiar flatness of feature, caused by want of adipose substance to cushion out their cheeks. Their stature is low....Their limbs slender, and playing badly and ungracefully. A very general bowing of the legs. Great numbers of girls and women walking lamely or awkwardly, with raised chests and spinal features. Nearly all have flat feet. ... Hair thin and straight – many of the men having but little beard, and that in patches of a few hairs, much resembling its growth among the red men of North America. A spiritless and dejected air, a sprawling and wide action of the legs, and an appearance, taken as a whole, giving the world but little assurance of a man, or if so 'most sadly cheated of his fair proportions'.

Such were the effects of working twelve hours or more a day, six days a week, in a temperature of 80° with high humidity, under the eye of a stern and sometimes vicious overseer who maintained a rigid code of discipline. That said, it is not clear whether their conditions of work were worse than those on the farm. A peasant farmer would have had a house made of mud and wood, with an earth floor; most farm work would have been done by hand and the farmers would be subject to years of famine when harvests failed.

Children, in particular, were poorly treated. This diatribe against the practice of employing small children was published in the *Leeds Mercury* in 1830:

> Thousands of little children, both male and female, but principally female, from seven to fourteen years of age, are daily compelled to labour from six o'clock in the morning to seven in the evening, with only – Britons, blush when you read it! With only 30 minutes allowed for eating and recreation. Poor infants! Ye are indeed sacrificed at the shrine of avarice, without even the solace of negro slaves; ye are no more than he is, free agents; ye are compelled to work as long as the necessity of your needy parents may require or the cold-blooded avarice of the worse than barbarian master may demand! Ye live in the boasted land of freedom and feel and mourn that ye are slaves, and slaves without the only comfort that a negro has. He knows that it is his sordid mercenary master's interest that he should live, be strong and healthy. Not so with you. Ye are doomed to labour from morning to night for one who cares not how soon your weak and tender frames are stretched to breaking.

Eventually the Christian conscience of the governing classes was touched. Despite opposition from business leaders in Britain, factory acts were introduced to limit the hours of work, initially for women and children in 1844 and then for men in 1850. This was the first of many direct controls introduced by governments to curb the excesses of the capitalist system.

For the first time the concentration of poor in the towns gave them some political power. As peasants or farmers, they were isolated and even when driven to rebel they were easily overwhelmed by the landowners. Combined in larger numbers they had much more influence and could, by using the strike weapon, occasionally win their arguments. Thus began a new powerful competitive organisation, the trades union, aiming to improve the conditions of work of the new working class. Workers paid a small part of their wage over to the trades union to look after their interests. The education and political nous of the old artisan class enabled them to become the first leaders and organisers of this new movement.

This was the start of a long political battle for the right of trades unions to negotiate with employers on behalf of their members, with the ultimate sanction of withdrawing their labour. The governing classes resisted. In the early nineteenth century in Britain, anti-labour legislation made it a breach of contract to strike. As the voting franchise extended, calls for legalising trade union activity gained increasing support. Trades unions were legalised in the 1871 Trades Union Act and the 1906 Trades Dispute Act absolved them from losses following a strike. Once the political battle for recognition had been won the trade unions embarked on a largely successful campaign to improve the lot of the working classes during the twentieth century.

In Britain the trade unions initially gave their backing to the Liberal Party, but with the extension of the voting franchise they came to support a new party with the workers interests at heart, the Labour Party. In 1906 the new Labour Party won 29 seats in Parliament.

In Europe, but not so much in the UK and the USA, the trade unions supported socialist parties that were inspired by the political philosophies of Karl Marx and Frederick Engels. Karl Marx (1818–1883) was a philosopher and historian, born in Trier in Germany in a wealthy, middle-class family. He moved to Paris in 1843 where he met his collaborator Friedrich Engels (1820–1895). Engels was the son of a cotton manufacturer from Wuppertal, also in Germany, who was sent at the age of 22 to Manchester to learn the trade. There he formed a relationship with Mary Burns, a poor Irish woman, who had been working in the cotton mills since she was a child. Mary introduced Engels to the grim reality of working-class life in England at the start of industrialisation. From this Engels was able to write his seminal work, *The Condition of the Working Class in England*, detailing the horrors of child labour, poverty and overcrowded accommodation. From this he naturally deduced that capitalism necessarily involved the exploitation of the working class.

Engels worked with Marx on his famous political pamphlet *The Communist Manifesto*, in which he wrote:

> Our epoch, the epoch of the bourgeoisie, possesses this distinctive feature: it has simplified the class antagonists. Society as a whole is more and more splitting down into two great hostile camps, into two great classes directly facing each other: Bourgeoisie and Proletariat.

Marx followed up with his great philosophical and historical work *Das Capital*. He believed that there was natural class conflict between owners of capital (who he labelled the bourgeoisie) and labour (the proletariat). Marx saw history as a class conflict. In the French Revolution, the aristocracy was overthrown by the bourgeoisie. He predicted the eventual downfall of capitalism and the triumph, by revolution, of the proletariat.

This adoption of a revolutionary ideal by European socialist parties understandably scared the rest of the population, just as the French Revolution had scared the aristocracy outside France at the turn of the past century. The effect was to delay full democratic development in Europe and raise the level of political antagonism. Eventually this was to have disastrous consequences in Europe with the rise of fascism.

History proved Marx wrong in his analysis of the new class structure of industrial societies. There were not just two classes the bourgeois and the proletariat. In Europe there were effectively four classes: upper, middle, working and the rest. The rest were farm labourers, clerks, shopkeepers and others who lived in the countryside and county towns. They did not identify at all with the working classes of the industrial towns. Their aspirations were more conservative and they did not have the same desire to overthrow the existing order of society. For want of a better term I shall call this the rural class. This class became important later as a focus of political resistance to socialism.

These four classes operated in a different social spheres to each other; the gap between them in terms of wealth, health and education was huge. This gap only began to close gradually with the extension of the voting franchise to the lower classes. It was the middle classes who were the main initiators of change; they aimed to wrest power from the aristocracy and saw the extension of the right to vote for all adult men as one of the main means to achieve this aim. In 1870 the USA extended voting rights to former slaves. In Europe progress was slower; kings, the Catholic Church and aristocrats clung to power as long as possible.

The growth of democracy in the nineteenth century went hand-in-hand with the development of political parties, another key competitive community in the evolutionary hierarchy. Most parties are held together by a common consensus across a broad variety of issues. At the start things were simple; you either broadly favoured the transfer of power away from the establishment (the church, aristocracy and monarchy) or you favoured the status quo. I shall call these two groups radicals and conservatives. This is the basis for the Liberal–Conservative split in England in the nineteenth century or between republicans and monarchists in the early days of the French Republic. In Germany, radicals supported the Progressive Party, which provided the opposition to the established order in the early days of the newly formed German Empire.

Conservatism is an enormously powerful force. Reaction to the threat of change will engender feelings of danger to one's community and fear of losing wealth and status compared to others. Conservatives dislike changes to the traditional structure of the community and want to preserve its culture. Radicals, in contrast, believe in a style of government that offers the chance for all to improve their position in life. They seek to restrict the power of the existing authorities to protect the establishment from change. Radicals want to ensure a fair playing field for all in education, the law and the economy. They are inherently capitalist in approach, in the Adam Smith style of capitalism, which opposes government interference with industry, but at the same time seeks to redress unfair competition and practices by industry itself. They believe in the change and benefits brought about by the technological revolution and that hard work should be rewarded.

These two broad approaches dominated the early days of democracy. Of course, the reality of government meant nothing was as black and white as this. Conservative governments would enact radical legislation and vice versa if they could gain political advantage. To give two examples: the second Electoral Reform Bill in England was enacted by a Conservative government in 1867; and Bismarck, leading a very conservative government in Germany, was the first to introduce state-sponsored old age and infirmity insurance in 1889.

Into this mix came the socialists with their message that the state had a duty of care to its citizens. They believed that capitalism should not exploit the working classes, and state planning of the economy was in many ways better than the capitalist approach. In Britain, the (socialist) Labour Party first came to power in 1923, but did not really get a chance to implement its approach until after the Second World War. In Continental Europe, the socialist message was more strident, with its Marxist view that revolution was necessary to implement change and allow the working class to gain power. In a democracy, parties threatening to cause revolution are extremely divisive. Democratic government allows for the election of parties to represent specific interests, such as the Centre party representing Catholics in the newly formed, protestant-led German Empire before the First World War, or the Irish Nationalists in the British parliament at the turn of the twentieth century. These represented minority views and did not threaten the voting majority. The socialists, however, with their aggressive promotion of working class rule were extremely threatening to those of conservative belief and caused a major crisis in democracy itself.

The final change to the evolutionary structure of industrial society was the change from local government being self-regulated bodies raising their own funds to become incorporated in government administration. This happened in Europe during the nineteenth and twentieth centuries. Governments reformed ancient administrative boundaries, establishing common voting systems and funding. A patchwork of local administrative bodies was swept away and one local government body coordinating all local public services was created. Government ministries now oversaw the operations of local government. There were considerable benefits to be had from this approach. Government funds could be directed to help improvements in specific communities. Nationwide standards of public health, housing, welfare, water and waste services and community governance could be imposed. Nevertheless, some local independence and initiative was lost in the change. Whereas in former times local corporations could represent people against government-imposed policies, now they all too easily became part of the government machine. An organisation that had largely been locally self-governing now became more restricted by government regulation. The local government community shown in the evolutionary structure of mediaeval society therefore changed to become another state institution alongside the military, education, the legal profession, the police and the civil service.

In truth, as transportation became cheaper and communication became faster, local communities became less important as a focus of allegiance than the broad region and the nation as a whole. Cultural practices that had varied according to locality gradually became common across a region. In the USA regional government had been enshrined in the Constitution through the definitions of state and federal rights and responsibilities. Regional government is thus shown as a new community in the evolutionary hierarchy of the industrial society. Initially, it existed mostly in the USA and the newly created British overseas dominions. Only later in the twentieth century would European regional administrations be created that evoked strong local feelings and support, such as the Basque and Catalan regions in Spain and the Northern Irish and Scottish regions in the UK.

Industrial societies in Europe and Japan

While the adoption of the structure of the industrial society evolved naturally in the USA and the UK, change in other countries proved much more difficult. While all countries industrialised and allowed companies to be formed and compete, they were less ready to adopt the social changes that came from secularisation and democracy. In particular, the aristocracy and the church were reluctant to give up their privileged position.

This was a problem right across Europe, even in the UK. We have seen that in 1832 the UK extended the voting franchise to the middle classes. Before 1867 only 20% of adult men had the vote. By 1885, following two reform acts in parliament, the better-off working class men had been enfranchised, resulting in about 60% of adult men having the right to vote. However the upper chamber, the House of Lords, was controlled by the aristocracy and had a power of veto over legislation until 1912.

In France the issue was not democracy, since the French had adopted universal adult male suffrage in 1848. The issue was the strong influence of the Roman Catholic Church on French political life. Republican politicians, supported by the anti-clerical middle classes, managed to pass laws secularising the provision of health and education. Priests were excluded from the administrative committees of hospitals and of boards of charity; lay women replaced nuns in many hospitals and a system of state schools staffed by lay teachers replaced a mixture of religious and private schooling. Nevertheless, church involvement in state affairs continued to be a major political issue until finally by an Act passed in 1905 France was declared a secular society.

In Germany, while giving ground, the monarchy still hung onto the reins of power. In 1848 the Prussian government yielded to popular pressure to form a constitutional style of government. However, it was one that allowed the old aristocracy still to maintain their influence. Voters were divided into three classes; in 1849, the first year of the elections, 5% voted in the first class, 12% in the second class and 83% in the third. The ballot was not secret and therefore open to pressure and manipulation. This system survived into the imperial period after 1870, ensuring that aristocratic rule in Prussia survived until the First World War.

In 1870, as a result of the political cunning of Otto von Bismarck (1815–1898) and the might of the Prussian army, a new German empire had been created, recognising the Prussian king as its emperor. In this new empire, however, many of the powers remained with the old constituent kingdoms and Prussia maintained the three-tier franchise which kept considerable power in the hands of the conservative landed classes. Although the German empire had a parliament (Reichstag) which was elected by full adult male universal suffrage, the Reichstag had no power over the executive, which was appointed by the chancellor who in turn was appointed by the Prussian king and emperor of Germany. Thus, Germany continued to experience authoritarian government and the practice of democracy was limited.

The Kingdom of Italy had a similar power arrangement to Germany, in which executive power belonged to the king and legislative power to elective assemblies. However, unlike Germany, it was practically impossible for an Italian government to stay in office without the support of parliament. In 1882 a system of proportional representation was introduced, giving voting rights to two million adult men.

After most adult men in the leading Western countries had the vote, one of the first dramatic effects was the spread of primary education to the lower classes. Up to now only the upper and middle classes had access to education. True, some churches had been providing training in the scriptures to working-class boys, which did enable some children to read. This, however, was the limit of organised working-class education. This changed dramatically in Europe and North America and soon all children were required to attend elementary school: by 1870 all states in the USA had free elementary schools, albeit mostly in the urban areas. In the UK in 1870 the Foster Education Act required all children between 5 and 10 years of age to attend school, though with some leeway at harvest time; in France the Jules Ferry laws established compulsory education for 6–13 year olds, again with provision for 15-day absences to take up farm or industrial work. In Germany the Volksschule was free and compulsory for boys and girls of primary school age.

Secondary education was still confined to the middle and upper classes and strongly biased towards the classics. Both Latin and ancient Greek had a dominant place in the curriculum. Class divides between the upper and middle class were also reflected in their schooling: such as the public school and grammar schools in the UK or the Gymnasium and Realschulen in Germany. Education in a Gymnasium included the study of Latin, classical Greek or Hebrew, whereas education in a Realschule was aimed at students not destined for university, but who would pursue a technical or office-based career.

University education was also very classically biased, and only slowly were science and engineering subjects added to the course options. Germany led the way in establishing technical colleges (Hochschulen) in engineering and other business-based subjects, which the French copied with the Grandes écoles. Perhaps as a result of the German initiative in education, it is noticeable that from the late nineteenth century onwards, German-educated scientists began to play a major role in scientific advance.

The nation-states of Belgium, the Netherlands and Scandinavia also developed a secular capitalist culture. The Austrian empire, with its mixture of German, Hungarian and Slav communities, found it difficult to embrace these concepts. The feelings of a communal bond in a nation-state are not necessary for secular capitalist countries to function, but if countries are not divided between many ethnic groups, ideas of equality and fair play are much easier to pursue with the support of the population at large. Dividing the empire in 1867 between Hungarian and Austrian kingdoms only served to increase nationalist divisions. Slav minorities in the Hungarian kingdom were even less favourably disposed towards their government.

That secular capitalism was not appropriate to Westerners alone became apparent when Japan adopted the culture in an amazingly short period of time. The static world of this rigidly controlled peaceful kingdom had been shattered in 1853 when Commodore Perry from the USA arrived in Japan carrying the stark message: agree to trade or face war. The Japanese, although isolated, were not without links to the outside world and were well aware of the fate of the Chinese after the opium wars. In 1858, without fighting, the Japanese agreed similar terms to the Chinese; eight ports were opened up to foreign trade and foreign consuls maintained jurisdiction over their own nationals. The only restriction on Western trade that the Japanese managed to obtain was to ban the import of opium. These unequal treaties undermined the authority of the shogun and he was eventually overthrown by a number of daimyo and samurai from south-west Japan in the Meiji Restoration of 1867–1868. The term restoration was used because the leaders of the rebellion were supposedly restoring the emperor Meiji to power. In reality it was a coup d'état and the emperor remained a figurehead. Real power resided with the leaders of the revolution who formed a ruling oligarchy. Some of their leaders had already travelled in the West and realised the military weakness of the Japanese. They were bowled over by Western achievements in architecture, education and industry. The new leaders drew the conclusion that if Japan was to survive it must embrace Western institutions, copy Western military hardware and industrialise. In a change that could only happen in a very ordered society, in the space of 25 years the country was economically and socially transformed. The class structure was overthrown. Around 280 daimyo domains were replaced by 72 prefectures with government-appointed leaders. Samurai privileges were eliminated. A new system of government was installed based on the constitutional monarchies of the West. It was a cabinet style of government with a prime minister and ministers of state in charge of the bureaucracy; all were responsible to the emperor. There was a constitution with an elected house of representatives and an appointed house of peers which had the power to write laws and approve budgets. Capitalism was seeded by setting up model enterprises and 1400 miles of railway were built. Literacy was improved by requiring all boys and girls to attend four years of elementary education.

There was always a nationalistic drive behind the Japanese changes. The ambition was to achieve the same military success as the West, reflected in the mantra of the reforms 'rich country, strong army'. A modern Japanese navy was established and a modern conscript army recruited from all classes, not just the samurai, as was the case before. The scale and success of these changes was truly astounding. They transformed Japan into a country that could take its place at the forefront of development alongside European powers. By 1904–1905 the Japanese were strong enough to take on and defeat Russia in war.

By adopting secular capitalism the Japanese were the one Asian country to escape colonisation in the late nineteenth century. Western powers tried to copy British colonial practice and indulged in a mad scramble for overseas dominions, irrespective of their economic value. Africa was rapidly divided between European powers; France, Italy, Germany, Belgium, Spain, Portugal and Britain all gained sizable African territories. By 1913 only Ethiopia and Liberia remained independent. Ethiopia had gallantly resisted an Italian invasion and Liberia was set up by Americans to repatriate some of its slaves and thus could be considered a US colonial development in which US ex-slaves governed the local native population.

The remaining 'available' parts of Asia were also acquired by European powers. France conquered the Vietnamese empire. The Netherlands consolidated its hold on Indonesia. The British founded Singapore and controlled Malaysia. Afghanistan and Thailand were the only two countries in south Asia and the Far East, apart from China, that were not controlled by Western powers or Japan; Afghanistan, having twice thrown out the British, and Thailand, having ceded territory to France in Indo-China.

After the First World War the Ottoman Empire in the Middle East was parcelled up between the British and French, leaving only Turkey and Saudi Arabia as independent countries. Thus in the whole world, only seven countries: Ethiopia, China, Turkey, Iran, Saudi Arabia, Afghanistan and Thailand had escaped Western or Japanese colonial domination at some time in their history.

Technical development and business

Secular capitalism was a political culture that both enabled and supported industrialisation. The pace of technical progress quickened as the laws of economics were free to operate. Successful companies and their technologies thrived. Failed companies ceased to exist. The fittest companies survived and began to drive material advances in society. Ideas developed by academics and enthusiastic amateurs alike were taken up and commercialised by companies. The development of electric power was typical. Whereas electric power was born from the scientific ideas of Michael Faraday (1791–1867), its practical success was achieved by corporations at the turn of the twentieth century.

Michael Faraday was one of the world's most famous hands-on scientists. He came from very humble origins. Faraday was born south of the Thames in London and at 14 was apprenticed to become a bookbinder. It was from reading books that he developed an interest in electricity. In 1812, at the age of 20, and at the end of his apprenticeship, Faraday attended lectures by the eminent English chemist Humphry Davy and eventually became his chemical assistant at the Royal Institution in London.

In the class-based English society of the time, Faraday was not considered a gentleman. When Davy set out on a long tour of the continent in 1813–1815, his valet did not wish to go. Instead, Faraday went as Davy's scientific assistant, and was asked to act as Davy's valet until a replacement could be found in Paris. Faraday was forced to fill the role of valet as well as assistant throughout the trip. Davy's wife, refused to treat Faraday as an equal (by, for example, making him travel outside the coach and eat with the servants).

However, his talent soon showed through. He was elected a member of the Royal Society in 1824 and became a director of the laboratory at the Royal Institution. Faraday had a broad interest in science but his most famous work was on electromagnetism. That an electric current produces a magnetic field had been discovered in 1821 by a Danish scientist, Hans Christian Ørsted. Faraday was able to demonstrate that a rotary mechanical motion could be achieved as a result of the interaction of an electric current with a magnet, the foundation of the concept of the electric motor. He also showed that with two intertwined coils of wire, electricity could be induced from one coil to the other, the foundation of the concept of the transformer. He had no mathematical education and others had to provide the mathematical basis to describe these phenomena. However, his concept that electromagnetic forces extended into the empty space around the conductor in a similar way to lines of flux around magnets was crucial for the successful development of the electromechanical devices in the remainder of the nineteenth century.

After the telegraph system, the first commercial uses of electric power were for lighting and tramways. To allow their use across a wide geographic area required both an economic source of power generation and the ability to distribute the electricity. It took until the 1880s before industry was able to implement the first practical systems.

The first city to use electric street lights was Wabash, Indiana; in 1880 a flood of light engulfed the town from four lights mounted on top of the courthouse. One of the original lights can still be seen on display at the Wabash County Courthouse. In 1881, a Siemens AC Alternator driven by a watermill was used to power Europe's first electric street lighting in the town of Godalming in the UK. The first regular electric tram service using pantographs or trolley poles, installed by Siemens, went into service in a suburb of Berlin, in Germany, in 1881.

However, in these early days, street lights, electric motors in factories, power for trams and lights in homes all used specialised generators with their own separate transmission cables, often using different voltages. It was known that if high voltages were used electricity could be transmitted over longer distances from one power source. In 1882 Thomas Edison (1847–1931), who invented the first practical light bulb, switched on the world's first large-scale electrical supply network that provided 110 volts direct current to 59 customers in lower Manhattan. In 1887 Nikola Tesla (1856–1943) a Serbian-American engineer, and probably the most important figure in the development of the use of electric power, filed a number of patents for an alternative system of alternating current (AC) power distribution. AC power distribution was safer and more efficient and eventually won out. By 1914 55 transmission systems each operating at more than 70 kV were in service in the USA. The First World War gave a further impetus to the development of electric power and large electrical generating plants were built to provide power for munitions factories.

It is worth noting that three companies that were instrumental to the start of electric supply are still with us today: Siemens from Germany, Westinghouse, who gained the rights to Tesla's patents and GE (General Electric) who claim Edison as their founder. By 1920 electric power was widely used in transport (trains and trams), lighting and in factories. However, its use in the home was still to fully develop.

The other source of power that was developed in this period was the internal combustion engine. The first engine was produced by Étienne Lenoir – a Belgian engineer working in Paris in 1858. It was a single cylinder two-stroke engine in which a mixture of coal gas and air was ignited by an electric spark. Despite being noisy, inefficient and prone to overheating, almost five hundred of these engines, which had only 6 to 20 horsepower, were built and used as stationary power plants.

The efficiency of the internal combustion engine was much improved by compressing the gas prior to ignition. This method was known as the Otto cycle after one of the two German engineers August Otto and Eugen Langen who formed a partnership to make the first efficient engine. Otto and Langen set up their company in the town of Deutz in Germany and within a few years made engines that were capable of producing power of up to 1000 hp. In 1884 their company, called eponymously Deutz, developed the carburettor and a reliable low voltage ignition system. This allowed the use of liquid petroleum fuel for the first time, which in turn made it feasible to use internal combustion engines in transportation. However, Deutz wanted to concentrate on stationary engines. This led to a disagreement with one of their managers, Gottlieb Daimler, who wanted to develop smaller, lighter engines for use in transport. Daimler left the company and took one of their other employees, Wilhelm Maybach, with him. In 1885 they succeeded in creating an engine that could power a motorised cycle. The world's first motorcyclist was Daimler's 14-year-old son Adolf.

Today Otto's company, Deutz, is still one of the largest manufacturers of heavy-duty vehicles in the world, and the successor to Daimler's company, Daimler-Benz, is one of the world's largest and most respected makers of luxury automobiles.

For heavy plant, a further improvement in engine design was made by Rudolf Diesel, a German engineer living in Paris. He showed that compression alone could ignite the fuel and air mixture of some petroleum fuels. The resultant diesel engine was yet more efficient than the petrol engine. In 1893 he produced his first engine and by 1896 he had improved it to a theoretical efficiency of 75%. (Steam power had an efficiency of only 10%). This spelt the end of a century of domination of the steam engine. Diesel's engines soon replaced steam power in industrial plants and pumping gear in mines and became the power source of choice for trucks and ships. Steam remained for many years predominant on the railways but, after the Second World War, diesel and electric engines proved to be more economical. By 1970 the steam engine had disappeared as a serious source of power and was to be seen only in museums.

The production of the internal combustion engine would not have been possible without the development of steel. Steel is an alloy of iron and carbon as well as other elements such as chromium, vanadium, tungsten and manganese. Compared with wrought iron, steel has better strength under compression and tension. However, its chemistry is complicated and although in 1855 Henry Bessemer invented a blast furnace that was technically capable of producing steel, it was not until the 1890s that steel of sufficient quality was available to replace wrought iron in the construction industry. It took the development of modern industrial organisations, using large-scale production, focusing on unit cost reduction and quality control before steel could be made reliably. The pioneer was a Scotsman by birth working in America: Andrew Carnegie (1835–1919).

Carnegie was not an engineer but he was a great businessman. When he discovered that the fundamental chemistry of steel-making was not fully understood. he employed technicians (chiefly German) as chemical engineers; firstly to understand, and then to control the steel-making process. He was the first major manufacturer to understand the importance of management accounting and insisted on knowing the costs of each stage in the steel-making process. He recognised that it is the productivity of labour, not the cost of labour, that is the most important determinant of profitability. He used measures such as output per person to explain the importance of efficiency improvements. Marketing theory was applied to the process of selling steel; he argued that 'The price in the market is not your affair – you must meet the price whatever it is'.

With this combination of technical ability and accounting and marketing skills he created the first recognisably modern business corporation run by professional managers. He was also the first business leader to adopt a policy of employing the best managers and rewarding them accordingly. Good managers became business partners. He said: 'I am after the winner. If he can win the race he is our racehorse. If not he goes to the cart'. He paid his managers the highest wages in US industry.

With his business leadership, between 1880 and 1900 US steel production rose from 1.15 million tons to over 10 million tons annually. Steel rails, which had cost $160 a ton in 1875, came down in price to $17 a ton in 1898. This is from Paul Johnson's *History of the American People*, p. 563:

> These enormous savings [in steel prices] worked their way through into every aspect of the economy, with consequential benefit to the public. No president, by miracles of administration, no Congress, by enlightened legislation, was capable of bringing comparable material benefits to all Americans in this way.... there was a practical logic in the admiration men like Carnegie inspired.

Andrew Carnegie was one of a new breed of businessmen who had amassed considerable wealth through their business activities – enough to give them a position of power equivalent to the top level of the aristocracy in European countries. Across USA and Europe they built themselves magnificent houses, designed by the best architects, furbished according to the most expensive fashions of the day and surrounded by huge, beautifully landscaped estates. Their children often married into the nobility or successfully oiled the wheels of government to gain an aristocratic title for themselves. In supposedly egalitarian America, they formed a new elite – equivalent to the aristocracy in Europe – setting fashions of the day. It is thought that the American eating custom of not jointly operating knife and fork in the left and right hand simultaneously, was initiated in New York at this time in the fashionable circle that collected around the Astor family.

Like Carnegie, many of these nouveau riche were single-minded, ruthless operators who knew business was about creating and exploiting market dominance for their product. Many were not too concerned about how this was achieved. They included men like Cornelius Vanderbilt, who created the New York Central Railway which ran from New York to the north-west towards Chicago. Vanderbilt earned his position of power not only by building and investing in railroads but by ruthlessly eliminating competitors or acquiring their businesses, by stock manipulation, unfair competition or influencing state legislators to manipulate state law in his favour. Similarly John D. Rockefeller acquired 80% of the world's oil refining capacity by unfairly wiping out his competition. According to the newspaper *New York World* in 1880 Standard Oil (his company) was 'the most cruel, impudent, pitiless, and grasping monopoly that ever fastened upon a country'.

Such men, known as robber barons, were simultaneously loathed, hated, envied and courted by the general public. Looking back, however, it can be said that such men improved standards of living by improving capacity and infrastructure, reducing prices and standardising and improving the quality of their product. Even when the US government had limited monopolistic behaviour by legislation, the companies they founded went on to grow and make increased profits well into the twentieth century. There was, however, another class of robber baron who created nothing, apart for money from themselves and chaos in the companies that they controlled. An example of this is the confrontation of Fisk, Drew and Gould with Commodore Vanderbilt over the acquisition of the Erie Railroad,

Daniel Drew began as a cattle-drover, selling cattle in New York. On the way to the sale he would feed the cattle salt and encourage them to drink water to increase their weight. This sharp practice was referred to as 'watering the stock' and was later applied to all assets sold at an inflated value. He teamed up with Jim Fisk, who began in finance by buying cotton in occupied areas of the South during the Civil War, then selling it to the North. They specialised in buying up poor stock cheap and dumping it on the unwary. Jim Fisk first coined the phrase 'Never give a sucker an even break'. In 1867 they went into partnership with Jay Gould, an expert on railway stock. Their tactics involved making the stock rise and fall until it reached a critical low point and then swooping to acquire all the stock.

The story of their acquistion of the Erie Railroad is taken up by Paul Johnson in *The History of the American People*, p. 550, and reads like an incident in the wild west:

> At the climax of the battle, Drew, Fisk and Gould, who had taken over the Erie HQ in New York, gathered up $8 million in greenbacks there, tied them in bundles, threw them into the back of a hackney cab, drove to the New Jersey ferry, crossed, collected an army of thugs, and fortified Taylor's Hotel on the Jersey City waterfront, renamed Fort Taylor, with their armed men and three cannon. They also had a shore patrol in four lifeboats, each containing a dozen gunmen. All this was to fend off the naval assault of the Commodore who, it was said, 'could be heard roaring from the New York shoreline.... That Fisk, Gould and Drew milked the Erie is undoubted. The once profitable railroad became bankrupt in 1877.

This would not be the last time that greed on Wall Street would ruin profitable enterprises and create misery and chaos.

As businesses grew, a flood of immigrants arrived in the USA drawn by the opportunities to improve their wealth and status. Historically it had been the British, afterwards the Germans then the Irish who had been the main sources of US immigration. Now people came from all over Europe to seize the opportunity of a new life with better working conditions, particularly Swedes, Norwegians, Italians, Poles and Jews from Russia. In 1890 the USA began to exercise control over the immigrants coming to the USA and built a reception station on Ellis Island in the harbour entrance of New York City. On the first day of opening three large ships landed and 700 immigrants were processed. Some who were sick or in poor health were sent back on the boat they arrived in. The rest entered America to find their friends and relatives – many going on to take a long journey by train to some distant part of the USA. They joined diaspora communities scattered across the country. These immigrant communities established their local church or synagogue where they could meet and use their local language. This is the first time that a nation was forged with such a diversity of ethnic groups. Amazingly, it worked. People learned English, respected the other communities and the Constitution, got on with their lives and were all proud to be called Americans. Compared with Europeans, the new Americans were wealthier and had far greater opportunities for self-improvement. Class divisions were largely absent; there were no kings or aristocrats.

Two groups, however, constituted an underclass: African Americans and Chinese. Black slaves were freed after the Civil War, but continued to suffer vicious discrimination in the South. In the North, the discrimination was less blatant but still effective, so that black people were obliged to live in poor ghettoes, such as Harlem in New York. The Chinese came over to work on the West Coast of the USA, firstly following the Californian Gold Rush in 1849 and later to work on the transcontinental railway. By 1860 the Chinese were the largest racial group in California. As hard-working, effective and cheap workers, they suffered racially inspired violence, particularly when some of the labour leaders blamed them for depressing wages. As a result in 1882 the USA passed the Chinese Exclusion Act to suspend Chinese immigration. After this, the USA effectively operated a 'whites only' immigration policy up to the end of the Second World War.

Health

This was the era when doctors finally realised that hygiene is important and that many diseases are caused by 'germs' or bacteria. Bacteria were first observed by a Dutchman, Antonie van Leeuwenhoek, as far back as 1676, using a single-lens microscope of his own design. However it was not until the nineteenth century that the science of microbiology advanced sufficiently to prove the relationship between specific bacteria and disease. Up until that time it was believed that infectious disease was spread by a miasma of noxious air coming from rotting organic matter. Two names stand out in the development of microbiology, Louis Pasteur and Robert Koch.

Louis Pasteur (1822–1895) was from the Jura region of France and the son of a tanner. In a splendid example of how an improving education system allowed the poor to rise through the ranks, he had become a professor of chemistry at the University of Strasbourg by 1848. He was the first scientist, rather than doctor, to make a major impact on disease management. Working on bacteria and yeasts, in 1859 he demonstrated that the fermentation process is caused by the growth of microorganisms, and that this growth is not due to spontaneous generation from noxious air. Using this knowledge he was able to provide useful information for brewers and wine-makers and introduce pasteurisation as a means of preserving the freshness of milk. It was not until 1870 that he turned his attention to infectious disease. His first major success came with anthrax. Anthrax is unusual in that the bacteria that cause the disease can be seen in the blood. Drawing from the work of Jenner on smallpox, Pasteur decided to try to produce a vaccine by attenuating (reducing its concentration) the bacterium and purposely infecting the animals with the attenuated bacteria. The success of the resultant vaccine led to his becoming an international celebrity. Pasteur also went on to create the first vaccine for rabies.

Although Pasteur had made some progress with identifying bacteria as the cause of infectious disease, it was Robert Koch (1843–1910) who, in his research into tuberculosis, finally proved there was a direct relationship between bacteria and disease; for this he was awarded a Nobel Prize in 1905. Robert Koch was the son of a mining official in the Kingdom of Hanover. He worked his way up from a district medical officer to professor of hygiene in Berlin University. He was the first professional microbiologist and pioneered important scientific techniques. In addition to tuberculosis, he identified the causative pathogens for cholera and anthrax. Using his methods, Koch's pupils found the organisms responsible for diphtheria, typhoid, pneumonia, gonorrhoea, cerebrospinal meningitis, leprosy, bubonic plague, tetanus and syphilis, among others.

The germ theory of medicine was now finally established and it became clear that cleanliness was very important in the hospital environment. To give one example, puerperal fever is a bacterial infection incurred during childbirth which led to septicaemia and death. For centuries one of its causes was the poor hygiene of doctors assisting at a birth. This had been reported as long ago as 1795, but now was finally believed, leading to an immediate improvement in hygiene at birth and consequently in the survival rates of mothers.

Another important example was the risk of infection during surgery. The first public demonstration of surgery under an anaesthetic (ether) had taken place in Massachusetts General Hospital in 1846. The effectiveness of chloroform was demonstrated a year later. This gave surgeons better conditions in which to operate, allowed them better opportunities to conserve tissue and made longer operations more practical. However, the longer time of the operation also meant an increased risk of infection. Thus, despite the used of anaesthetics, patients' survival rates did not improve until antiseptic surgery was pioneered by the British surgeon Joseph Lister (1827–1912) in 1867. He used carbolic acid (phenol) as a general antiseptic to reduce the risk of infection from bacteria. Antiseptic surgery was soon replaced by aseptic surgery, the aim being not to kill contaminating bacteria but to exclude them in the first place. Asepsis excludes bacteria by sterilising equipment, instruments and dressings, the surgeon's hands and the patient's skin. For the first time in history invasive surgery could now be accomplished without a high risk of death resulting from infection.

The final triumph of this period was the development in 1909 of the first antibiotic against *Treponema pallidum*, the bacterium that causes syphilis, by a German Paul Ehrlich (1854–1915). Western medicine was, at last, entering an era where its ministration was more likely to have a positive than a negative effect on the patient.

Military developments, the First World War and its consequences

States had always made war one to another, often for the flimsiest of reasons. Outwardly it seemed, therefore, nothing extraordinary when the events following the assassination of an Austrian aristocrat by a Serbian nationalist caused Russia, Italy, France and Britain to enter a war against Germany, Austro-Hungary and the Ottoman Empire. However, it was to prove the bloodiest European war yet undertaken.

During the nineteenth century weapons of war had increased in potency, with breech-loaded rifles, machine guns and improved artillery. Rifles, which caused the bullet to spiral on firing, were more accurate than muskets. Although they had been in existence since the eighteenth century they were slow to load. Hence, muskets were still the preferred weapon of infantry in the Napoleonic wars, allowing a comparatively rapid, imprecisely aimed volley to be fired against an enemy charge. During the nineteenth century breech-loaded rifles from a cartridge were developed. Now accuracy and speed was possible. In the early part of the twentieth century, soldiers were trained to shoot accurately over long ranges. British Lee-Enfield rifles were able to fire distances of up to 1.6 km. Individual shots were unlikely to hit, but a platoon firing repeatedly could produce a beaten ground effect similar to light artillery.

Even more effective at producing beaten ground was the machine gun. The Gatling gun, patented in 1861 was the first gun capable of high speed firing and was used in a limited way in the American Civil War. It was hand-cranked and fed by a belt of bullets. The first self-powered gun was the Maxim gun, invented in 1884. It used the recoil power of the first bullet to load the next and had a much higher rate of fire. The gun was water-cooled to prevent overheating. This design was widely adopted and derivatives were used by all sides in the First World War.

The killing power of artillery also developed in a dramatic way. In the nineteenth century artillery had been divided into light and heavy, depending on the weight of solid shot fired. Light guns, deployed at battalion level, were usually 4–6 pounders (referring to the weight of shot or shell), whereas heavy guns were 8–12 pounders. By the end of the nineteenth century this had all changed and three different types of gun were in use, the field gun, the howitzer and the mortar.

The long-barrelled field gun was rifled for accuracy and capable of quick firing, as its recoil system bounced the barrel back into the firing position. It used a high explosive shell with a strong steel case, a bursting charge and a fuse. The fuse detonated the bursting charge which shattered the case and scattered hot, sharp case fragments at high velocity. Most of the damage was caused by shell pieces rather than the blast itself.

Howitzers were developed to be used under cover or against hidden targets. These fired heavy shells on a high trajectory through a short barrel and were originally designed to destroy fortifications. However, as the war developed, their high angle of attack made them increasingly useful against trench defences. Heavy howitzers (200–400 mm) could fire shells weighing over 900 kg over 18 km.

Another high-trajectory gun was the mortar. The projectile was dropped into its broad, stubby barrel and fired by a pre-loaded explosive charge. By the end of the war some of these guns were capable of firing shells up to about 2 km.

Against all this firepower common soldiers in the First World War hid in trenches and surrounded themselves with barbed wire. The horse-mounted cavalry was no longer an effective military force. On the western front, after the initial dynamic phases a stalemate was reached. All battles seemed to follow the same pattern; the pounding of enemy trenches with artillery shells was followed by a suicidal charge through barbed wire. On the basis that they were expected to lead from the front, a British Second Lieutenant around the time of the battle of the Somme could be expected to survive less than a month. The glory of battle was totally absent; bravery almost certainly resulted in death.

Naval warfare technology also changed dramatically during the twentieth century. As naval gunnery improved in line with the development of artillery on land, an entirely new sort of warship emerged. The development of high-explosive shells made it necessary to use iron armour plate on warships. Guns with a longer range of accurate firing meant that ships could engage in battle at a distance, rather than the broadsides of Nelson's day. The ultimate ship for this new form of battle was launched by Britain in 1906: the Dreadnaught, the first battleship to be equipped with only large guns and capable of speeds of over 20 knots. An arms race with Germany developed, fuelled by patriotic fervour on each side. However, during the First World War the use of dreadnaughts was not significant to the outcome. It was another new sort of ship that proved most effective: the submarine. Submarines had a long history. In 1620 the first submersible was recorded; in 1775 the first submersible to use screw driven propulsion was made; in 1873 the first non-human propelled submarines were developed. However, it was in 1896 that the Irish inventor John Philip Holland designed submarines that, for the first time, made use of internal combustion engines on the surface and electric battery power when submerged. This proved to be a breakthrough. His designs were used in the Russo-Japanese war (1904–1905) and thereafter submarines were bought in large numbers by all the navies of the world. Armed with new self-propelled projectiles called torpedoes as well as the traditional gun, German submarines threatened Britain's supply lines by attacking its merchant fleet. They were remarkably successful, sinking over 5000 allied ships. In contrast, the big dreadnaughts only engaged in a small number of inconclusive battles.

The total number of military and civilian casualties in the First World War was over 37 million, the deadliest conflict so far in human history. Deaths in battle amounted to around 4% of the total population in France, Germany, Italy and Austria-Hungary, with a further 5% wounded. Despite this there was little pressure for peace from any country until 1917 when there was a mutiny among French troops and a revolution in Russia. It appeared that the human will to win and the patriotic fervour of the common people overcame all common sense; mutual destruction was pursued whatever the cost.

Political cultures at the end of the First World War

Proof that war accelerates change can be seen in the aftermath of the First World War. The old church and state cultures that had been holding onto power ceased to exist. The empires of Germany, Russia and Austria were disbanded. Their emperors abdicated, aristocrats lost their titles and the church lost power. Outside Europe the Chinese emperor abdicated in 1912 and the Ottoman sultan resigned in 1922 following the Turkish War of Independence. USA, France, Italy, Japan and the UK, which adhered to a secular capitalist political culture, had triumphed militarily. All the other political cultures suffered setbacks. Orthodox Christianity was replaced by communism in Russia. Neo-Confucianism clung on in China but lost its political leadership. Only a handful of Islamic states remained independent and the Latin Catholic cultures suffered revolutions and regime changes. Progress in each of these five cultures is discussed briefly below.

Secular capitalism of Europe, Japan, North America and Oceania

The USA had clearly emerged from the war as the greatest economic power in the world after a mere 150 years of independence. As winners in the war, Great Britain and France remained huge colonial powers but, as a result of the war, their countries were considerably weakened. Key to the development of industry was the joint stock companies which were now evolving to become huge organisations. This was particularly true of the USA and Germany; names like United Steel, GE, Krupps, Siemens and Esso were already widely known across the world.

Using gross domestic product (GDP) estimates we can begin to gauge how the British lead in industrialisation had been reduced by its rivals in the Western world and Japan. Figure 9.2 shows the percentage of world GDP held by the Western powers and Japan. It is extracted from data by Angus Madison in an OECD publication, and is based on the modern configuration of states, so must be treated with caution. Its end year is 1913, before the First World War. It shows that the USA had now become the most important industrial power in the world with an economy twice the size of the second largest, Germany. Both the USA and Germany had overtaken Britain in industrial output. As the economies of Europe would have declined during the First World War, the American lead would be even higher by 1920.

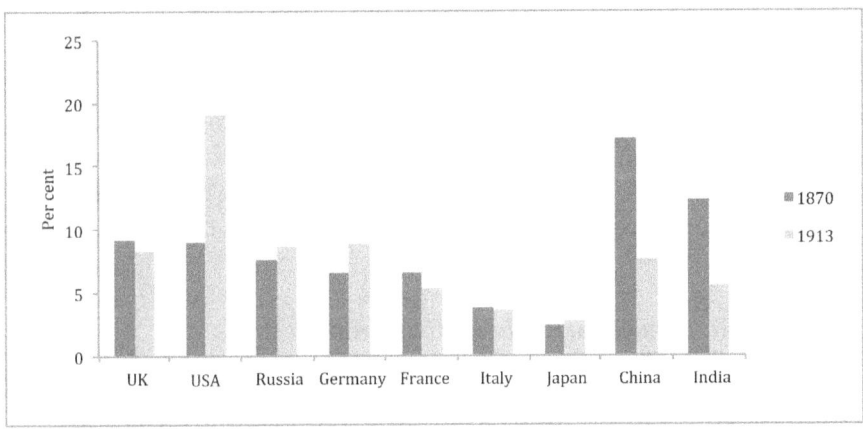

Figure 9.2 Share of world GDP 1870–1913

The USA had also overtaken the UK in terms of GDP per capita – a measure of wealth per person in a country. Figure 9.3 shows the extent to which the rest of the world was catching up with the UK. In 1870 the average wealth of the British was over 3.5 times the world average and that of the USA was 2.8 times. By 1913 the situation had completely reversed. Despite their strides towards industrialisation, Japan and Russia remained some way behind the other nations in terms of industrial development and wealth.

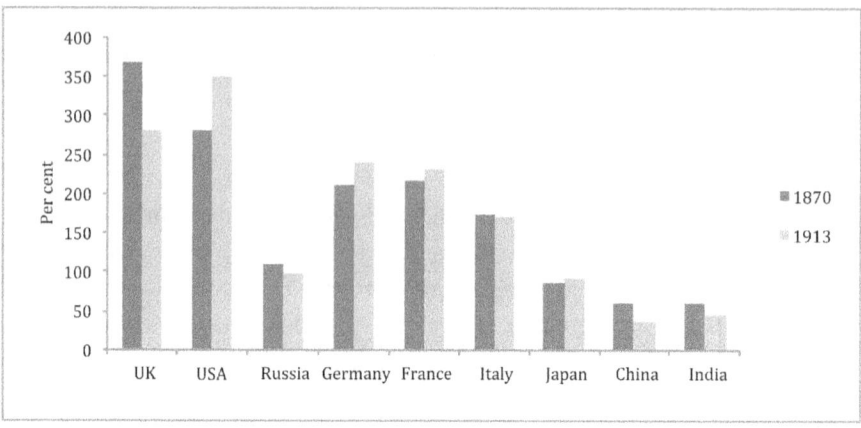

Figure 9.3 GDP per capita compared to world average

After the war, secular capitalist democracies were set up in Germany and the new states in Eastern Europe. It seemed as if secular capitalism would become the culture of Europe from now on. However, as we shall see, secular capitalism cannot easily be imposed on a country. It has to have strong positive backing from the population at large. New political philosophies, like communism and fascism, were to become alternative political systems of government in many European states in the next period.

Latin Catholic cultures in Iberia and Latin America

In these countries the papacy still had a considerable influence in the running of the state. Both in Iberia and Latin America, this was a period when monarchists, republicans, the army, the Catholic Church and the landed elite struggled to gain power and no political culture established primacy. As a result there was a succession of competing systems of government including Catholic monarchies or dictatorships, republican dictatorships and Catholic oligarchies, interspersed with periods of genuine democratic government. In truth, neither the ruling elite nor the electorate was yet ready to support secular capitalism. As a result, no long-term stable political culture emerged that satisfied all parties. This situation would also be true of other new states born in the twentieth century, as the old empires contracted. I refer to these states in future as developing states.

Of the Latin American states, only Mexico, Brazil and Argentina managed to attract sufficient investment to advance their economies substantially. In Mexico, the dictatorship of Porfirio Díaz (1876–1911) achieved some political stability, establishing rule of law and modernising the economy.

Días was an astute politician who built a national base of supporters. During this period, the country's infrastructure was greatly improved, thanks to increased foreign investment from Britain and the USA.

However, while the economy gained strength, resentment grew among the peasantry and working classes against the land owners and capitalists. In the period 1910–1929 Mexico entered a period of armed revolution which grew increasingly broad-based, radical and violent. The revolution sought far-reaching social and economic reforms by strengthening the state and weakening the conservative forces represented by the Church, the rich landowners and the foreign capitalists. Different strong men fought bitterly for control of regions; millions of people were uprooted; many died or fled to the USA.

In Brazil in 1889, the empire was replaced by a nominally democratic republic. In reality the country was controlled by landowners. The country's economic development was spearheaded by the cultivation of coffee and rubber. Between 1890 and 1930 over four million migrants arrived from Europe and another 200,000 from Japan. Most went to work on the coffee estates to the south of Brazil but a few stayed in Sao Paolo and began to develop Brazil's industries.

In Argentina, after a long period of in-fighting, the boundaries of the modern state finally emerged in the period 1850–1880. The local Indians were brutally suppressed and the agricultural potential of the state began to be exploited, particularly for wool and beef. This was helped by British investment in the railways and the development of refrigerated ships, so that Argentine beef could be sold in Europe. The Argentinian economy blossomed and about six million immigrants came from Europe between 1890 and the start of the First World War. However the development of their political culture was slow. Power remained in the hands of the landed elite almost up to the First World War.

The communist culture of Russia

The Orthodox Christian Russian leadership had been unable to adapt to the challenges of the new industrial world. This is explained by Orlando Figes in A *People's Tragedy – the Russian Revolution* 1891–1924, p. 14:

> But the last two tsars and their more reactionary supporters – in the gentry, the Church and the Rightist political circles – were at best ambiguous towards the idea of 'modernisation.' They knew, for example, that they needed a modern industrial economy in order to compete with the Western powers; yet at the same time they were

deeply hostile to the political demands and social transformations of urban social order. Instead of embracing reform they adhered obstinately to their own archaic vision of autocracy. It was their tragedy that just as Russia was entering the twentieth century they were trying to return it to the seventeenth. Here then, were the roots of the revolution, in the growing conflict between a society rapidly becoming more educated, more urban and more complex, and a fossilised autocracy that would not concede its political demands.

The Orthodox Christian empire of Russia did not survive the First World War. In the end this old-fashioned absolute monarchical organisation was not up to providing the leadership necessary to win a protracted military conflict. Revolution was perhaps an inevitable outcome and the new communist party was single-minded enough to seize power. The emperor was ultimately executed and his ally, the Russian Orthodox Church, suffered an almost complete fall from its dominant position in Russian society.

The Islamic states of the Middle East

The Islamic empires had notably failed to keep pace with the technical and political developments of the West. Before the First World War, the area controlled by the once powerful Ottoman Empire had been reduced to just Turkey and the old Fertile Crescent of the Middle East. In 1908 army reformers forced the government to adopt a system of constitutional monarchy. It failed to stem the tide of defeat, which continued as the Ottoman Empire backed the losing side in the First World War. In 1922 the last Sultan abdicated and the lands of the Ottoman Empire were divided among the winning allies. Only Turkey survived as an independent nation.

The Qajar Empire of Persia also diminished in size, having ceded territory in the north to Russia during the nineteenth century. In 1906 the Shah was forced to adopt limitations on his government and rule as a constitutional monarch. Again, this did not stop Western aggression. In 1907, after the discovery of oil by the British, Russia and the UK divided Iran into spheres of interest. Eventually, in 1925, a military strong man, the Prime Minister Reza Shah Pahlavi led a coup-d'état and the last Qajar shah was forced into exile.

The neo-Confucian culture of China

Before the First World War China had had to endure continual humiliation as Western nations and the Japanese queued up to demand concession after concession. China had to cede yet more trading rights to foreign powers with privileged legal status. Concessionary rights for eight nations were established in Shanghai and six other ports. The Japanese and Russians were more aggressive still. Russia seized the territory beyond the Amur River and founded Vladivostok. Japan, having acquired much Western military technology, was easily able to defeat the Chinese army. Their army seized Taiwan, invaded China, and then established a colony in Korea.

For the Chinese government the world had turned upside down. The imperial government was unable to respond to the external threat. Totally stuck in its Confucian traditions, the combination of internal rebellion and Western and Japanese aggression brought the empire to its knees. In 1912, almost 2300 years of imperial rule in China was finally brought to an end. The emperor abdicated and a new unstable political era started as China attempted to deal with the modern industrial world.

Summary: situation at the end of the First World War

The technological revolution had begun to transform the world. The countries in the van of this development were in the West and Japan. Their citizens were now several times better off than the other centres of civilisation. Large parts of Asia and most of Africa became their colonies.

The world population had risen to 1.9 billion people, an increase of 0.8% per annum (doubling every 90 years). With a flood of immigrants increasing US population, America continued to grow at the rapid rate of 1.7% p.a. Despite the losses in the First World War, Europe still grew at 0.8% p.a. Figure 9.4 shows that for the first time population growth rates over 0.5% p.a. were seen in Asia and the rest of the world. In 1900 the largest city in the world was London, with 6.5 million, followed by New York, with 4.2 million. All the largest cities in the world were now in Europe or USA, except for Tokyo in Japan, with 1.5 million. For the first time none of the world's top ten cities were in China or India.

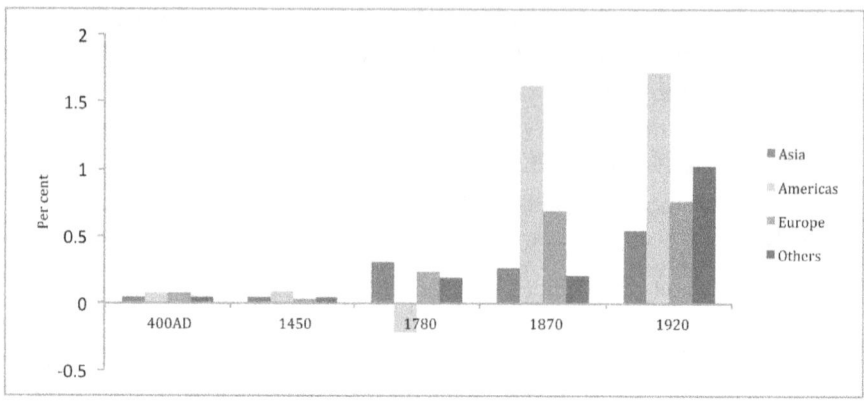

Figure 9.4 Population growth rates to 1920

One of the most important reasons for the high rates of population growth was an increase in life expectancy. Figure 9.5 shows that in some Western countries it exceeded 50 years for the first time. Increasing industrialisation and wealth had a positive effect on the health of the population. I suspect a large part of the increase in life expectancy was due to the increased proportion of middle classes working in less physically stressful occupations and experiencing improved living conditions. The life expectancy for the working class would not have improved so significantly. China and India still showed unchanged low levels of life expectancy of around 20–30 years. The low life expectancy in Russia was due to the huge loss of life during the First World War, followed by the horrors of the communist revolution and the subsequent civil war.

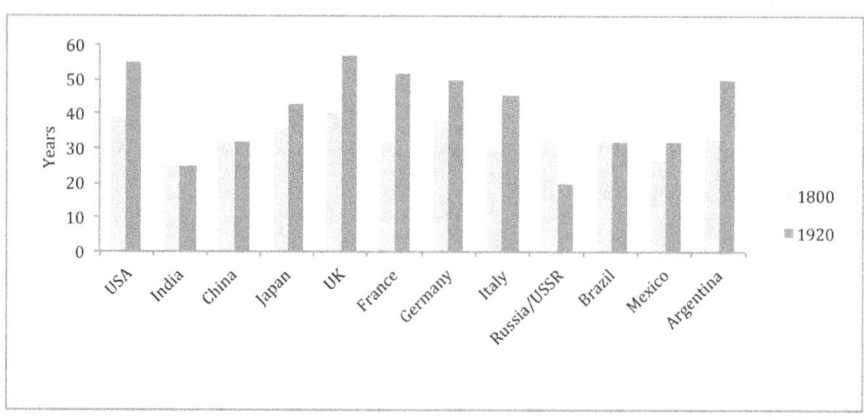

Figure 9.5 Life expectancy 1800–1920

The old mediaeval society structures of the West had been swept away. In these newly industrialised countries absolute monarchs, the privileged aristocracy and subsistence peasant farmers were now largely a thing of the past. A new evolutionary structure had emerged for the industrial society. In this world the dominant political culture was secular capitalism which had been pioneered by Britain and the UK and extended across Europe as a result of victory in the First World War.

However, change always brings a reaction. The cultural and relational changes initiated by the technological revolution led to resistance from those parts of society that had lost out. They would initiate new political cultures which would compete with the secular capitalist culture of the West during the rest of the twentieth century.

CHAPTER 10:
THE CHALLENGE TO SECULAR CAPITALISM (1920–1950)

With France, USA and Britain triumphant after the First World War it looked for a while as if the secular capitalist political philosophy would be adopted by most advanced countries. The benefits of the technological revolution had been spread to an increasingly high proportion of the population as **mass production** techniques brought the price of many goods down to a level that could be afforded by the ordinary people. For 10 years after the war, secular capitalist countries embraced new technologies as their economies expanded. **Buses** and **cars** sped up transportation. Information could be distributed much faster by **radio** and **telephone**. For the first time, ordinary people began to share in the wealth of the nation. The consumer society was born; all classes acquired significant material possessions and had more leisure time. Aided by medical advances such as **antibiotics**, there was a general improvement in life expectancy and the quality of life.

All this progress came to a shattering halt with the Great Depression, when banks and businesses failed and a large percentage of the workforce found themselves unemployed. The faith in the capitalist approach had been shattered and two new political cultures, **communism** and **fascism** took hold. Extreme nationalism once again reared its ugly head and another war became inevitable. The Second World War would be truly global and the killing power of military **aeroplanes**, **tanks** and the **nuclear bomb** would make it the most destructive war yet.

Developments in secular capitalist countries

Companies led the way to material progress. Mass production was pioneered by Henry Ford (1863–1947) in making the Model T car in the early part of the twentieth century. Mass production is the assembly of an item (a car for example) in a flow line from pre-manufactured parts, whereby a number of workers each repeat the same assembly operation as different cars pass in front of them. The specialisation of labour involved in concentrating on one repeatable process achieved huge increases in efficiency. Three innovations: standardised parts, improved materials and small electric motors made mass production possible. Standardised parts, manufactured to close tolerances, were developed in the USA at the end of the nineteenth century. They were pioneered in the US armament industry in the civil war and used by such companies as Singer to manufacture sewing machines in the 1890s. High strength steels increased the robustness of the assembly and the use of tungsten carbide in drilling and cutting allowed for more rapid processing. However, according to Henry Ford, the real breakthrough came with the development of the small electric motor. Previously, machinery had been driven by a cumbersome system of powered belt drives that had to be fixed in one location, but now the electric motor allowed powered tools to be moved to the point of assembly.

The story of the success of Henry Ford is one of the great triumphs of capitalism. Henry Ford was from a farming background in Michigan. He trained as a machinist in Detroit and had a spell working and learning from the inspirational entrepreneur, Edison. He built his first petrol engine in 1892 and a decade later launched the Ford Motor Company. His great idea was to build a reliable car which was cheap enough to be used by the ordinary people. As a child he knew all about rural isolation and believed the internal combustion engine, combined with cheap petrol, was the solution. His vision was that, by gradually increasing economies of scale, a mass-produced car could be made available to all.

His success is outlined in *The History of the American People*, p. 619 by Paul Johnson:

> The result was the model T of 1908. Farmers (and their wives and daughters) found they could handle it and even learn to 'fix' it on the rare occasions when it needed attention....
>
> Ford marketed the first Model T at $850 in 1908, when he sold 5986. By 1916, when he sold 577,036, he had got the price down to $360. By 1927, when the series was discontinued, he had sold over 15 million of these autos, and made them and their rivals standard equipment for the American family. Ford gave the entire, vast nation a mobility it had never had before – and which in time spread throughout the world.

In America, Ford had used the internal combustion engine to make personal transport more convenient and faster for ordinary people. Developments in Europe were slower and car usage extended only to the middle classes. Most people used mass transportation. The railway remained the dominant form of long-distance land transport. However the bus, also powered by the internal combustion engine, began to challenge both the tram in urban areas and the railway in the countryside. Buses proved a cheap and convenient form of transportation as they did not need the capital outlay of laying track or the ongoing expense of its maintenance.

The internal combustion engine also enabled the development of air transport. The Wright brothers, Orville and Wilbur, made the first powered flight in 1903 in North Carolina. After that, aeroplane speed, range and reliability slowly increased to allow them to play a critical role in reconnaissance during the First World War. However, it was not until the 1930s that their construction evolved from low-powered biplanes made from wood and fabric to sleek, high-powered monoplanes made of aluminium. The payload gradually increased so that passenger travel was possible; the Douglas DC-3, introduced in 1936, was a harbinger of things to come. It had a capacity of 20–30 passengers and could achieve 190 mph. With only three refuelling stops, flights could cross the USA in approximately 15 hours. In contrast, a train journey from Los Angeles to New York, with a stop in Chicago, would take about 3 days.

Ocean liners continued to be the principal form of intercontinental travel; their only competitor was flying boats. Using these planes Imperial Airways offered a London to Sydney service in 1936 that took 11 days; a journey by liner would typically take 7 weeks. In reality only the rich and government officials could afford to fly such distances; mass air transport still remained some way off.

Although transportation of humans over long distances was slow by modern standards, information transmission approached modern speeds with the development of the telephone and the radio. The telephone evolved from the telegraph with the development of a microphone, transmitter and receiver. This allowed people to converse over distance, greatly increasing the amount and usefulness of the information exchanged. Many were involved with its creation, but the clear commercial winner was the Canadian Alexander Graham Bell (1847–1922) and his backers, the American Telephone and Telegraph Company (AT&T) incorporated in 1885. As with electric supply, the key to its success was to establish a distribution network incorporating telephone exchanges for routing the calls. Initially these exchanges were manually operated, but in 1891, Almon Brown Strowger, an undertaker in Kansas City, Missouri, patented the stepping switch, a device which led to the automation of telephone circuit switching. AT&T was chartered to build and operate the original long-distance telephone network. Building out from New York, AT&T reached its initial goal of Chicago in 1892, and then San Francisco in 1915. In a way that was previously unimaginable, telephone conversations could take place between people thousands of miles apart with scarcely a pause in the response.

Mass information exchange became even easier with the coming of radio transmission. This was possible only after the scientific discovery of electromagnetic radiation. One of my great heroes, James Clerk Maxwell (1831–1879), developed the theory of electromagnetic radiation. Maxwell was a Scot. He studied Faraday's magnetic lines of force and was able to show that the speed of the propagation of these fields of force was approximately the speed of light. Further, he conceived them as a wave, with a frequency of oscillation and a wave length defined by mathematical equations. Through a great leap of imagination, he realised that light was a particular form of electromagnetic radiation. This was the start of modern physics, and of the notion that the world was defined by mathematical laws which could scarcely be imagined in practical terms. There was no mechanical explanation of how an electromagnetic wave propagates. Unlike a sound wave it was not dependent on oscillating particles; and it could pass through a vacuum. Nevertheless, electromagnetic radiation could be measured and its properties predicted using mathematics. This was enough to enable scientists and engineers to develop amazing new technologies, such as the radio, television and radar, all based on the transmission of electromagnetic radiation.

Although Maxwell published his equations in 1864, the existence of electromagnetic waves was only proved by the German physicist Heinrich Hertz in 1886. Thereafter, many people were involved with the development of suitable transmission and receiving devices for electromagnetic radiation. However, it was Guglielmo Marconi (1874–1937) who was to provide the dynamism to push forward the early development of radio. Marconi was a great Italian inventor and businessman who proved that, even at the dawn of the twentieth century, a talented amateur could still make a major breakthrough. Working from home he developed a transmitter and receiver capable of working over 2 km. Finding no support for his ideas in Italy, at the age of 21 Marconi went to London. There, with support from the British Post Office, he arranged a series of demonstrations for the British government and in 1897 Marconi sent the first ever wireless communication 6 km over water to the other side of the Bristol Channel. The first commercial uses of radio were for ship-to-shore and ship-to-ship communication, using Morse code. Radio was famously used during the sinking of the Titanic in 1912, firstly between the sinking ship and nearby vessels, and then in communicating the list of survivors to shore stations.

In 1900 a Brazilian priest, Landell de Maura, first demonstrated the wireless transmission of the human voice and a network of enthusiastic amateurs began communicating with each other across the world. To enable the development of mass commercial broadcasting, governments had to license a specific radio frequency exclusively to one broadcaster. The KDKA station operated by Westinghouse in Pittsburgh was the first licensed radio station. Radio's ability to provide mass entertainment was soon demonstrated in the airing of comedy and music shows, sports broadcasting (providing commentary on the world heavyweight title bout between Jack Dempsey and Georges Carpentier in 1921) and news broadcasting (bringing the results of the 1920 presidential election). The stage was set for the golden period of mass radio broadcasting in the 1930s and 1940s. People could now be entertained and informed by the sound coming from a small box in their own home.

More rapid transportation and increased speed of information flow inevitably makes societies more cohesive and efficient. Secular capitalist countries maintained and extended their lead over the rest of the world in the 1920s. There was an unprecedented expansion in world trade. Figure 10.1 shows it doubled between 1918 and 1927 – a very high rate of increase of 8% per annum. Increase in trade is a good indication of a general increase in wealth. It shows that suppliers are increasingly specialising in their production of goods and services and thus becoming more efficient and effective. This general increase in wealth after the First World War was led by the USA and enjoyed by most secular capitalist countries. Stock values kept rising and there was apparently unlimited opportunity for continued economic growth in the capitalist countries of the West.

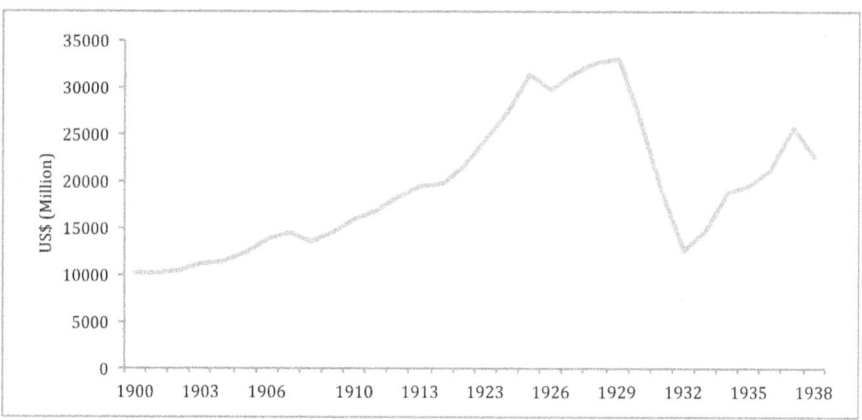

Figure 10.1 World trade 1900–1938 (excluding the war years)

This new growth challenged the old social order. Although in Europe the old monarchical style of government had been replaced and the aristocracy had lost its privileged status, class was still important. Although there was equality before the law in European secular capitalist countries, because of the education system there was no equality of opportunity. Speaking very broadly, the wealthy could afford to send their children to university, the middle classes could afford to keep their children at school to attain a certificate of education and the poor barely managed to achieve literacy and numeracy. The four classes I have outlined, upper, middle, worker and rural operated in different spheres using, in Britain, different ways of speaking, and going through different school systems and having different cultures and lifestyles. Nevertheless, capitalism forced change. Before the war in Europe the richest 10% of the population owned 90% of the wealth of the nation. After the war middle class managers and professionals gained sufficient income to acquire significant capital in their own right. However, according to Thomas Piketty in his book *Capital in the Twenty-First Century*, the proportion of wealth held by the top 10% had declined to only 80% by 1930. Society had become more equal but it was a long way from being egalitarian.

Nevertheless the purchasing power of the working, rural and middle classes gradually increased. Dramatic improvements were made in their material well-being and quality of life. This was most obvious in the quality of the new dwellings being built. Up to the the end of the First World War, most people were housed in buildings that had no internal plumbing and were damp and overcrowded. Now around most towns in Western Europe and the USA, neat suburbs appeared with substantial houses or flats containing separate bedrooms, bathrooms, kitchens and living rooms. The houses and flats were clean and dry with adequate provision for heating and with hot water on tap. All had interior plumbing with no need to visit the privy at the bottom of the garden or to share facilities with other families. The houses were lit at the touch of a switch. Kitchens were equipped with gas or electric cookers, replacing the old ranges heated with coal or wood. Baths were permanent fixtures, and could be filled directly from a hot water tank, unlike the tin baths of earlier years. It would be several decades before such housing was available to everyone but a step change in the overall quality of the housing stock had been made.

Housework was still a time-consuming job and with rearing children it remained the principal task of women. Vacuum cleaners, refrigerators, washing machines and other helpful gadgets for the housewife were only commonly used after the Second World War. Doing the household laundry, in particular, was still a time-consuming manual operation. Shopping for food was largely a daily chore, but for the first time tinned and powdered foods began to reduce the necessity to shop every day. Ready-to-wear clothing could be bought in the shops; so clothes no longer had to be made at home or by a local seamstress. With increasing money available, the new department stores tempted people to buy luxury goods. Houses were furnished with modern furniture, sofas, dining tables and sideboards, all of which would have been beyond the reach of earlier generations of ordinary people.

People had more leisure time. Annual holidays to the seaside became common at the start of the century. For most people, travelling to the coast was still largely undertaken by train or coach. Only richer families could afford to travel in their own car. If a car was too expensive, a motorbike was possible, perhaps with an attached sidecar to take the children. With electric lights common, people looked for entertainment in the evening. Other than just visiting the local pub or bar, they could stay in and listen to the radio to be entertained by drama, news and music. Many went to the cinema at least once a week. It began to become common to have Saturday afternoon at leisure in addition to Sunday. Families would take excursions into the surrounding countryside by bus or train. The middle, working and rural classes may not have acquired much capital but their wages were sufficiently high for them to become the most important purchasers of goods and services.

Industries now focused on supplying ordinary people, rather than just the rich. They advertised on radio and in the newspapers. Fashion was important. Keeping up with the Jones became a critical motivation for purchase. Credit terms were extended to buyers. People were encouraged to buy cars, furniture and other durable goods on hire purchase. The 'never-never', as it was known, allowed people to pay for goods by instalments over several months. Houses were bought with mortgages on the same principle. The availability of credit in society began to become even more important to the functioning of the world economy.

Attitudes to women also changed. Before the First World War in the USA, ex-slaves were allowed to vote but women could not. Indeed, only a handful of countries had allowed women the vote. New Zealand was the first in 1893, followed by Finland, Denmark and Norway. Extending the voting franchise to women had been strongly resisted by the male establishment. However, the important work undertaken by women during the war changed public perceptions; with men away at the front, women took over male roles in munitions factories and shipyards, driving vehicles and working on the land. In the USA and Britain it was accepted, after the war, that women had gained the right to vote. In Britain in 1918, women over 30 were enfranchised; in 1928 this was extended to those over 21, the same qualifying age as men. In the USA sexual discrimination on voting was ended in 1920. However, in truth, the role of the sexes had not changed in a fundamental way. After the war women returned to a largely domestic role and men reclaimed their jobs in transport and in the factories. It would be many decades before sexual discrimination in the job market would be addressed. Indeed the consensus in France was that women should resume their role in the home; voting rights for French women were not conceded until 1945.

At the end of the First World War, with the collapse of the Austro-Hungarian Empire and the contraction of Russia, a host of new European states appeared: Czechoslovakia, Poland, Finland, Estonia, Latvia, Lithuania, Hungary, Austria and Yugoslavia. The winning allies, USA, Britain and France, all democracies, encouraged the new states to copy the democratic style of government. Indeed, all these new states started off as constitutional monarchies or republics. However the practices of democracy and capitalism had had little time to take root and soon the manner of government in many of them changed.

The inter-war years were a proving ground for two very different alternative philosophies of government, communism and fascism. These became particularly attractive after the crisis of capitalism caused by the Great Depression. This began with a collapse in the price of shares on Wall Street in October 1929. The value of stocks fell by about a third. This was a straightforward correction of investor over-exuberance and stocks soon stabilised at a level achieved at the end of 1928. However, there was a subsequent credit crisis which resulted in bank failures. These reached a critical mass in November and December 1930 when 608 US banks failed, including the Bank of the United States (in total 9000 US banks failed during the 1930s). A downward spiral resulted when businesses were unable to borrow as debts were foreclosed, which resulted in more business failures. With the expansion of trade in the 1920s, countries had increased their interdependence. However, in the 1930s the world did not have any financial institutions that allowed countries to cooperate for their own mutual benefit. Countries increased trade barriers to protect their own economies and this reduced world trade, which resulted in more business failures. Figure 10.1 shows that trade fell below the levels achieved before the start of the First World War. The results were catastrophic for the economies of all countries in the West, leading to a collapse of industrial production and a huge rise in unemployment, as demonstrated in Table 10.1. Massive unemployment and a return to the poverty levels of the past dented people's confidence in secular capitalism. Governments with some history of democracy such as Britain, the USA and France held on, but many other countries turned to fascism or communism.

	United States	Great Britain	France	Germany
Industrial production	-46	-23	-24	-41
Wholesale prices	-32	-33	-34	-29
Foreign trade	-70	-60	-54	-61
Unemployment	607	129	214	232

Table 10.1 Percentage change in economic indicators 1929–1932

Jerome Blum, Rondo Cameron, Thomas G. Barnes, *The European World: a History.* 2nd edn. (1970) p. 885

Fascism

Fascism started in Italy after the First World War as a wave of violent unrest among the rural and middle classes in reaction to socialism. Many hundreds of socialists were killed. Political leadership was provided by Benito Mussolini (1883–1945) who formed his fascist party, the Partito Nazionale Fascista (PNF) in 1921. Mussolini was an ex-socialist who resigned from the party in protest against its anti-war policies. His fascism was a mixture of the egalitarian policies of the socialists combined with revolutionary nationalism. The guiding principle was the mobilisation of men and women from all classes in support of the state. Fascists regarded the middle classes as soft and idle. They saw themselves as representatives of those who worked, a new elite ready to create a nationally inspired Italy.

In 1922 the Fascist Party had a quarter of a million members and had effectively taken over the administration of law and order in many rural areas. However, fascism was still not a force in the Italian parliament; it had won only 35 seats in the 1921 elections. Mussolini came to power by a combination of political manoeuvring and the threat of force. At this time the main parties in the Italian parliament were the conservatives, liberals, catholics and the socialists. No party dominated and governments were formed from unstable coalitions. In the autumn of 1922 plans for a fascist march on Rome were laid. The government felt it could not rely on the army and the police to defend the institutions of power. Under the threat of force, conservative and liberal politicians saw fascism as less threatening than socialism and appointed Mussolini prime minister on 29 October 1922. For a while the formalities of democratic government were observed but in 1925 Mussolini banned all political opposition, abolished freedom of the press and established a fascist dictatorship in Italy.

In Germany the Weimar Republic, formed after the First World War, was always fragile. The Kaiser had abdicated at the time of the armistice and before the end of the peace settlement, leaving the existing Reichstag leaders to pick up the pieces. It was the tragedy of the Weimar Republic that it had to face humiliating peace terms, dictated by the winning allies, which included the loss of German territory and war reparations (financial penalties). Germany lost their overseas colonies, as well as Alsace and Lorraine to France and much of their eastern territories to Poland. The reparations were excessively penal, being fixed in 1921 at 132,000 gold marks, which were to be paid at a rate of 2000 million marks annually, plus 26% of her exports. As the alternative was renewed war, the government had to accept the peace terms. However, the nationalist parties blamed the government for being weak.

What followed was a national disaster. The German government could not keep up the reparation payments. In retaliation, in 1923 French and Belgian troops occupied the Ruhr, the heartland of German industry. The German government adopted a policy of non-cooperation; coal mine owners were ordered not to hand over coal stocks and civil servants were instructed not to obey French instructions. A general strike was declared in the Ruhr area. The government continued to pay the Ruhr workers' wages and salaries and extended credit to factories and mines. However, these additional costs, taken together with the loss of tax revenues and export earnings, ruined the nation financially. Government debt could be financed only by printing money. The story of what happened next is from *The Weimar Republic 1919–1933*, pp. 35–6 by Ruth Henig:

> Within six months, the German currency had collapsed completely. In August [1923], a dollar cost 4.6 million marks [whereas in July 1922 it had cost 439 marks]. Three months later, in the worst of the hyperinflation, it cost an almost unimaginable 4,000 billion marks….To enable wages to continue to be paid, and commodities to be purchased, 133 printing offices and 1783 machines were by this stage churning out paper notes for the Reichsbank. Over 30 paper factories were being employed to full capacity. Everyday food items such as loaves of bread, joints of meat and vegetables were costing millions of marks, or enormous piles of paper notes.

This catastrophic event, wiping out savings and reducing people on fixed incomes to penury, was an enduring humiliation for the German people. It reinforced the alienation from the government of both those on the left and the right. At all post-war elections the communists and nationalists won significant proportions of the vote. Governments were made up by fragile combinations of liberal, catholic and socialist parties.

In the event the German government managed to overcome the 1923 crisis but when it was hit by the Great Depression the coalition could not agree on a policy of reducing social benefits and the German chancellor began to rule by emergency decree, overruling parliament. In the 1932 elections the Nazi party gained 37% of the vote and became the largest party in the Reichstag. And in 1933 Adolf Hitler became chancellor; the Weimar Republic was dissolved and Germany became a fascist state.

Adolf Hitler (1889–1945) was one of the three most destructive individuals in the twentieth century, together with Stalin and Mao. All were self-obsessed people with a strong vision and their absolute control of a one-party police state enabled them to ruthlessly pursue their ideas irrespective of the suffering caused. As a direct result of their policies millions of people died. Hitler was born in Austria, joined the German army in the First World War, where he reached the rank of Gefreiter, generally translated as corporal, and received the Iron Cross. Appalled at the loss of German territories after the First World War, he joined right-wing political movements and took part in a failed armed rebellion in Bavaria in 1923. While serving a prison sentence after his capture and trial, he wrote his political statement, *Mein Kampf.* He identified the causes of German problems as, firstly, a Jewish conspiracy for world domination and, secondly, the Weimar government, particularly the socialists and communists. He was clear that he wanted to abolish the democratic system of government, believing it to be corrupt and exploited by opportunists. His book touched a popular nerve, and when he became the chairman of the Nazi party, he increased its popular appeal by staging increasingly theatrical oratorical performances, in which he would whip the audience into a frenzy of nationalistic fervour. After his appointment as chancellor, he did exactly as he said he would. Within months, press freedom was suspended, non-right-wing parties dissolved themselves, trade unions were banned and Jews were debarred from state employment. The civil service was purged and the Nazi party's paramilitary arm, the SS, became a sort of parallel administration, in which the law was increasingly ignored and people were subject to arbitrary beating, imprisonment in concentration camps, or execution. Hitler was obsessed with recovering the territorial losses and creating Lebensraum (living space) in the east. This led directly to the Second World War.

Germany and Italy were not the only countries to reject democracy. Liberals were unable to prevent the conservative backlash following the failure of capitalism in the Great Depression. One by one, democratic European countries reverted from democracy to dictatorship. Spain became fascist after a civil war in 1936. The new countries in Eastern Europe were particularly prone to nationalist issues because of the democratic demands of ethnic minorities. Hungary, Romania and Latvia installed fascist governments. Poland and Yugoslavia introduced nationalist dictatorships.

Most importantly, in the 1930s the Japanese also stepped back from representative government. Strong nationalist sentiments gained ground among the military leaders. The Japanese army invaded Manchuria in China in 1931 without the permission of the government. When parliament failed to censure the military leaders it was clear the Japanese parliament had lost control. Thereafter, the military combined with bureaucrats and senior government officials to effectively run the country. Even more military aggression became inevitable as nationalist instincts were exploited by the nation's new leaders.

Communism and Russia

In 1917 after three years of war Russian citizens and soldiers at the front had had enough; they wanted peace. The czar, Nicholas II, was a weak leader and the logistics of running such a huge conflict as the First World War had proved beyond the czarist government's abilities. Russia was still largely a pre-industrial society, where aristocrats lived off their estates, while the overwhelming majority of the population were peasants. The one religion, Russian Orthodox Christianity, was a powerful supporter of state order. There was a parliament (the Duma) but all executive power lay with the czar. In some large cities, such as St Petersburg or Moscow, there were factories and a middle-class elite, but overall Russia was predominantly an agrarian society.

In February 1917, short of bread, the citizens of St Petersburg took to the streets and the soldiers mutinied and refused put down the rebellion. The workers and soldiers organised a council, or soviet, to represent their interests. The St Petersburg Soviet agreed the basis for a transition to a democracy with representatives of the Duma. The bourgeois and liberal aristocratic leaders of the Duma took the reins of power, the czar abdicated and Russia became a republic. For one brief moment a genuine democracy was formed and most of Russia rejoiced. However, the new government did not end the Russian participation in the war and food shortages continued. Worse, having abolished the police and with the army in mutiny, the government had no way of enforcing its orders without obtaining political agreement from the workers' soviets. The working classes gained in confidence and soviets were formed across all Russia. The peasants took control of the estates and land was forcibly redistributed in advance of any authorisation from government. Peasant assemblies effectively took over law and order in their own areas. The revolution progressed but the government was not in control.

In April 1917 Vladimir Lenin (1870–1924) returned from exile after 10 years. Lenin, born to a wealthy Russian middle-class family, had taken an interest in left-wing politics after the death of his brother, a political activist. Lenin had been exiled to Siberia once and had spent the previous 10 years in Switzerland, immersed in Marxist politics and theory. On his return, he immediately took control of the socialist Bolshevik party and began to plan the seizure of power. Lenin believed that some people are more politically aware than others and should assume the responsibility for leading society to socialism. He created a secretive and disciplined party which took its instructions from the top. Bolsheviks came to accept Lenin as leader and teacher.

According to classic Marxist theory a worker-led state could only occur after the bourgeoisie had seized power from the monarchy and created a capitalist state. Russia was still an overwhelmingly peasant society and therefore was not ready for a communist revolution. Such theories did not deter Lenin. While other socialist parties sought compromise and consensus, the Bolsheviks acted. They gained power because they were professional revolutionaries dedicated to the overthrow of government. Brilliantly and ruthlessly led by Lenin, though in a minority overall, the party and Lenin were able to seize the reins of power of central government in October 1917.

Once in power the bloodletting really began. The first targets were the rich and educated, labelled bourgeois. This is from *A People's Tragedy: The Russian Revolution: 1891–1924*, pp. 524–5 by Orlando Figes:

> [Lenin] stressed that the 'proletarian state' was 'a system of organised violence' against the bourgeoisie: this was what he had always understood by the term 'Dictatorship of the Proletariat.' Licensing popular acts of plunder and retribution was an integral part of the system, a means of terrorising the bourgeoisie into submission to the Proletarian State.... The archaic plunder of the bourgeois, Church and noble property was legitimised by Bolshevik decrees of revolutionary confiscation and taxation, which the local Chekas [state police] then enforced through the arrest of ... hostages. The mob trials of bourgeois employers, officers, speculators, and 'enemies of the people' were institutionalised through People's Courts and the crude sort of 'revolutionary justice' they administered.

In 1918 this degenerated into a full Red Terror in which anyone deemed to be counter-revolutionary was subject to arrest, imprisonment, torture and execution by the cheka. Those who were especially at risk were members of other socialist parties, but beyond that, the pattern was arbitrary. This is from the same book, p. 643:

> Arbitrary arrests were particularly common in the provinces, where the local Cheka were very much their 'own men' pursuing their own civil wars of terror. But the principle urged by Lenin – that it was better to arrest a hundred innocent people than to run the risk of letting one enemy of the regime go free – ensured that wholesale and indiscriminate arrests became a general part of the system.

The Red Terror was essentially a brutal attempt to force people to submit to the will of the party, irrespective of its rationality. The next to feel the ire of the Bolsheviks were the peasants. Throughout the early years of the revolution grain supply was critically short in the cities. People left the cities to be nearer the source of food. This is a description written in *My Disillusionment in Russia* of her return to St Petersburg by Emma Goldman in January 1920, having last seen it in the 1880s:

> It was almost in ruins, as if a hurricane had swept over it. The houses looked like broken old tombs upon neglected and forgotten cemeteries. The streets were dirty and deserted; all life had gone from them. The population of Petrograd [St Petersburg] before the war was almost 2 million; in 1920 it had dwindled to five hundred thousand. The people walked about like living corpses; the shortage of food and fuel was slowly sapping the city; grim death was clutching at its heart. Emaciated and frost-bitten men, women, and children were being whipped by the common lash, the search for a piece of bread or a stick of wood. It was a heart-rending sight by day, an oppressive weight by night. The utter stillness of the large city was paralysing. It fairly haunted me, this awful oppressive silence broken only by occasional shots.

The cause of the grain crisis was the peasant's reluctance to sell grain for paper money. Grain prices remained controlled while inflation was rampant. The peasants could make more profit from grain by using it to fatten cattle or make vodka and sell it in the black market. With the pressure on prices peasants switched from growing grain and cultivated crops not under state control. The government's response was to set up food brigades from the towns to descend on villages and forcibly remove all the grain they could find. The result was anarchy in the countryside with peasants in revolt and fighting armed townspeople.

The Bolsheviks retained power only because of the incompetence of their enemies. There was a moment in 1921 when, with the peasants in rebellion and grain supplies short, the Bolsheviks lost control of their industrial and military supporters. Strikes broke out all across Russia and the naval base at Kronstadt mutinied. The opposition leadership failed to seize the moment and the revolt was crushed. In the end the Bolsheviks survived by reintroducing some form of capitalism. Peasants were allowed to keep their surplus production once they had paid their taxes. Individual small trade initiatives were tolerated. This New Economic Policy was introduced in 1921 and by 1924 some sort of normality had been reached, in which an ordinary person could live without risk of starvation. Capitalism in Russia was not dead yet and it rescued the Soviet regime.

The communists also made a determined attempt to break the hold of religion on people's lives. They saw religion as offering a palliative for the oppressed workers; Marx called religion the 'opium of the people'. Communism was actively promoted as a philosophy of life as an alternative to the Orthodox religion. The aim was to change Russian society into venerating the state rather than worshipping God. In their ideal world there would be just one class and everybody would be motivated to work for the good of all. This was summarised in the slogan 'from each according to his ability to each according to his need'. In this new philosophy of life Marx was seen as a prophet and Lenin the high priest.

A systematic campaign against the Church started shortly after the start of the revolution. Church property was nationalised. Priests were characterised as living off the backs of the peasants and plotting the return of the csar. Communist propaganda ridiculed the Church; its relics were exposed as fakes and its miracles were mocked. Communist festivals were held in competition with key religious events such as Easter. Instead of christening their children, parents were asked to promise to bring them up as good communists. Alternative versions of wedding and funeral services were devised. To back up the battle for hearts and minds, in 1922 Lenin ordered all valuable items to be removed from churches.

The political power of the Orthodox Church was broken. However, the attachment of ordinary people to the Church and its rituals was not so easily changed. People still preferred to bury their dead rather than cremating them, as had been ordained by the communists. Red weddings failed to supplant their religious equivalent. Local priests returned, supported by voluntary donations and grants of land from peasant communes. Religious belief cannot be so quickly changed by an appeal to rationality. In the end Orthodox Christianity survived and maintained a place in the hearts of ordinary Russians.

The communist revolution had been a human catastrophe. Something like 10 million people had died from the civil war, the terror, famine and disease. Lenin and the Bolsheviks had clung to power irrespective of the human cost. They had created a police state run by a privileged, corrupt elite which had many parallels with the aristocratic regime it had replaced. The cheka replaced the csarist secret police; the new aristocrats were the communist party officials with their dachas in the countryside. In place of the supposed godlike veneration of the peasant for the czar, Lenin was now subject to public idolisation. At the bottom the peasants still scratched a subsistence living in the challenging Russian climate.

In 1924 Lenin died and was replaced by an even bigger monster: Joseph Stalin (1878–1953). Son of a cobbler and a housekeeper in modern-day Georgia, he joined the Bolsheviks in 1903 and was a key follower of Lenin during the revolution. In 1922 he became general secretary of the communist party. Utterly ruthless, with a peasant-like paranoia and cunning, he exercised the black arts of power politics to eliminate his political rivals one by one. Through famine, political execution, death in concentration camps (or gulags) and the forced exile of ethnic groups, his regime was responsible for the deaths of an estimated 10–20 million people. This excludes deaths in the Second World War, for which Stalin was partly responsible.

Nevertheless, Stalin transformed Russia into a major economic and military power. In 1928 he introduced the first of his five-year plans. Each five-year plan dealt with all aspects of development from industrial production to education and welfare. The emphasis varied from plan to plan, but generally the key elements were the generation of power, the output of capital goods, the development of agriculture, and the strengthening of the military. The aim was to be able to match the economic and military challenge of secular capitalist and fascist nations. The implementation of the plans was an intense national effort which was single-mindedly and brutally enforced. In 1926 78% of workers were employed in agriculture. But by collectivising the farms and demonising the more successful of the farmers, Stalin ruined the agricultural sector, driving peasants into the towns to become the new industrial workforce. Millions died, especially during the 1933 famine. However, having driven the standard of living of workers down to a level unmatched in modern history, the new Union of Socialist Republics (USSR) was able to invest a considerable proportion of its wealth in education and science. With millions of workers being trained, the country gradually acquired the skill base necessary for sustained growth. For example, the number of graduate engineers increased from 47,000 in 1928 to 289,900 in 1941. From an incredibly low base the industrial sector was transformed. If you accept Russian figures as correct, over the period of the first two five-year plans (1928–1937), coal output rose from 24.4 to 96.3 billion tonnes, steel production from 4 to 17.7 tonnes and electricity output rose sevenfold. The USSR, had managed to avoid the effects of the Great Depression by a new state-directed approach. It appeared to the outside world that there was an alternative to capitalism and that state-directed economies were an alternative way to prosperity.

More deadly military technology and the Second World War

The Second World War was fought in a totally different way to the First; improved tanks and aeroplanes made for a more dynamic movement than the static war of attrition that characterised the previous conflict. The tank, invented at the end of the First World War, had developed dramatically in speed and firepower during the interwar years. Military leaders were able to use the tank as a modern version of cavalry, providing a highly mobile attacking force. Attacks could now be made with mass tank formations; the barbed wire and trenches of the First World War were no longer a suitable defensive shield. Tank technology was a decisive factor in the war and the German panzer tanks were superior to the British but not as good as the Russian T-34.

The aeroplane, which had been used for reconnaissance in the First World War, became faster and was capable of carrying large bombs weighing a ton or more. These high explosive bombs had the potential to destroy virtually everything built above ground level. Bomb targeting, however, was very imprecise. The initial purpose of bombing was to disrupt the supply network of the enemy by hitting bridges, ports or railway yards. Targeting was so poor, however, that civilian targets were almost always hit during the raids. Soon, much of the bombing was aimed at the indiscriminate destruction of cities. Ordinary citizens were now in the front line. War between countries became total war with all citizens involved and the country's industry, farming, mining and construction all focused on survival.

In order to detect threatening bombing raids, radar was invented. This was another application of electromagnetic waves. Radar used pulsed radio waves that could be detected when they were reflected from an aircraft. The British developed the first practical radar system in 1934 and its deployment, together with effective fighter planes, allowed the British to survive the aerial war against Germany in the Battle of Britain (1940).

At sea, airpower was also important. Aeroplanes firing torpedoes were a major threat to the now old-fashioned battleships. Flat-topped aircraft carriers capable of acting as a runway for landing and launching aeroplanes at sea were developed between the wars. These extended the range of engagement between enemy ships. Before a battleship could close on an enemy aircraft carrier it would have received several lethal sorties of aircraft. The battleship was no longer queen of the sea. The largest naval battle of the Second World War, the Battle of Midway, was fought between USA and Japanese aircraft carriers.

Military intelligence was also important. German U-boat submarines were faster and more effective than in the First World War. Acting in wolf packs they were initially very successful, causing an unsustainable rate of loss to British merchant shipping. Using Polish intelligence on German coding methods, British cryptographers were able to decode German military radio communications. Decoding, speeded up by the development of the first automated computers, was sufficiently rapid for shipping convoys to be redirected away from the wolf packs and for the U-boats to be intercepted and destroyed.

As a result of the global spread of this more deadly technology the Second World War resulted in the biggest loss of life in human history. The war was truly worldwide in scope, with the two deadliest areas of conflict being between Russia and Germany, and Japan and China. The bombing and the mobility of the war meant that more civilians than ever were caught in the conflict. By some estimates, 22–25 million military personnel and 38–55 million civilians died; some 8–10% of Germans and an astonishing 10–15% of citizens of the Soviet Union perished.

In addition, Adolf Hitler exterminated 6 to 7 million Jews, Gypsies and other 'undesirables' in the holocaust. The unluckiest nation was Poland, caught in the cross-fire between Germany and Russia; over 5 million Poles died. The war ended in the largest destructive attack yet conceived, the dropping of the H-bomb on two cities in Japan.

China, Japan and Communism

The early twentieth century was a disastrous period for China. Having failed to combat the power of the Western nations and the newly acquired military expertise of the Japanese, the imperial system of government finally collapsed in 1912 with the resignation of the 12-year-old Emperor Puyi. A republic was proclaimed in Nanjing, and Yuan Shikai (1859–1916) was offered the Presidency. Yuan Shikai was an old-style military warlord whose probable aim was to start a new imperial dynasty. He had been appointed to command the imperial army in 1895 and was the officer most directly responsible for the creation of China's first modernised army. Yuan gained significant political influence and the loyalty of a nucleus of young officers. The Qing court relied heavily on his army, due to the proximity of its garrison to the capital. Having defeated his former imperial masters he turned on the Chinese republicans and established military rule. On his death in 1916, central Chinese government fell apart and regional warlords seized power. There was no move to democracy and no significant industrialisation apart from the Western enclaves on the coast. The peasants still provided most of the state income and opium smoking was still endemic.

In 1927 one regional warlord, Chiang Kai-shek, established a semblance of central control from his capital at Nanjing and made outwardly ambitious plans for modernisation. However, the Japanese invasion of Manchuria in 1931 and the subsequent all-out war with Japan in 1937 gave the new government no chance to establish its peace-time credentials.

The Japanese military in the Second World War demonstrated what can happen when an aggressive nationalist military organisation controls the government. The Japanese government appeared to lose all sense of humanity and strategic direction. Having already consolidated their control of Manchuria, there was no strong political reason for the invasion of the rest of China. However a mad nationalism prevailed and, with their technological superiority, within a year they had conquered most of the north-east of China. The Chinese then changed their tactics and tried to drag out the war as long as possible with the aim of exhausting the enemy so the war became a stalemate. Having invaded the most populous nation on earth and become tied down in China in a war of attrition, the Japanese then proceeded to declare war on the imperial powers: the British and French. Not content with this, they then followed this up with a surprise attack on the US naval base at Pearl Harbour. This was effectively a declaration of war on the richest nation on earth, having already taken on the most populous nation and the nations with the largest empires. They might have beaten any one of these three sets of powers, but to take on all three at the same time was suicidal. In addition, they imposed a harsh imperial regime on the occupied countries, ensuring that native resistance would be strong. Their strategy was flawed, their battle tactics brutal and their occupation vindictive. The extract below is from Justin Wintle's *The Rough Guide Chronicle*, China, p. 371 describing the Japanese attack on the Chinese capital Nanjing:

Japan's army in China had as its slogan 'Loot all! Burn all!, Kill all!. Entering a city abandoned by Chiang Kai-shek and its defending officers, its troops perpetrated a slaughter that stands out even among the dismal catalogue of 20th century atrocities. All told, up to 300,000 Chinese were butchered, the majority of them non-combatants, but also 80,000 unarmed soldiers. All the usual cruelties were present: gang rape, genital mutilation, sexual enslavement and torture. Their hands tied behind their backs, victims were used for 'bayonet practice.' Just as disturbing was the way that the 'rape' [of Nanjing] was initially celebrated in Tokyo's newspapers. Japanese officers, bragging the numbers they had killed, were congratulated. Only an outcry by the world press persuaded the Japanese government to censor such reports.

At the end of the war in 1945 there were three centres of power in China: the Japanese in the north-east; the Nationalist forces in their capital Chongqing in Sichuan province in the south-west and the communist forces in Shaanxi Province in the north-west. There, the communists had managed to recover their military strength, after having been encircled and almost annihilated by Nationalist forces. The 6000 mile Long March from southern China to Shaanxi via Sichuan in 1934–1935 had depleted their force from 100,000 down to 7000 but they had survived. With the surrender of the Japanese in 1945, it was a straight fight between nationalists and communists, for the control of China. It would take another four years of civil war before the Chinese would experience peace, when in 1949 Mao Zedong declared the People's Republic of China. As in Russia, it was peasant support against the exploitive ruling classes that made the difference. And, as in Russia, the peasants were soon to be cruelly disappointed. The communist system of government was a one-party state with an all-powerful leader supported by its centrally controlled judicial system. Allied to a fanatical belief in Marxism this proved to be a very nasty combination of arbitrary dictatorial rule. As with Stalin in Russia, the communist leader Mao Zedong joined the pantheon of the most destructive humans in history.

When the Chinese communists came to power, China was still largely a peasant country, with about 85% of the population living in the countryside. As with the USSR, China's priority was industrial growth outlined in a five-year plan which started in 1953. Monopolies were declared on all agricultural products and food was diverted to the towns to support growth. Mirroring Russian practice and in pursuit of greater agricultural output, the Chinese communists broke up the old village structures, replacing them by cooperatives to which peasants were obliged to contribute labour, land, tools and animals. It was a communist axiom that larger units would lead to greater productivity. The communist officials zealously enforced the change and by 1956 an amazing 92% of peasant farmers were members of collective farms. The pace of centralisation increased further in the subsequent years; the government nationalised land and equipment while farmers were paid only for their labour. Ignorant government officials dictated the farming strategy, with predictably disastrous consequences. No lessons were learned. Reporting negative results was taken as criticism. Under Mao's misguided leadership they plunged on and in 1958 launched the perfectly misnamed Great Leap Forward. This involved setting up communes by pooling the agricultural and small-scale industrial activities of a rural area. This eliminated any private land ownership and created a completely collective existence for its members, with very long working hours and the almost complete abolition of privacy. By the end of 1958, 25,000 communes had been established, each averaging 5000 households. Mass canteens served food to all commune members, eliminating home cooking. Instead of cash payments, work points were allocated, which were exchanged for food. Despite falling output, false production figures were reported to mollify senior communists, who, despite the evidence of their own eyes, continued to believe the approach would be successful.

With favourable weather conditions a good harvest was produced in 1958, but thereafter farm and industrial output began to decline catastrophically. In terms of lives lost, the famine that resulted from the Great Leap Forward was the worst disaster of the twentieth century. It is generally accepted that at least 30 million people died. The government insisted that industrial workers and, of course, communist cadres were given priority for food. Starving peasants were left to die, victims of a false ideology and a government machine that had no effective feedback to warn of failure. It was not until later, after small plots of land were handed back to private ownership and the big communes were broken up, that output began to recover. The new communist government in China had got off to a disastrous start and the agonies of the Chinese people continued.

Nuclear power

It is ironic that the greatest scientific advance of this period resulted in the most destructive weapon yet devised – the nuclear bomb. From the end of the nineteenth century a number of brilliant scientists had delved deeper into the nature of matter and learned of its potential for creating huge amounts of energy. In 1900 Max Planck (1858–1947), a German professor, proposed that matter must be made up of oscillators that could change energy levels in very small amounts, which he called quanta. It was a first practical indication of atomic structure.

Planck set the stage for the German-born Albert Einstein (1879–1955) to make his contribution to theoretical physics. In 1905, while Einstein was working as a clerk in the Swiss patent office, he produced three papers that have ensured his place in the pantheon of great physicists alongside Newton and Maxwell. In the first paper he showed that light itself was made up of the same energy quanta as observed by Planck and could be considered as a stream of very small particles, now known as photons. In the second he proved mathematically that the random motion of miniscule particles suspended in water is due to the random motion of discrete water molecules. This was the first real proof of the existence of atoms.

The third was his paper on special relativity. The paper was a result of one of his famous thought experiments. Einstein was concerned about how the laws of nature were affected by relative movement between the observer and the object being observed, especially when the speed of movement approached the speed of light. The speed of light was believed to be constant.

By imagining that the laws of nature remain the same no matter how fast you are moving, Einstein developed the ideas of Newtonian physics so they could be applied on an astronomical scale. Unexpectedly it also led to his most famous equation $E = mc^2$; proving that mass and energy are related and that matter itself contains enormous amounts of energy.

At around the same time, the great New Zealand experimental physicist Ernest Rutherford (1871–1937) began work on heavy elements that emitted natural radiation, such as uranium and thorium. He identified two types of radiation, alpha and beta, by their effect on other particles and later confirmed the existence of a third type, gamma. In reaching these conclusions he noticed that each type of radiation emitting matter decayed in a regular fashion, having a standard time to reach half its activity. This simple observation has revolutionised mankind's ability to date past events accurately. Elements with long half-lives, such as carbon 14, which has a half-life of 5730 years, can be used to date findings in archaeology. Other, even longer lived isotopes such as uranium-235 from igneous rock (half-life 700 million years) have been used to date earlier events on the earth. We now even have a date for the creation of the Earth at 4.5 billion years ago, which is 750 million times earlier than the famous estimate of 4004 B.C. by Archbishop Ussher in the seventeenth century.

In 1907 Rutherford identified the particle involved in alpha radiation as the nucleus of the helium atom. By observing how the path of alpha radiation is deflected after hitting gold foil, Rutherford was able in 1911 to propose the first accurate model of an atom, in which a positively charged nucleus containing almost all the atom's mass is surrounded by low mass, negatively charged electrons. This picture of an atom as mostly charged space in which electrons are somehow contained within it but at a distance far from the nucleus (thousands of times the radius of the nucleus itself) was completely unexpected. In his book of science for children *The Magic of Reality* (which should be read and understood by all adults), Richard Dawkins gives an analogy for the astonishing amount of space between carbon nuclei in a diamond crystal. If you scale up the carbon nucleus to be the size of a football, then the neighbouring carbon atom nucleus would be 15 kilometres away.

In 1913 the great Danish physicist Niels Bohr (1885–1962) proposed a solution to the mystery of how electrons were contained in the atom by supposing that electrons were constrained to fixed orbits around the nucleus and could move only from one orbit to another by giving or receiving a fixed quanta of energy, the same quanta Planck had observed a decade earlier. The classical Rutherford–Bohr model of the atom as a nucleus surrounded by electrons arranged in a limited number of fixed orbits was born. However, the nature of the nucleus was disputed until in 1932 the Englishman James Chadwick proposed that the nucleus was composed of charged and uncharged particles of the same mass, now known as protons and neutrons. Each element was now characterised by having a specific equal number of protons and electrons. Hydrogen has one, helium two, all the way up to californium 98, which is the heaviest element occurring in nature. Each element could have several isotopes with differing numbers of neutrons in the nucleus. Some of these isotopes are unstable and will decay to a more stable form by releasing radiation.

The discovery of the neutron gave scientists an insight into how to break up the atom. Alpha particles used in the gold experiment by Rutherford are positively charged and are repelled by the nucleus. The neutron has no charge and therefore when a stream of neutrons is directed towards an atom there is no force preventing them from impacting the nucleus. A neutron hitting a nucleus results in nuclear fission – the breakup of the nucleus to form other particles and energy. Some isotopes release neutrons and energy when they decay. These neutrons will penetrate the nucleus of other atoms in the isotope and release more energy and neutrons. It was quickly realised that if an isotope was sufficiently active in producing neutrons and sufficiently densely packed, then the nuclear fission resulting could create a chain reaction and produce a huge explosion, as its mass was converted to energy as predicted by Einstein. One isotope capable of creating this chain reaction was found to be the isotope uranium-235, which makes up 0.72% of naturally occurring uranium.

In August 1939 Einstein signed a letter addressed to President Roosevelt by worried scientists, which warned of the potential development of extremely powerful bombs of a new type. It urged the USA to take steps to acquire stockpiles of uranium ore and accelerate research into nuclear chain reactions. The war acted as a stimulus to bomb development. The USA took up the challenge and concentrated enough money, project management and scientific know-how in the Manhattan Project to develop a nuclear bomb. By 1945 they were ready and the first atom bomb was dropped on Hiroshima in Japan on 6 August, followed by the second over Nagasaki on 9 August. Within the first 4 months of the bombings, the effects killed 90,000–166,000 people in Hiroshima and 60,000–80,000 in Nagasaki, with roughly half of the deaths in each city occurring on the first day. On the day of the explosion 60% died from flash or flame burns, 30% from falling debris and 10% from other causes. During the following months, large numbers died from the effect of burns, radiation sickness and other injuries, compounded by illness. An overall US estimate was that 15–20% died from radiation sickness, 20–30% from burns and 50–60% from other injuries, compounded by illness. In both cities, most of the dead were civilians.

The human desire and ability to injure fellow beings had now reached a critical new level of horror. Just two atomic bombs killed at least 200,000 people and within 10 years thousands of bombs were held by competing powers and the even more powerful hydrogen bomb had been developed. Competing sides now had the potential to destroy all an enemy's cities and towns and pollute their land with such high levels of radiation that it would be unsafe to cultivate for many generations to come. It was clear that humans now had the ability to destroy themselves in war.

Improvements in healthcare

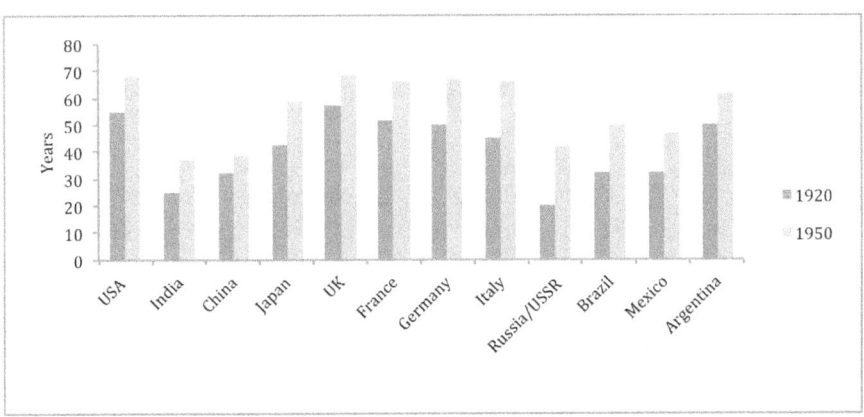

Figure 10.2 Life expectancy 1920–1950

Figure 10.2 shows that in the period to 1950 a large increase in life expectancy occurred in all countries of the world and it exceeded 60 for the first time in Western European nations and the USA. Life expectancy in South America improved dramatically, with Argentina also exceeding 60 years. Russian life expectancy remained low, showing the effects of the Second World War, but it was higher than during the communist revolution, when it had fallen to 20. Life expectancy in India and China remained below 40 but even these countries showed significant signs of improvement. How had all this come about? Table 10.2 shows the causes of the gain in life expectancy by disease type in England and Italy between 1871 and 1951, when life expectancy improved from 41 to 68 years in England and from 34 to 66 years in Italy.

Causes of Death	England		Italy	
	Gains (years)	% of gain	Gains (years)	% of gain
Infectious Diseases	11.8	42.9	12.7	40.1
Bronchitis, pneumonia, influenza	3.6	13.1	4.7	14.8
Diseases of the circulatory system	0.6	2.2	0.8	2.5
Diarrhoea, enteritis	2	7.3	3.4	10.5
Diseases of infancy	1.8	6.5	2.3	7.3
Accidents	0.7	2.5	0.5	1.6
Tumours	0.8	2.9	0.4	1.3
Other Diseases	7.8	28.4	7.7	24.3
Total	27.5	100	31.7	100

Table 10.2 Causes of the increase in life expectancy in England and Italy (1871–1951)

Source: G. Caselli, 'Health transition and cause-specific mortality', in R. Schofiels, D. Reher and A. Bideau, eds, *The Decline in Mortality in Europe* (Oxford: Oxford University Press, 1991)

It can be seen that the most important reason for the increase in life expectancy was the reduction in the number of deaths from infectious disease. This was mostly an outcome of increasing personal wealth, which had resulted in improved housing conditions, better sanitation and a healthier diet. However, towards the end of the period there were major medical discoveries in the field of antibiotics that further reduced the danger of infectious diseases. In 1936 sulphonamide, the first of the new sulphur antibiotics, was proved effective against the blood infection, streptococcal septicaemia. Penicillin was discovered in England at the start of the Second World War. However, it was not clear how it could be mass-produced. Amazingly, after a worldwide search, a mouldy cantaloupe melon found in a market in Peoria, Illinois was found to contain the best and highest quality penicillin. The discovery of a natural source and improvements in fermentation technology enabled the USA to produce 2.3 million doses in time for the invasion of Normandy in the spring of 1944.

In 1943 streptomycin was isolated and proven to be the first drug effective against tuberculosis (TB). Mass production started in the early 1950s. Finding a cure for TB was very important. With the exception of malaria, TB killed more people than any other infectious disease. In the nineteenth century, it killed an estimated one in four of the adult population of Europe and as late as 1918 one in six deaths in France were still caused by TB. Improved living conditions and the quarantine of those with the disease had greatly reduced its impact by 1950 and eventually the use of antibiotics almost eliminated the disease from the richer countries.

The other major reason for the increase in life expectancy was the decline in the rates of infant mortality. For the first time the rate of infant deaths was brought under control in the West. As an example, in 1850 there were 160 deaths per thousand births in England and Wales but by 1950 the number of deaths had declined to 30 and by 2000 it was around five. The principle reason for this was the better living conditions. However also, in this period, there were advances in the professional standards of midwifery and much greater hospital support for difficult births, both of which contributed towards the decline in infant mortality.

Political cultures after the Second World War

This period had seen a major clash of cultures between the secular capitalist culture of USA, France and Britain, the communist regime of the USSR and the fascist and nationalist cultures of Italy, Germany and Japan. Fortunately for humanity, the main nations that had caused the conflict, the brutal regimes of Germany, Italy and Japan, were defeated. The field was left free for the two remaining political cultures: secular capitalism and communism. The defeated nations were financially ruined and had to be nurtured back to economic health by the victors.

Italy, Japan and West Germany fell into the secular capitalist sphere of influence, and East Germany and East Europe fell within the communist sphere.

The secular capitalist culture of most of West Europe, North America, Japan and Oceania

This group includes the victors of the Second World War: the USA, the UK, France, Belgium, the Netherlands, Norway, the former British colonies (Canada, Australia and New Zealand) and neutral countries like Sweden, Switzerland, Finland and Austria. After the war secular capitalist systems were set up in the defeated nations of Germany, Italy and Japan. Lessons were learned from the First World War and this time the defeated nations were supported to enable them to recover economically. As a result secular capitalism took root and prospered.

At the end of the Second World War the USA was not only the most important world power, but the only one with its economy intact. While Britain and France emerged on the winning side and still controlled huge empires, they were manifestly weaker than before the war; soon their colonies would start to gain independence.

The Communist culture of USSR, East Europe and China

The USSR had emerged from a traumatic period of revolution and war at the head of its own empire of communist states. As leader of the communist world the USSR was set to challenge the USA in the decades to come. Religious practice had been positively discouraged in communist countries. This did not mean that religious feelings disappeared. Russian grandmothers still crossed themselves and prayed for their family. Chinese families still maintained a corner of the house where they venerated their ancestors. In China, as well, non-Western folk medical traditions, like acupuncture and herbal medicines, continued and were encouraged by the new regime. However, at a political level religion had no influence. The new equivalents of gods were Marx and Lenin and the new high priests Mao and Stalin.

In communist countries a new division arose between party officials and the rest. Party officials retained a privileged status. In times of famine, they remained well fed. The trappings of power were acquired by the new leaders: big cars, nice apartments and, just like the old aristocracy, dachas for their private use.

Islamic monarchies

These monarchies include Iraq, Egypt, Jordan, Saudi Arabia, Libya and Iran. Islamic states were now at their lowest ebb, with little or no influence on the world stage with their rulers heavily influenced by the British and French governments.

Developing countries

In addition to Latin American countries, this group now included Turkey, Syria, Lebanon from the old Ottoman empire, some-time European monarchies: Greece, Spain and Portugal and the ex-British colony, South Africa. These countries continued to try to modernise at the same time as reconciling the conflicting interests of their power groups. There was a huge range of unstable political cultures: modernising dictatorships and monarchies, reactionary dictatorships and monarchies, democracies favouring specific racial groups and reactionary or modernising rule by one party. To illustrate the range of political cultures a brief summary is given below of progress in Turkey, South Africa, Spain, Mexico and Brazil.

After the First World War, Turkey was occupied by the allies and Greece, with the objective of bringing it under colonial control. Mustafa Kemal (who later called himself Ataturk), a military commander who had distinguished himself during the Battle of Gallipoli, launched the Turkish War of Independence and by 1922 had expelled the occupying armies. He was the first leader of an Islamic country to attempt to separate state and religion on Western lines in order to create a secular state. He closed religious convents and dervish lodges (dervishes had a role similar to that of friars in mediaeval Christianity) and introduced a new civil and penal code, quite separate from shar'ia law. He also banned the wearing of the fez, introduced the Roman alphabet and gave women equal rights to men. Ataturk was essentially a strong man running Turkey as a one-party state with a modernising agenda. He forced the population to change their ways of life and, in doing so, set up a tension between old and modern values that has continued in Turkey up to the present.

South Africa gained self-governing status from the British in 1909. Once independent the ruling white elite immediately began the process of giving themselves an elevated status, which eventually grew into the Apartheid philosophy. A total of 87% of the land was reserved for white ownership and the native Bantu peoples and the Indian immigrants were not permitted to vote. Democracy was only exercised among the white minority.

Spain went through a prolonged period of institutional instability, with a constitutional monarchy from 1876 to 1923, a dictatorship supported by the monarchy from 1923 to 1930, a republican government from 1931–1939; after a civil war which started in 1936, a reactionary military dictatorship led by General Franco lasted until 1975. Three power groups squared up to each other. The republicans favoured land reform and giving land to the peasant farmers, and were strongly anti-clerical. The right-wing groups believed in the Catholic tradition, stability and hierarchy. The third element was the Basques and Catalans who wanted independence from the rest of Spain. In the event the right wing triumphed and a regime in the tradition of a Catholic monarchy was restored.

In 1934, in Mexico, after years of revolution and conflict the Party of the Institutionalised Revolution (PRI) formed a modernising one-party state which remained in power for 71 years. The PRI agenda was broadly socialist and supported peasant and worker organisations and a programme of nationalisation. The party was successful in unifying the country; for the first time everyone in Mexico was content to accept their Indian heritage. The old Spanish landowners had lost out and land was redistributed on a large scale.

Brazil also began to reach some degree of national consensus when the revolution of 1930 brought to an end the oligarchy of the coffee plantation owners. From 1930 to 1945 Getulio Vargas led a popular government, which gradually assumed dictatorial powers. He was a nationalist with strong working-class support. Oil, steel and electricity companies were nationalised and a health and social welfare programme was put in place. The ruinous coffee subsidy was abolished, industry encouraged and agriculture was diversified away from just coffee and rubber. Sao Paolo became the biggest industrial centre in South America. Vargas brought Brazil into the Second World War on the allied side, but fighting for democracy abroad and retaining a dictatorship at home proved contradictory. Presidential elections were resumed in 1945. Despite progress, Brazil still had not found a stable political culture

Summary: situation in 1950

At the end of a traumatic period, when the social effects of the technological revolution were being absorbed, two political cultures dominated the world stage: secular capitalism in the West and Japan and communism in East Europe and China. In the West the lives of ordinary workers had been transformed and their health and life expectancy was significantly better than any time in the past. Nevertheless, in Europe, class divisions remained strong, reinforced by different standards of education for each class. In communist countries class distinctions had been eliminated but the workers and peasants remained poor. A new state-directed style of economic management was initiated. In this approach the government set production targets and individual and business freedom to compete was eliminated.

The remaining independent countries in Europe, Asia and Latin America had struggled to find a stable form of government that could embrace change.

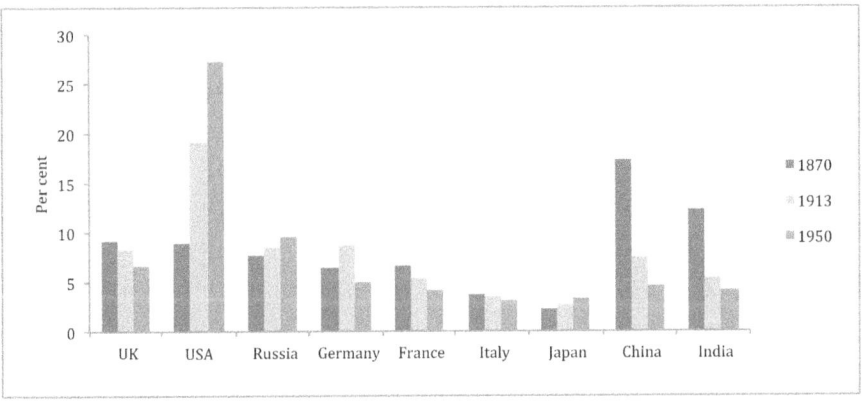

Figure 10.3 Share of world GDP1870–1950 (per cent)

Figure 10.3 shows that by 1950 the USA economy was by far the most important in the world with a GDP three times that of its nearest challenger, the USSR. India and China continued to decline in importance, having failed to make sufficient progress towards industrialisation. Although the USSR was second in importance economically, its wealth per capita remained low, well below Britain, France and Italy.

The world population had reached 2.5 billion with a population growth rate of 1.04% p.a., the fastest yet (doubling every 67 years). Figure 10.4 shows that population growth in the Americas continued at the high rate of 1.7% p.a., increasing America's share of world population to 13%, a fantastic increase from 1780, when it represented only 2% of the overall population. Europe's population, on the other hand, after the devastations of two world wars only grew at 0.4% p.a.

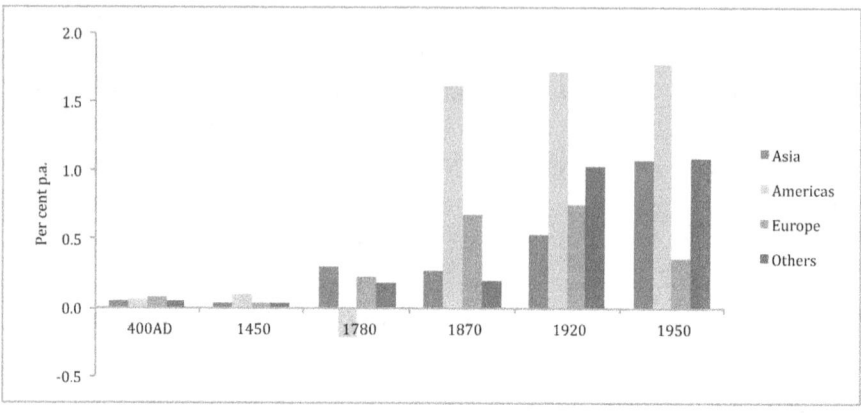

Figure 10.4 Population growth rates to 1950 (per cent per annum)

Cities continued to increase in size. New York was now the largest city in the world, with a population of 12.5 million, the first city to have a population of over 10 million. This was followed by London with 8.9 million and Tokyo with 7 million.

CHAPTER 11:
THE SUCCESS OF SECULAR CAPITALISM AND THE END OF EMPIRE (1950–1990)

During the 40 years after the war technology transformed the lives of ordinary people living in secular capitalist countries. There were no major wars in Europe and North America and there was fairly continuous economic growth. This is the age of my youth when there was progress on all fronts. Everything seemed possible.

Improved communications enabled companies to be managed as international businesses. Managers could call any country in the world from a phone on their desk. Hard-copy business information could be sent quickly by fax. Slide-rules and cumbersome electromechanical adding machines were replaced by simple, cheap **electronic calculators**. By the end of the period most office workers had access to a computer terminal. Businesses relied on **computers** for their accounts and taking orders. Working conditions improved in factories and new systems of **quality manufacture** supplied goods to new standards of reliability and affordability.

People had more leisure time and with improved transportation were able to travel further afield. Most people had a car and with the new **motorways** journeys of a few hundred miles could be made quickly and conveniently. With the development of aircraft powered by **jet engines**, air travel became commonplace. Holidays were taken in distant destinations where sun could be guaranteed.

Home life changed as well, and as housework became easier more women had the chance to work. The drudgery of the weekly wash was reduced with the help of the **washing machine**. People shopped at supermarkets, where they brought goods from a huge range on display. **Refrigerators and freezers** meant that families could shop once a week and the food would remain fresh. Cheap plastic replaced glass and paper for packaging; **plastics** could be formed into virtually any shape to make a robust, lightweight casing for any consumer item. Many new consumer goods, from biros to vacuum cleaners, made home life easier and more pleasant. New **man-made fibres** were woven into fabrics which were light, durable and easily maintained.

In Europe socialist governments came to power and instituted the **welfare state**, which provided health care that was either free or affordable for all. Medical interventions became increasingly effective as huge developments were made in **diagnostics** and **pharmaceuticals**. On retiring, at the latest age of 65, people could expect to live at least another 10 years in relative comfort.

Socialist governments also introduced state-controlled monopoly industries for power generation, transport and other infrastructure industries. In secular capitalist countries there was **free secondary education** and attaining a university degree was made more affordable. Society became more egalitarian; **the right of women to have equal job opportunities** was recognised. Class distinctions diminished. New techniques of **birth control** gave couples the opportunity to enjoy sex without the need to worry about its consequences. Scientists developed nuclear technology for peaceful uses and **nuclear power** produced electricity for many countries of the world. The code of life, **DNA**, was discovered. This was an age of astonishing progress when people became noticeably healthier and lived longer more fulfilling lives. If you live in the West, looking back, this may well come to be seen as the golden age.

Secular capitalism and new technology

Apart from the USA, all the main combatant countries had emerged from the Second World War in debt, with a destroyed or dilapidated infrastructure. Within 25 years there had been a dramatic transformation and all secular capitalist countries had reached new standards of affluence. No countries advanced faster than the two defeated nations, Germany and Japan. In Japan over the 23 years from 1950 to 1973, GDP expanded by an average of more than 10% p.a. Their car companies such as Toyota, Honda, Nissan grew to be major world players and Sony, Toshiba, Panasonic and others grew to dominate the newly developing consumer electronics industries. Part of the reason for this was a manufacturing technique, variously known as just-in-time, Kanban or quality management.

Traditional mass production had aimed to maximise volume throughput. Workers were encouraged to work as fast as possible and were often paid according to the number of units produced, in a system known as piecework. Stocks of components were maintained at a high level to avoid any stoppage in production. While the price of mass-produced items fell using this system, the quality of the product was often poor and any change to new models was very expensive due to stock write-offs. An American statistician W. Edwards Deming (1900–1993), an advisor attached to the US army during their occupation of Japan after the Second World War, had a significant influence on the early development of quality management. In the new system, quality was paramount, stocks were kept low and a huge emphasis was placed on getting the job right first time and eliminating scrap. Workers were encouraged to suggest improvements, rather than just follow instructions blindly. The resulting products were quicker to develop, more reliable and, astonishingly, cheaper. This Japanese manufacturing technique became the model for assembly operations during the latter part of the twentieth century. Japanese cars, motorcycles and electronics became market leaders across the world. The most dramatic illustration of this is in cars sales in America; from a start of zero in the 1950s Japanese cars gained a reputation for reliability and value for money and gradually increased their sales until, by 2008, the Japanese outsold their American rivals in their home market.

The accelerating rate of technological development after the Second World War enabled companies to develop, manufacture and market products on a global scale. With improved transport links and printed matter instantaneously transmittable by fax over telephone lines, managing international companies from a central location became practical. Whereas companies exporting goods would once have been content to work through distributors or agents, they now formed local sales companies. This allowed companies to control their own destiny. Companies with bases in different countries merged to improve their total impact. Multinational companies were born and thrived. Food companies like Kraft and Unilever took over local brands and established them on a worldwide basis. Manufacturing companies like Toyota or Honda established assembly plants outside Japan to improve their flexibility in local markets. Technology companies like IBM and Microsoft began manufacturing parts in such places as Taiwan or Malaysia in the Far East, where cheap, reliable labour was available. Outsourcing became fashionable and many assembly jobs were transferred to low-wage economies. Energy suppliers like Shell, BP or Esso applied their skills to the exploration, processing and distribution of oil and gas to make cheap energy available to the world. Chemical giants like DuPont, Bayer and BASF developed plastics with its huge range of uses, replacing ceramic, metal and natural fibres in many of their applications. Pharmaceutical corporations like Glaxo and Pfizer developed drugs that were effective against a large range of ailments.

The whole of the Western world developed and thrived. No longer was economic success limited to a few countries. Broadly speaking, all countries with a stable government, a stable currency, legal protection for the operation of multinational companies and an educated and responsible workforce could participate in the general economic advance. Most countries that benefitted had a secular capitalist political philosophy. Authoritarian regimes and those dedicated to a directed state economy did not benefit in the same way. The world had divided into three: the secular capitalist world consisting of the USA, Canada, Western Europe, Japan, South Korea, Australia and New Zealand; the Communist world: USSR, China, East Europe, North Vietnam and Cuba and the third world, made up of modernising democracies like India or Turkey, military dictatorships like Egypt or Chile and African countries that were struggling to form viable nation-states. It was clear that only the secular capitalist world was thriving.

The car was one of the main symbols of this technological advance, but it took state investment in roads on a scale that mirrored and exceeded the investment in railways in the nineteenth century before car transportation achieved its full potential. If you travel to Africa and experience what it is like to travel on unpaved roads, you can get a feel for what it was like in Europe at the turn of the twentieth century. A magic material called asphalt transformed road transport. Asphalt (or bitumen) is a sticky black, highly viscous form of petroleum. Mixed with an aggregate it is known as asphalt concrete (blacktop, tarmac or macadam). Applied to the surface of roads from the end of the nineteenth century, it provided a smooth, water-resistant and durable coating to the road surface which dramatically improved the speed, comfort and convenience of travelling by car. Asphalt was first used in a significant quantity in Europe by exploiting naturally occurring deposits found in France. In Britain coal tar, a by-product of coal gas production, was a widely used alternative. Nowadays most asphalt comes from oil sands, like those found in Canada. As the growth of car usage took off, so did the demand for tarmac-covered road. As a result, in Britain not only the main roads, but also most minor roads were paved by the start of the Second World War, and today 94% of the roads in the USA are covered with tarmac.

Despite the better roads, long-distance travel by car before the Second World War was relatively slow and sometimes dangerous. Roads were not designed to cope with the speed and volume of car traffic; they were narrow, two-way roads that passed through town centres. It took another tranche of investment for the car to achieve its true potential. The two-lane highway with high speed junction layouts, known variously throughout the world as motorways, thruways and so on were first conceived as autobahns in Germany and built in the 1930s, primarily for defence purposes. By 1940 a high speed road network of almost 400 miles criss-crossed Germany. The Americans started their interstate highway system in the 1950s and by 1992 had completed some 47,000 miles of road linking all the states. The British opened the M1 from London to Birmingham in 1968 and then went onto complete most of their network by 2000. By this time the main part of the whole West European motorway network had been finished. This revolutionised transport for medium distance journeys of a few hundred miles. Personal mobility increased dramatically, enabling great changes in lifestyle. New out-of-town shopping and leisure facilities were built, all with convenient on-site parking. Workplaces also had to accommodate the car.

Company headquarters no longer had to be in the centres of large cities. Employees could commute to work from a greater distance, giving them greater job mobility. Suburbs spread out as mobility increased and living in the country and working in the town became a possible life style. The car gradually became a necessity of life as more people lived away from city centres, depending on the car for transport.

Trucks also totally changed the freight industry. No longer did heavy freight have to be transported by rail with multiple transit points. Now haulage could be door to door at a fraction of the previous cost. Figure 11.1 shows that truck and car usage mushroomed, with over 500 million car registrations by 1990.

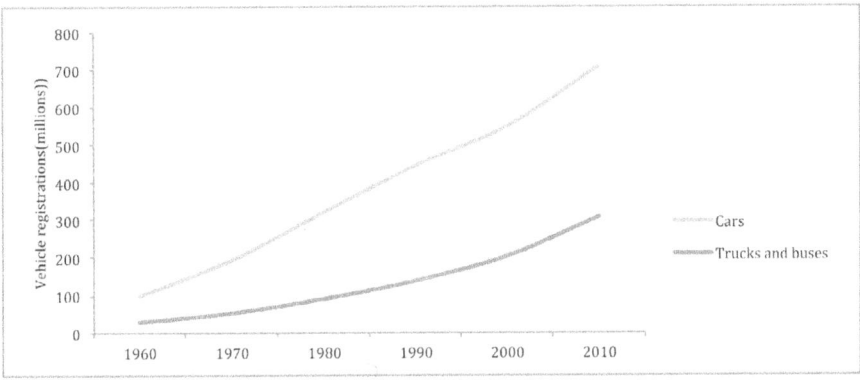

Figure 11.1 Trend of worldwide vehicle registrations, 1960–2010

1. Stacy C. Davis, Susan W. Diegel and Robert G. Boundy (2011) *Transportation Energy Data Book: Edition 30.* Office of Energy Efficiency and Renewable Energy, U.S. Department of Energy.

After the war air transport became the preferred method of travel for long distance journeys. Initially, planes were propeller driven but later jet engines increased the speed and range of air travel. Jet engines had been independently developed by two engineers, Frank Whittle in the UK and Hans von Ohain in Germany during the late 1930s. Both Germany and Britain developed jet-powered military planes at the end of the Second World War, but not soon enough to significantly affect the outcome. In the late 1950s the de Havilland Comet and the Boeing 707 were the first commercial jet aeroplanes capable of flying directly across the Atlantic. They transformed international air travel. There were over 300 million passenger flights in 1970, over half of which originated in the USA. International air flight became cheaper as jet engines developed and soon world travel became affordable for the average Westerner. British holidaymakers, who mostly went to British resorts before the war and popularised Spain in the 1970s, could now take their vacation in distant places like the West Indies. Between 1970 and 1990, the number of air flights increased threefold. The world's most exotic locations now became holiday destinations for ordinary people.

There were huge changes in sea transport as well. The shipping container transformed the way that general cargo was transported at sea. An intermodal shipping container is a steel box, 20 or 40 ft long, which can be carried by sea, road and rail. For centuries cargo had been stored in any shape or size in the hold of ships. In the early twentieth century cranes lifted the heavy goods from the hold, but considerable manual effort was involved in transferring them to land-based transport. Using the container, all the handling processes at the docks could be mechanised. The traditional role of docker virtually disappeared. The container was commercialised in the 1950s and international standards for containers were published by 1970. There are now over 17 million shipping containers in the world.

Industrialisation and the growth of cities created a need for cheap high density housing. Up until the end of the nineteenth century, housing could only be made of stone, brick or wood. Over the period from the nineteenth century onwards the construction of housing and office buildings was transformed. Concrete could not be used by itself for housing because, although it is strong under compression, it is weak under tension; any swaying in a tall building made of concrete alone would lead to catastrophic failure. However, at the end of the nineteenth century it was found that embedding steel bars in concrete allowed the composite material to improve its tensile performance. The result, reinforced concrete, became the material of choice for large buildings from then on. Allied to the invention of the lift, electric lighting and cheap steel, reinforced concrete enabled urban architecture to be transformed. A safe lift mechanism had been first demonstrated at the 1854 World Fair in New York by Elisha Otis (1811–1861), a mechanic from Vermont. He cut the supporting rope on a raised lift, and to the crowd's astonishment it fell only a few inches. Soon, with the ready availability of electric power, lifts that could reach the top of the tallest buildings in a very short time had been developed.

The first skyscraper pierced the sky in Chicago; in 1892 the Masonic Temple, reached 22 storeys in height. Thereafter, New York led the field in innovation with the 47-storey Singer tower in 1908 and the 57-storey Woolworth Building in 1913. These were the first icons in the glitzy face of building design, which would develop into a global race for bragging rights to claim the world's tallest building.

However, the workhorse of urban development was the apartment block. Concrete buildings proved cheap to manufacture and effective at meeting the housing needs of fast-rising city populations. The aesthetics were poor and a limited access to outside space failed to meet the needs of children, but concrete high-rise apartments soon became the face of city development throughout the world. Offices, hospitals, factories and municipal buildings were all built using the same materials, possibly with some decorative glass or cladding to diminish the monotony of the urban landscape. Concrete and steel were also used in building the tunnels and bridges that allowed metropolitan areas to expand. Great conurbations built entirely of these materials arose, with 10, 20 or even 30 million people living in hundreds of square miles of concrete jungle.

Developments in transportation and construction were based on improvements in old technologies. However, it was developments in a new industry, electronics, that was to truly differentiate the post-war years from the centuries before. There were two key applications: television and computers.

The development of television technology began in the inter-war years. The first continuously moving image was shown by the Scottish inventor John Logie Baird in 1926 with a scan rate to 12.5 frames per second – just fast enough to give an illusion of a moving image. However, his electromechanical technology was soon surpassed by an alternative electronic system using a cathode ray tube (CRT), a vacuum tube with a source of electrons at one end, which can be directed to bombard a fluorescent screen at the other. This was also first demonstrated in 1926, when a Japanese inventor, Kenjiro Takayanagi, displayed images on a CRT television with a 40-line resolution at Hamamatsu Industrial High School in Japan. His prototype can still be seen there.

Several inventors worked on improving the technology, so that by 1936 the BBC was able to introduce the world's first regular, high definition system using 405-line per second broadcasting. For the first time events could be watched remotely; it was not necessary to attend an event to visually experience it. For example, in Britain in 1953 many people gathered around 9 inch flickering screens to watch the coronation of Queen Elizabeth as it happend. Television usage took off after the Second World War, starting in the USA. In 1947 Motorola introduced the VT-71 television for $189.95, the first television set to be sold for under $200, finally making television affordable to millions of Americans. While only 0.5% of US households had a television set in 1946, 55.7% had one in 1954, and 90% by 1962. By the early 1970s 90% of West European households also had a television. With the advent of colour transmission from 1967, television had achieved its current dominance as the medium of choice for entertainment, news and sport.

Automated electronic computing was driven by the increasing need to undertake complex calculations. Most famously, Colossus, the world's first programmable computer, was made by the British to help with breaking the German Enigma coding system during the war. However, it was not until the development of the transistor that modern computing became possible. Transistors use the electric properties of semi-conducting materials such as silicon, germanium or gallium arsenide to act as an electric switch or an amplifier. In the early 1960s manufacturers began to produce computers that could be used to automate routine business processes, such as order entry, stock allocation, financial accounting, invoicing and payroll. At first programmes were written in languages specific to each manufacturer, but in 1959 the first industry standard language was developed – COBOL. By modern standards computers were physically big with central processing units, 24-inch disc drives and noisy line-printers occupying a large air-conditioned room.

It looked for a while as if the usage of computers would be confined to large companies and research establishments. However the potential of computer power was transformed by the miniaturisation made possible by the invention of integrated circuits (chips or microchips). These are formed by depositing a pattern of trace elements on a thin substrate of semiconductor material; while additional materials formed the connections. In the early days of integrated circuits, only a few transistors could be placed on a chip. This increased over time to millions, and today billions of transistors can be contained on a chip. Using chips, the first mini-computers became available for small businesses, then the first hand-held electronic calculators appeared and soon after, micro-computers could be actually placed on an individual's desk. In 1981 the IBM personal computer with a built-in floppy disc was introduced and rapidly became the industry standard. The operating system was provided by a small company, Microsoft. In time, Microsoft would become an industry giant dominating not only the software used for micro-computer operating systems but also the three main applications, report writing, spreadsheet calculation and presentation. The computer was no longer just a big calculating machine; it was transforming the way in which individuals conducted business.

The other major new technology of the post-war years was nuclear power. After the Second World War and the proven destructive potential of the atomic bomb, the state-sponsored scientific community attempted to find peaceful uses for atomic energy. The concept was to generate electricity using a normal steam turbine but with heat generated from a nuclear reactor. The reactor would have to be carefully controlled to create heat without the risk of it exploding.

The Russians were first off the mark, when, in June 1954, the Obninsk plant generated about five megawatts of power fed into the local power grid. The British followed in 1956 with the world's first power station to produce electricity in commercial quantities from Calder Hall in Cumberland. The USA followed with its Shippingport (Pennsylvania) reactor in 1957.

Figure 11.2 shows that installed nuclear capacity initially rose from less than one gigawatt (GW) in 1960 to 100 GW in the late 1970s and 300 GW in the late 1980s. France and Japan, having no source of natural oil, were the heaviest investors in nuclear power. After 1980 worldwide capacity stabilised and in fact more than two-thirds of all nuclear plants ordered after 1970 were cancelled. The economics of nuclear power had proved to be non-competitive compared with existing conventional methods of power generation. Uranium is a fairly common element on the earth's surface, and the high cost of nuclear power is not due to the cost of the fuel but to the upfront capital costs of building the new reactor, the expense of de-commissioning it afterwards and the disposal of the nuclear waste. Safety and regulatory concerns drove these costs up following serious accidents which released nuclear material into the environment, principally at Three Mile Island (USA) in 1979 and, worst of all, at Chernobyl in the Ukraine in 1986. Thirty-one deaths among the reactor staff and emergency workers are directly attributed to the Chernobyl accident. The Chernobyl forum (a group of UN agencies) estimated the eventual death toll was 4000 of those exposed to the highest levels of radiation; mostly as a result of radiation-induced cancer.

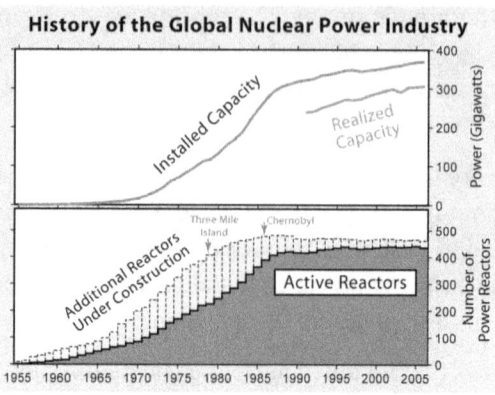

Figure 11.2 Use of nuclear power (top); active nuclear power plants (bottom)

The data are collected in Wikipedia from data from the International Atomic Energy Agency, principally 'Nuclear power reactors in the world' with additions received by direct communication.

New technology transformed the role of women in society. Up until the end of the nineteenth century a married women's role was centred in the home. Cooking, cleaning, caring for children and sick relations, fetching water, making and mending clothing had occupied the main part of a woman's life from Neolithic times onwards. Most widows, spinsters and young women had to work but tended to be employed in low-paid roles in domestic service or as textile workers in factories. As the education of women improved, more roles with more responsibility were available to them as teachers or nurses. When service industries expanded in the early twentieth century women were employed as shop assistants and waitresses. The plum professional and managerial jobs, however, remained almost exclusively the preserve of men. Upper and middle-class married women almost always stayed at home. Their role involved organising the home, managing the servants and bringing up the children. The triviality and boredom of the existence of middle and upper class women at the time is apparent in the great novels of the day.

After the First World War it became more difficult to obtain domestic servants and middle-class women undertook many of the domestic roles themselves. This was possible due to technical innovations as described above. After the war three innovations made the domestic role easier: the automatic washing machine and tumble dryer took away much of the hard physical drudgery of the weekly wash; the use of refrigeration allowed shopping to be made once a week, rather than daily; and he development of central heating eliminated the dirty job of maintaining a coal fire and reduced the level of dirt and grime in the house. Taken together this reduced the workload of the housewife sufficiently to allow her to undertake at least part-time and often full-time employment.

The result of these three time-saving innovations was a major transformation in the patterns of work of men and women. It now became possible for both men and women to work, greatly increasing household income and wealth and contributing to the autonomy and self-realisation of women who could do so. Table 11.1 shows the percentage of women in the workforce in the major secular capitalist countries. It shows a dramatic change by 1990 with (except for Italy) over half, and in some countries over two-thirds of women in employment.

Country	1955	1965	1975	1985	1995
France	45.8	46.1	51.1	56	59.4
Germany	48.5	49	50.8	52.9	61.7
Italy	27.4	31	34.6	41	43.3
Japan	61.3	55.8	51.7	57.1	62.2
UK	45.9	51	55.1	60.5	66.6
USA	38.3	44.4	53.2	64	70.7

Table 11.1 Women in the work force as a percentage of women of working age, 1955–1995

OECD, Manpower Statistics, 1950–62, Paris, 1963: OECD, Labour Force; Statistics, 1959–70, Paris, 1972: OECD, Labour Force Statistics, 1965–85, Paris, 1987: OECD, Labour Force Statistics, 1969–80, Paris, 1982: OECD, Labour Force Statistics, 1978–98, Paris, 2000.

This liberation of women from their traditional duties allowed them to play a full role in the economy and further enhanced the wealth of secular capitalist countries. 1950–1990 was a period of unsurpassed economic advance. Figure 11.3 shows that by 1990 the average individual in major secular capitalist countries was around five times wealthier than the world average. In the past this level was only ever achieved in Britain and the USA. The industrial success of Japan was clear, even overtaking the USA in the measure of GDP per capita. In contrast, the communist countries of Russia and China and the directed socialist government of India declined in economic importance, creating the largest gap in wealth between rich and poor states the world has ever seen. It was not necessarily that the average Chinese or Indian peasants were any worse off than they had been in the past; it was just that the benefits of industrialisation had failed to reach them. It was clear that secular capitalism as a political philosophy had surpassed any of its rivals in achieving economic success.

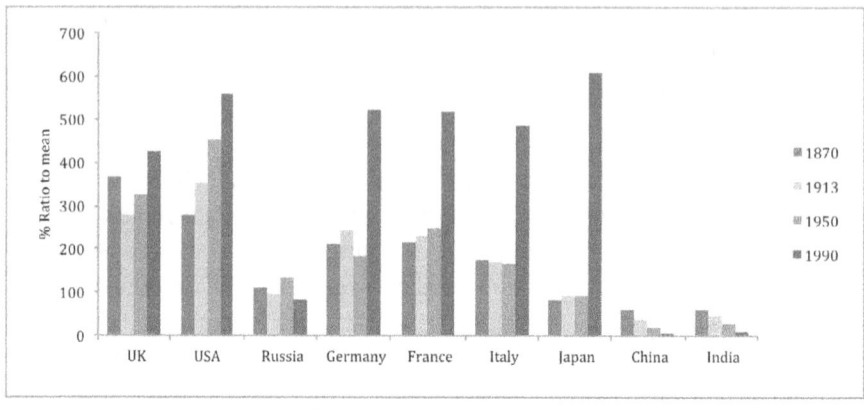

Figure 11.3 GDP per capita by major country (1870–1990) compared to the world average

Changes in the secular capitalist political culture

The beliefs, individual rights, rules of commerce and method of governance intrinsic to the secular capitalist political culture all developed further during this period. The fundamental belief in a future of increasing wealth and health based on the application of rationality, science and technology remained the same.

Developments in evolutionary theory added further important beliefs about racial equality. In 1953 the Englishman Francis Crick (1916–2004) and the American James Watson (1928–) first identified the famous double helix of DNA as the means by which the code of life was passed from one generation to the next. It was revealed as beautifully simple in structure but incredibly complex in operation. It was made up of just four linked compounds (bases) arranged in two strands of an interlocking helix and was common to all life-forms from bacteria to plants and animals. The most complicated life-forms had the longest strands of DNA. The human genome (the DNA code of life for humans) contains 3 billion base pairs; if human DNA was the thickness of a cotton thread it would be 90 kilometres long. Long sequences of base pairs, known as genes, form the code to produce proteins. Proteins enable the life-form to function by performing a vast array of specific tasks. The whole workings of the system of protein creation, the way that DNA replicates itself during cell division and the fertilisation process itself are all part of one fantastic chemical machine which operates with astonishing speed and reliability.

With the discovery of DNA, the science of genetics was now truly launched. It had begun with work by Gregor Mendel (1882–1884) an Austrian working at an abbey in Brno in the modern-day Czech Republic.

Mendel's work was ignored at the time and only rediscovered after his death. Mendel was interested in hybridisation; how different variants of plants were passed onto the progeny of the hybrids. He discovered that in cross-fertilising two plants with different characteristics, say tall and short, the progeny were either tall or short (not somewhere between) and that the proportion the plants with the two characteristics was split 3:1. The new discovery of DNA explained Mendel's findings perfectly. In genetics alternative forms of the same gene are called alleles. In this case there are two alleles: one will generate tall plants and one short. One of these alleles will be dominant, say the tall one. When the alleles pair up in the DNA molecule, if the dominant one is present it becomes the active gene. Thus in three out of the four possible pairings of the alleles, the plant will be tall.

Evolution was now perceived as the result of random mutations in DNA, passed on from parents to children. Most random changes in DNA would have either a negative effect or no effect at all, but occasionally a mutation would result in a physical advantage to the life-form. Then the natural selection process would result in the progeny of the changed life form becoming more successful.

Darwin's theory now had no scientific rivals; the code of life had been discovered. Geneticists moved on to determine the precise sequence of DNA of different life-forms. They could now demonstrate how close humans and apes are in their DNA makeup, confirming that apes and humans had the same ancestor. They could also show that humans were one species which had evolved from one small group of ancestors in Africa. The fascist theory of superior races was no longer credible. As a consequence, two more fundamental **beliefs** were added to the secular capitalist political culture: culture: Homo sapiens is one species that evolved from African ancestors and DNA is the code of life that passes from parents to children.

Individual rights changed considerably as a result of changing attitudes to women and homosexuals. As women gained more financial independence through direct employment they began to claim the right for equal treatment with men. The suffragette movement in the early twentieth century had been successful in gaining voting rights for women and equality for them in law. In the 1960s a new second-wave feminism movement broadened the debate to a wide range of issues. Two of the most important were a drive for equality in the workplace and a focus for reducing violence against women. The intellectual background came from the French writer Simone de Beauvoir (1908–1946) who in her book *The Second Sex* argued that, just because women are capable of getting pregnant, lactating and menstruating did not justify their unequal treatment in a patriarchal society. She argued that the male-centred ideology currently accepted as a norm needed to be challenged and its myths exposed. Second wave feminism was largely successful in the Western world in overcoming the assumption of the superiority of male leadership. There followed a whole series of laws and medical changes which improved the lot of women. In employment, legislation in Europe and many US states gave women the right to equal pay with men for the same job. Discrimination against women in companies, clubs and societies was made illegal in Europe. In law, rape was treated as a serious crime and the legal processes were made more sympathetic to women. Divorce became easier to obtain. If their wives were unfaithful, men could not claim custody of their children by right and they were required to pay child maintenance after a divorce. In practice, the custody of the child usually remained with the mother. In 1960 the first oral contraceptive pill was developed in the USA, giving women control over pregnancy for the first time. Abortion became widely available on demand in Europe and the USA, as long as the foetus was relatively young. Exact conditions varied from country to country, with states where the predominant religion was Catholic being the most stringent. A revised list of rights for secular capitalist citizens would include:

> equality of all individuals in the eyes of the law no matter what their race, **gender, sexual orientation**, class or religion
>
> freedom to pursue the religion they choose
>
> **the right to divorce on the basis of mutual consent**
>
> freedom from arbitrary arrest and imprisonment
>
> freedom to have and to publish one's own opinions, subject to libel laws
>
> **freedom of adults to pursue the sexual activities they desire so long as no undue violence is involved (that is, heterosexual and homosexual acts between consenting adults are allowed)**
>
> **freedom for women to abort a foetus early in pregnancy.**

The two most contentious changes concerned abortion and homosexuality. Women were now no longer driven to seek life-threatening back-street abortions. Consenting adults could no longer be sent to jail for committing homosexual acts. Established religions had great difficulty accepting these changes; divorce, contraception and abortion were and still are specifically opposed by the Roman Catholic Church. Churches were exempted from gender discrimination laws and many retained their policy that only men could become priests.

In the Western world, women were now as well educated as men and were widely accepted in the workplace and society as the equal of men. Nevertheless, the employment patterns of men and women remained completely different. Women dominated such professions as nursing and teaching; which could be seen as an extension of their old domestic role. Despite legislation, senior managerial and professional positions were still dominated by men. Women's role in society was still governed by genetic differences and cultural influences. It was apparent that the act of childbirth meant that there would always be overall differences in career opportunity, motivation and ambition between men and women. In secular capitalist countries, however, if individual women had the skill and motivation they now had the opportunity to achieve success in professional roles which had been unavailable to them 50 years earlier.

While equality of opportunity was being sought for women, the battle for equality of opportunity for all classes was also being fought. Attention focused on the different access to education of each class. After the war, free secondary education was extended in most countries to all children up to around 16 years of age. This was later extended to 18. In Europe the state provided economic support to parents sending children to universities. In Britain, educational selection at the age of 11, which benefitted the middle classes, was largely abolished. The result in Europe as a whole was much greater social mobility and class became less important in defining one's status.

In terms of **commerce**, many European countries experimented with socialist ideas of centralised planning. After the war, European socialist parties differentiated themselves from the communist parties by their commitment to true democracy and from their older conservative and liberal rivals by advocating centralised planning and controls and improved welfare support for the poor.

Centralised planning implemented by socialists in Europe was not like the communist conception of planning. It was perceived as a cooperative national venture with trade unions and business people directly involved in decision-making. National targets for growth and development were set. The language it used was that of macroeconomics: relating national output, measured as GDP, with price inflation, unemployment and economic efficiency. This model perceived the economy as a vast clockwork machine in which government investment at one end resulted in jobs and growth at the other. Price and wage increases needed to be contained by agreement. The model implicitly acknowledged that the government was somehow in charge of economic activity and that by a few tweaks of the economic controls the economy could be made to chart a steady course for growth. Organisations were set up to control prices and wage increases. Ailing industries were rescued to preserve jobs. Socialist parties were also in favour of state monopolies, known as nationalised industries, particularly for utilities, telecommunications, mining and transportation. They believed this eliminated unnecessary investment in competitive infrastructure, as having one supplier enabled the country to plan rationally for the overall good.

In the post-war years, this model worked reasonably well. The infrastructure that had been destroyed in the war was rebuilt and the wide availability of consumer goods benefitted all. However weaknesses in this approach began to appear in the 1960s and 1970s. Britain was worst affected among the European powers. The power of the trade unions began to have a negative effect on growth; by resisting change and focusing on short-term gain for workers they prejudiced the long-term good of the companies that employed them. One of the worst examples was the car manufacturer British Leyland, which launched its famous Mini in the 1960s. The company became a byword for strikes, poor quality manufacture and weak management and eventually collapsed; leaving Britain as the only major power without its own mass car manufacturer.

Lack of competition meant nationalised industries were slow to change and could not adapt to customers' needs. For example, in Britain in the 1960s, only one style of telephone was available and there was a waiting list several months long for installation; showing that Britain was way behind the USA in its vital telecommunications infrastructure. Trade unions in nationalised industries were particularly powerful, holding the company and by extension, the country to ransom. Unions naturally resisted redundancies. In the most extreme case the militant leader of the Miner's Union, Arthur Scargill, claimed that the National Coal Board should keep pits open even if their coal reserves had been exhausted.

In the 1980s British Prime Minister Margaret Thatcher finally wound back the clock on the socialist approach. The pretence at centralised planning of Britain's industrial development ceased. Price and income controls were phased out. Tripartite talks between government, the unions and industrial leaders no longer took place in Whitehall. Nationalised industries were broken up and privatised and market forces left to do their brutal work. After years of poor performance British industry was decimated; many uncompetitive companies went to the wall. The power of the trade unions to negatively affect the whole economy for the short-term good of their members was broken at the same time. The population returned to the old capitalist work ethic, recognising that companies have to be competitive to survive. Foreign companies appreciated the change and eventually Britain again became a profitable place for investment. For example, in Sunderland Nissan set up a factory which became the most efficient car producer in Europe. The British economy recovered and its path to industrial growth was restored.

Other European countries recognised the same symptoms, albeit on a lesser scale. Soon the British approach was copied across Europe and, with France lagging behind, the socialist centralised planning approach gradually died.

Socialism had one great success however, the establishment of a social support system for its citizens. This is a system that ensures a minimum standard of living for the old, sick and unemployed, by providing pensions, unemployment benefit and access to health care for all. It is funded by insurance contributions and a redistribution of income from rich to poor. The system can operate either with universal benefits for all (rich or poor) as in the Nordic countries, or as an extension of previously existing employment based systems, such as in Germany, or as a targeted benefit system, as in Australia.

The British system, known as the welfare state, is a mixture of universal and targeted benefits. It was established in the late 1940s following the publication of the Beveridge Report in 1942. William Beveridge (1879–1963) recommended that the government should tackle the five giants of want, squalor, ignorance, idleness, and disease. He argued that to tackle these problems the government should provide adequate income to people, adequate health care, adequate education, adequate housing and adequate employment. It proposed that all people of working age should pay a weekly National Insurance contribution. In return, benefits would be paid to people who were sick, unemployed, retired or widowed.

The socialist approach to care and development for all the population led to a number of support services being made available to the population at large. All countries developed a level of support for the unemployed and the old, in the form of pensions, unemployment insurance and disability provision. In Europe the right to free health-care advice and affordable treatment became common and the health of the poor as well as their quality of life improved dramatically. The USA, however, having never had a socialist government, was very wary of any initiative seen to be socialist. While there was some financial support for the poor, they had no access to affordable health care. As a result America, which invests more in health care than any country in the world, has a poorer record on overall health care than most countries in Europe. Table 11.2 compares recent statistics for the USA with three European countries and shows life expectancy, under-five death rates and maternity mortality are all worse in the USA. The latest effort in the USA to overcome this issue, Obamacare, remains deeply controversial.

	USA	UK	Germany	Sweden
Life Expectancy at birth(years)	79	80	81	82
Under five death rates(per 1000 live births)	8	5	4	3
Maternity mortality ratio(per100,000 live births)	21	12	7	4

Table 11.2 Key health statistics for the USA, UK, Germany and Sweden

Source: UN World Health Organisation report for 2012

Before the First World War state spending in most countries had been less than 10% of GDP. Most of the budget was spent on the armed forces, law enforcement, primary education and administration. During the twentieth century, as the welfare state developed and state education expanded, state spending in economically advanced countries rose to 35–55% of GDP. The state now had enormous powers of patronage and state agencies now employed a significant proportion of the population. Each state agency developed their own organisational, technical and cultural memes. The various state institutions had to compete among each other for state funding. Health care began to require ever-increasing financial resources, more people wanted to go to universities, adding to education costs, new highways needed to be built and maintained and so on. Teachers, the medical profession, the police, and all state agencies had to make their case to the country and to government to gain their share of the funding. State institutions now became an even more important element in the evolutionary structure.

The main **commercial** tenets of secular capitalism remained unchanged throughout the twentieth century: the freedom to trade legally without government interference and the right to form limited liability companies and trades unions. Secular capitalist countries also encouraged the growth in international trade by recognising international patent and copyright law and complying with international trade agreements.

In terms of **governance** the basic principles remained the same. All secular capitalist countries were democracies and governments were periodically subject to election by all adults. The vote was extended to women and all adults from 18 or 21 years of age. Most nations succeeded in keeping corruption out of government practice, but all had relapses, with resulting political scandals. The American system of a legally enforceable Bill of Rights was adopted in Europe. This made it possible for individual Europeans to challenge government decisions in the courts on the basis of an infringement of their individual liberties.

It was remarkable that the cultural values of secular capitalism were shared across many countries, many of which contradicted the local religious culture. The secular capitalist political culture had emerged as part of a general rational, humane and analytical view of the world. In former times religion had determined the cultural values of society. Historically, religions had been able to adapt to new circumstances and remain relevant. But now all religions struggled to include women's rights, homosexuality, sex before marriage, divorce, abortion, science and evolution in their core beliefs. It was astonishing that the secular capitalist culture had developed on an international basis without any religion specifically promoting it. Indeed these secular capitalist values were rarely explicitly stated. Most citizens were happy to be part of a secular capitalist culture and be part of a religious congregation at the same time; for example, many Catholics were prepared to practise contraception. This reflected the decreasing importance of religion in the day-to-day life of citizens. In the West church attendance had begun to drop; Sundays had ceased to be a day of rest for religious contemplation but instead became a day for shopping and attending sports events. Christmas became increasingly an orgy of present-giving and overeating without any strong religious overtones.

In the 40-year period after the war, the secular capitalist political philosophy had served democratic countries remarkably well. It had enabled peoples' lives to be completely transformed, allowing most a longer-lived, more comfortable and fulfilling existence. This new generation took all this comfort and convenience for granted. The collective memory was short but humans had come a long way from the two hundred years ago when most people were peasants worn down by a hard subsistence lifestyle. Secular capitalism was a successful cultural set of memes honed by experience to exploit the opportunities of the technological revolution. It had given those who lived in secular capitalist countries the greatest health and freedom from want that has ever been available in the 200,000 years of mankind. It is astounding that these values are not more explicitly treasured and worrying that these cultural values are not taught with the highest priority from generation to generation.

Outside secular capitalist and ex-communist countries, however, religious influence remained as strong as ever. Religion remained on the second level of the evolutionary hierarchy of the industrial society and was an important focus for community allegiance. It is in this period that Samuel P. Huntington wrote his book *The Clash of Civilisations and the Remaking of World Order,* predicting that, with the collapse of the cold war between Communist countries and the West, the new fault lines between conflicting communities would be drawn up between what he called civilisations. By civilisations he meant groups of people that were broadly united in their religious affiliation. Thus the dividing lines between Hindu, Shi'a Muslim, Sunni Muslim, Jew, Buddhist, Catholic, Orthodox and Protestant Christian would open again. In this opinion he was remarkably prescient. Many of the recent global conflicts have at their heart an antagonism between communities divided by their religious beliefs.

The end of empires and immigration

The middle of the twentieth century saw the gradual dismemberment of Western empires, resulting in many new independent states in Africa, Asia and the West Indies. The British empire, which had reached its largest extent in the 1920s, was by the end of the century reduced to a few scattered islands.

Trade, plunder and the lure of gold had been the initial motivations of the Western seaborne empires; land settlement happened afterwards, almost by accident. The British were lucky in North America and Australia in that the local natives were susceptible to Western diseases and unable to match Western military power. Ethnic cleansing was pursued vigorously; the natives were quickly moved off the prime land and Europeans were able to take over and exploit the natural resources of new continents. After the USA gained independence in 1783 in a military rebellion, the British realised that direct government from Great Britain over these new colonies with their independent and entrepreneurial populations could not be maintained indefinitely. They developed a method of government known as dominion status, by which the king remained the head of state, with limited power, but the dominion was otherwise self-governing. Canada gained dominion status in 1867, Australia and New Zealand in 1907. This created three new resource-rich nations with ample land available for British and European emigration.

After the dominions gained their independence the jewel in the crown of the British Empire was India, the subcontinent that had been acquired by the East India Company and later absorbed into the empire. The British, however, ran it on the cheap, relying heavily on the acquiescence of the natives. There were only 1250 British civil servants in India in the inter-war years running a country of some 400 million people. For most of its history India was a geographic concept rather than a state. The area occupied by the British had never before been controlled by one country. There were about 25 major spoken languages written in several scripts and three major religious faiths, and parts of the country were still ruled by Indian princes. It did not appear to be one country at all.

It took the organising genius of one man to engage with the peasantry and push forward the idea of a self-governing Indian nation. Mohandas Karamchand Gandhi (1869–1948) was a UK-trained lawyer from Gujarat who found his vocation while supporting the political rights of Indians in South Africa. There he developed the concept of the power of truth (satyagraha) combined with non-violent action (ahimsa) which became the bedrock of his strategy for political protest.

Gandhi was bought up by a devout mother in the traditions of the local Hindu community devoted to Vishnu. There were Jain influences on the religion of this area and from an early age Gandhi absorbed several characteristic cultural practices: a compassion for sentient beings, vegetarianism, fasting for self-purification and mutual tolerance among individuals of different creeds. On leaving India to study law at University College in London he took a vow to continue to be a good Hindu by maintaining abstinence from meat, alcohol and promiscuity. In 1893 he travelled to South Africa, where direct experience of the racial discrimination practised against the Indian community allowed him to develop his natural ability for social activism and for leading oppressed communities. It was here for the first time that Gandhi led a non-violent protest against injustice. His idea was based on the conviction that, if a cause were just or true it would have the power to win over opponents as long as the opposition was not directly threatened. In South Africa the heavy-handed response of the police to peacefully protesting civilians caused a huge public reaction and led to Gandhi's first political success.

On returning to India, Gandhi again started campaigning for the oppressed. He had an immediate impact on the independence movement. India's future first Prime Minister Nehru said, 'Then Gandhi came.... Get off the backs of these peasants and workers, he told us, all you who live by their exploitation; get rid of the system that produces poverty and misery'. His first success was with the poor farmers producing indigo and other cash crops in Gujarat. There he set up an ashram – a term usually used for a religious community – for his followers to gather information on the sufferings of the local villagers. It was in this period he was first addressed as Mahatma (great soul).

In 1921 Gandhi became a leader of the Indian National Congress and began a policy of non-cooperation with the British government. A key aspect of this policy was the promotion of Indian woven cloth. During this time Gandhi adopted the simple clothes of a religious ascetic and started to wear the dhoti, a hand-woven piece of cloth wrapped around his waist. In the words of Churchill he became 'a seditious middle temple lawyer now posing as a fakir ... striding half naked'. The dhoti confirmed Gandhi's identification with ordinary Indians and their traditions and was simultaneously threatening to the Western standards of the British authorities.

In response to Ghandi's non-violent campaigns (which often involved violence from more ardent nationalists), the British used force, imprisonment and their old tactics of divide and rule. The Government of India Act (1935) was designed to offer some measure of self-government to India at the same time as emphasising the diversity of the electorate and thus binding India as part of the British Empire indefinitely.

By this Act eleven Indian provinces achieved self-rule. At the centre two federal assemblies would exercise home rule for India, controlling all policy except finance, defence and foreign policy. Separate seats in the Federal Assembly were reserved for princes, Muslims, Sikhs, untouchables and women to ensure that the diversity of Indian interests was represented. Although the federal organisation never functioned, as the princes refused to cooperate, the Congress Party took control of eight of the eleven provinces. Membership of the Congress Party grew from 0.5 million in 1935 to 4.5 million in 1939.

By the time the Second World War was over, the Congress Party had become so strong that the move for independence was unstoppable. However, the idea of one India stalled, as the concept of Hindu politicians in power proved too threatening for the Muslim community. Muslims insisted on a separate state for the provinces for which they were in the majority. A boundary commission hastily divided the Punjab and Bengal, so that Muslim communities there could join the newly named Pakistan, and Sikh and Hindu communities could join India. Inevitably the division left some Hindu and Muslim families on the wrong side of the political divide. On independence in 1948, the inter-community tensions that had remained mostly under control for centuries broke out in an orgy of violence. The worst hit was Punjab, where the Muslim community was outraged at the extent of the land awarded to the Sikhs and Hindus. Everywhere, majority communities assaulted and expelled minorities in acts of savagery of which only the human animal is capable. According to Piers Brendon in *The Decline and Fall of the British Empire 1781–1997*, p. 41:

They roasted babies on spits, impaled infants on lances, boiled children in cauldrons of oil. They raped, mutilated, abducted and killed women, sometimes hacking off the penises of their dead husbands and stuffing them in their mouths. They subjected men to frenzied cruelties, burning them alive in their houses, stabbing them in the streets, butchering them in hospitals, strangling them in refugee camps.... Some of the worst massacres took place on the railways ... often they left nothing but coaches of corpses, which arrived at their destinations, in charge of Eurasian engine drivers, with gore seeping from every aperture.

Altogether some 500,000 people were killed during one of the greatest mass migrations in history. Around 4 to 5.5 million people crossed the border in the Punjab to seek refuge. About half a million Hindus from Sind migrated to India and an equal number of Muslims from the rest of India to Pakistan. In addition almost a million Hindus from east Pakistan crossed into India. The birth of the new nations of India and Pakistan was marred by ethnic and religious tensions that have endured to present times.

After the Second World War, the impotence of the West European powers in face of the military force of the Japanese led to a massive decline in their prestige and authority in Asia. Subject nations flexed their muscles and the weakened West European nations simply did not have the power or the will to maintain their authority. In 1948, in addition to the British withdrawal from India, Burma became independent and Sri Lanka was created from the old British colony of Ceylon. The British held on in Malaya a little longer, defeating the local Chinese communists before withdrawing in 1957. In 1949 the Dutch lost control of Indonesia after failing to defeat a guerrilla war for independence. In 1954 the French withdrew from Indo-China after military defeat by the communist North Vietnamese. Four new countries were created, North and South Vietnam, Laos and Cambodia.

The boundaries of these new Asian countries had been created by the colonial powers and often did not have a history of dominance by a particular language or ethnic grouping. As in India, ancient ethnic rivalries soon caused communal violence. Well-known examples are of the Tamils versus the Sinhalese in Sri Lanka and the Burmese versus the Mon, Shan and other minority groups in Burma. The process of nation-building still had to take place, involving civil strife, military force and the establishment of an elite. In several countries a military strong man assumed power. Despite these difficulties, the new countries largely remained intact and by 1990 some of them, Malaysia and India in particular, were beginning to thrive as democracies.

Once a country has been conquered by another and the spoils of war taken by the winners, there are four reasons to maintain power: to take over the land by ejecting the natives, to draw income from the country in the form of tax and trade, to replace the local ruling elite with placemen from the invading country, and to prevent retaliation in a future war. After the Second World War, when the important states like India had gained their independence it gradually became clear to the colonial powers that none of these four benefits was available to them in practice. Maintaining a colony no longer made economic sense. Native resistance prevented more land settlement. Tax income barely covered the cost of administration. When small arms became readily available after the Second World War, armed native rebellion threatened the colonial elites. Once the Asian colonies had gone, the Western powers saw no profit in holding onto the remaining African and central American colonies. They were all largely poor countries which would bring more trouble than it was worth if it was necessary to hold on to them by force. Nonetheless, there was one last act to play. White European settlers in Kenya, Rhodesia and South Africa had ambitions to maintain control over their colonial settlements. However, unlike the position of North American colonists in the seventeenth century, disease had not decimated the local population and the natives had found the resource and ability to fight back. The Mau-Mau uprising in Kenya eventually led to Kenyan independence in 1964. The whites in Rhodesia declared independence in 1965, but eventually, after insurgent activity and international pressure they had to concede power to the black majority in 1979. It took longer to change South Africa, the biggest and most prosperous country in sub-Saharan Africa, but even there, by 1994 the nation had its first black African President, Nelson Mandela.

By 1990 the Western European colonial empires had largely ceased to exist and many new states had been formed. World power, which formerly was monopolised by a few states, was now shared between a large number of independent entities. In Africa these new states were artificial creations, formed by Europeans drawing lines on maps, with no attention paid to the languages spoken or the religious or ethnic groups that straddled these lines. All these countries contained mixtures of tribal groupings speaking many languages. Take Kenya for example; there are four major native ethnic groups using over 30 languages among them. Groups of hunter–gathers still survive in remote areas. Farmers come from three different language groups: Cushite-speaking peoples from the Ethiopian highlands, Kalenjin peoples from the north-west speaking Nilotic languages and Bantu peoples from the west and south. Arabs once colonised the coast, converting some natives to Islam. Swahili, a Bantu language with elements of Arabic vocabulary, became widely used both as a langauge in the coastal areas and as a lingua-franca across Kenya as a whole. After 1600 further waves of pastoral Nilotic-speaking people came down from the north-west including the Samburu and Maasai, who were the dominant political force at the time of the British colonisation. The British arrived and settled in Highland areas and with them came Indians, predominantly Punjabis and Gujaratis, adding further to the ethnic mix. Many Kenyans speak three languages: Swahili, English and their own native tongue. Christianity is now widely practiced in the central areas and the old tribal religions are disappearing. Christianity is, of course, divided into a number of Catholic, Protestant, charismatic and African indigenous sects. Islam is still practiced on the coast.

Inevitably, the leaders tend to favour their own ethnic group. Democracy is practised in name only, with widespread corruption and vote rigging. Its first leader, Jomo Kenyatta, was a typical African leader. Educated by missionaries and then taking a degree in London, he became a national leader during the Mau-Mau uprising. On assuming power, he rapidly turned Kenya into a one-party state, with himself as its charismatic leader. Soon every shop had to display his picture and stories of his actions dominated all the news bulletins. Rampant corruption became the norm and his own Kikuyu tribal group dominated power.

This pattern of one-party dictatorship, corruption and vote rigging was common across all Africa. Given the diverse nature of the states, maybe this is the only structure that will work until the nations become more mature. Certainly most states have managed to maintain central power and only Somalia and the Democratic Republic of Congo can be classified as failed states, with substantial parts of the country reverting to chaotic, violent control by local militias. Measured by population growth, the area thrived. The population of sub-Saharan Africa, according to the UN, was 186 million in 1950 and had grown to an astonishing 515 million by 1990.

During this period that change in lifestyle resulting from technological development began to transform the developing world, particularly in south-east Asia and Latin America. As communication and transport improved, the isolation of peasant and tribal communities in the developing world lessened. As trade and industrial technology spread people became wealthier and more efficient. As the benefits of advances in medicine became significant, people lived longer. And as wars became less frequent, more young men survived to play their part in the world's future.

However, those seeking economic advance had another option, emigration. Until the 1950s immigration to Europe was relatively small scale. Usually, after a few generations the immigrant population was absorbed into the mainstream. Thus, European states had a relatively stable gene stock and the populations had a limited range of cultural norms. For example, in the face of religious persecution by Louis XIV in the seventeenth century, some half a million Huguenots (French Protestants) emigrated from France to other protestant countries. When they first emigrated they tended to settle in their own communities and pursue their trades and build their own churches. As an illustration, about 50,000 Huguenots settled around Spitalfields in the East End of London between 1670 and 1710. They were expert silk weavers, thrived and built seven churches in the area. At first the Huguenots kept their own distinct identity, speaking in French and attending their own religious services. As with many immigrant groups, their churches acted as a community centre in providing welfare to the poor and support to new arrivals. Over time the Huguenots became assimilated into English society; their names were anglicised, their churches became Anglican and, when the silk industry began to decline, the Huguenot community started to disperse. Today there remains no trace of the Huguenot culture in British life, save for a few French-sounding surnames. Their genes have survived however; it is estimated that three centuries later a quarter of London's population will have some Huguenot ancestry.

Until the twentieth century in Europe, most immigrant communities were eventually totally absorbed into the mainstream community. The major exceptions were the Jews and the Romani, (Gypsies) who retained their own culture. Both communities were persecuted and the Jews were deliberately kept in separate ghettoes until the Enlightenment. The economic success of Jews in Germany, in particular, led to widespread resentment, which was manipulated by the Nazi party and ultimately ended in the tragedy of the Holocaust.

The nature of human community bonding means that immigrant communities are always at risk of rejection by the dominant culture. For example, when they first arrived, the Huguenots, like all immigrant communities, were resented for taking British jobs. The immigrant community, like the Huguenots, may eventually be absorbed. However, the Jewish experience in Europe up to the 1950s shows that there was always a risk of racial intolerance to communities that maintained their identity.

American immigration, where from the 1850s onwards Europeans from all countries emigrated, showed the same pattern as the Huguenots in Britain. Initially, communities congregated in specific areas focused around their own church. Swedes largely went to the mid-West and settled around their Lutheran churches; Germans were some of the earliest inhabitants of Pennsylvania with their Lutheran, Calvinist, Amish and Mennonite churches;

Italians tended to congregate in large cities with strong Catholic community ties and so on. Cross-community intermixing started early, particularly between Protestant north European and Anglo-Saxon communities. The Catholic Irish and Italian American communities retained stronger specific identities. However, all these communities consider themselves American. Indeed, American Italians fought against Italy in the Second World War with no major identity problems. Jews, as another major distinct community, have not faced the same problem of discrimination as in Europe. My guess is that this is because all immigrants have been able to identify and unify behind the same secular capitalist principles that are outlined in the American Constitution. This commitment to the Constitution, symbolised by the pledge of allegiance to the flag, while placing hand on heart, allowed all Americans from every sub-community to consider themselves as part of the same nation-state.

This did not necessarily apply to the African American population, whose ancestors were slaves, or to the Chinese who originally worked on the railroads. The Latino community, immigrants from Central and South America, continued to speak Spanish and were not wholly integrated either.

The period 1950 to 1990 has seen huge increases in emigration from all racial groups and cultures to the wealthier countries in Europe, North America and Oceania. There has also been a rise in the acceptance of multiculturalism and a strengthening of the laws against racial discrimination. All nations were seeking to avoid the same devastating impact of racism as occured in Nazi Germany before the war.

Starting with the USA, the 1964 Civil Rights Act restored the federal government's power to prohibit racial discrimination. Banned, at last, were the hated visible signs of racial discrimination in the South: the separate seats on buses and in restaurants, jobs reserved for whites only and the blatant gerrymandering of elections that had made the deep South of the USA more similar to South Africa than any modern secular capitalist country. A year later, in 1965 the Immigration and Nationality Act abolished the system of national origin quotas for immigration, which had effectively restricted immigration to those who were from Europe. Almost at once the pattern of immigration changed, with most immigrants coming from the Americas and Asia: Mexico, China, Vietnam, the Philippines and India were now the major countries of origin of the new immigrants. Completely new communities were created, meeting at churches, temples and mosques. The ethnic nature of the USA was changing; it was becoming a mixture of all ethnic groups, not just Europeans. While Europeans accounted for nearly 60% of US immigrants in 1970, they accounted for only 15% in 2000.

The same pattern was found in Europe. In the economic boom times of the 1960s there was a labour shortage for manual jobs. Across northern Europe immigration was encouraged. Gastarbeiter (literally guest workers) were employed in Germany, large numbers of whom came from Turkey. France received immigrants from its ex-colonies, including Algeria. British immigrants came mainly from the Caribbean, Africa and the Indian subcontinent.

This welcoming of other cultures as immigrants into Europe was something quite new. Europe now had a considerable population of Muslims, Hindus and Sikhs with a quite different culture from the Protestant and Catholic Christian traditions. As a result there were lots of small communities who, as the Jews of old, clustered around their places of worship, tried to maintain the language and traditions of the old country and to live a quite separate existence from the European mainstream. Of course there have been resulting racial tensions. Governments tried to minimise the inter-ethnic friction. Racial discrimination was made illegal across the subcontinent. Multiculturalism was the banner under which this proceeded, a sort of live and let live for each culture. It appeared that Europeans, while rejecting the the 'ein volk ein reich' culture of the Nazis have gone in the opposite direction, without clearly thinking through the risks and how to overcome them. Europe is not the USA, which has been a traditional melting pot of communities; nor have many of the immigrant cultures the same long history of democracy and respect for a written constitution. Unlike America, the values of European culture in the secular capitalist tradition have not been made explicit. Cultures clash, often in trivial ways: should Sikhs wear helmets when riding motor bikes, when this would mean removing their turbans; should Islamic women be allowed to cover their face in situations where facial expression is an important part of communication?

For a modern state to thrive without too much discord there has to be a broadly accepted set of values. These values naturally occurred in the nation-state; they built up over time in a population that shared the same background and language. In a multicultural state this is much less clear and the values of different religious and ethnic groups can be quite different from the mainstream. The hope is that values will be absorbed by the immigrant community and inter-community marriage will gradually break down cultural barriers. However, it is not clear that this will happen. Religions and faith-based schools are adept at instilling their own values into their communities, which may be quite different from those in the nation as a whole. Like the Catholics and Protestants in Northern Ireland this can lead to the perpetuation of inter-community hatred.

Both Europe and the USA are committed to a multi-ethnic society. To function smoothly this society will have to unite behind common values. Currently, these values stem from the political culture of secular capitalism. I believe the signs are good that, over time, humans from many backgrounds can come together as one community. The success of British athletes in the London Olympics involved many sportspeople from mixed marriages and minority backgrounds and was celebrated by the whole British nation. However, it also has to be recognised that there are dangers, particularly if religious community cultures become very entrenched. Unfortunately, it will take several generations before we know whether the multi-ethnic society we are creating is a success.

The collapse of communism

From the 1980s onwards, it was clear that the communist states of China, USSR and East Europe had failed to thrive in the same way as the West. Communist central planning was successful in building technological infrastructure: electric power plants, roads and railways and steel plants. In this, the USSR was particularly successful, catapulting it from a minor power to one of the two greatest powers in the world between 1920 and 1950. However, central planning could not match the spectacular success of capitalist countries in delivering all the intricate and, some might say, wasteful, requirements of an economy driven by consumer demand. In particular, communist countries never embraced the just-in-time revolution in quality. Too often meeting production targets led to the production of large quantities of goods, like tractors, for which demand was limited and which had the double disadvantage of being unwanted and unreliable. The Russian-built car brand, the Lada, was a joke in the West and was famed for its unreliability. When Mikhail Gorbachev (1931–) came to power in the USSR in 1985 his prime aim was to improve the Soviet economy. He soon realised that this could not be done without moving away from the central planning approach. In 1985 he introduced a perestroika initiative which, according to Gorbachev, was the 'development of democracy, socialist self-government, encouragement of initiative and creative endeavour'. In doing so, he felt it necessary to champion local initiative against the central bureaucracy. However, when totalitarian states step back from rigid control, the results are not always what were intended. His initiative led directly to the collapse of the communist system in the USSR and East Europe. Communist regimes in East Europe had been established by Soviet military might after the war. The USSR was effectively a colonial power maintaining one-party dictatorships sympathetic to its grand designs. In the past, all attempts to break away from this system, in East Germany (1953), Hungary (1956) and Czechoslovakia (1968), were suppressed by the Russian military. Gorbachev's new policy meant encouraging nationals to challenge the existing dictatorship; in doing so he indicated that the USSR would no longer maintain its interests by force.

The result was an immediate flight from communism. In November 1989, East Germans freely crossed the Berlin Wall, the hated symbol of the division of Germany. By 1990 all the regimes in communist East Europe had been replaced and free elections held. Poland, Czechoslovakia, Hungary and East Germany, with their ancient links to the West, quickly established a democratic system of government and a capitalist approach to economic management. East Germany was reunified with the rest of Germany after a gap of 45 years. In Bulgaria, Rumania and Albania, with no particular democratic tradition, the old communist party elite clung to some of the reins of power under the auspices of renamed parties.

The abandonment of communism in the USSR was a historical accident. Devolution gave more power to the 15 constituent republics of the USSR. Nationalist movements triumphed in elections in the Caucasus and the Baltic States and the charismatic, drunkard Boris Yeltsin (1931–2007) became President of the Russian Federation. An attempted coup by the military failed in 1991. Yeltsin, organising the resistance, suspended the Communist Party in Russia and shortly afterwards withdrew Russia from the federation. With the demise of the USSR, 15 new independent states were created. The new Russia was considerably smaller than the old Russian empire. Yeltsin replaced Gorbachev in the Kremlin and soon the central organisation of the Soviet state was dismantled. Russia started its sudden, inadequately prepared and ultimately catastrophic embrace of capitalism.

In China Mao Zedong died in 1976. Leader of China since the Communist Party's victory in 1949, he had behaved with an absolute callousness, indifferent to the effect of his policies on the lives of ordinary people. His personality, formed by long years of opposition to government, proved unable to successfully lead a great nation. He was vindictive to anyone who tried to argue with his approach and ruthless in the elimination of any perceived alternative power or authority. He once boasted that the first Emperor Qin Shi Huangdi had buried 460 scholars alive but that 'we' have buried 460,000. In his famous catalogue of disastrous initiatives (the Great Leap Forward, the Hundred Flowers campaign, and the Cultural Revolution) some 40–70 million people were killed by starvation, elimination or incarceration. His leadership team, as with Stalin in Russia, proved submissive and pliant to his wishes. So strong was his hold on power that no opposition was possible.

This is from Jonathon Fenby: *The Penguin History of Modern China*, p. 526:

> Backed by the immense cult of personality, the charismatic,
> narcissistic Son of Heaven, who thought himself capable of changing
> human nature through his mass campaigns, could demand complete
> loyalty to the cause of revolution as he chose to define it. Nobody
> and nothing could be excused from utter dedication and readiness
> to contribute whatever was demanded. Private life meant nothing.
> People were a blank sheet of paper, mere numbers to be used as the
> leader saw fit. Maoist autocracy reached heights of totalitarianism
> unparalleled by Hitler or Stalin, accompanied by massive hypocrisy as
> the leader who preached simplicity, morality and proletarian values
> had his favourite fish flown up from Wuhan, dallied with a succession
> of young ladies, had rarely used villas built for him at great cost,
> and raked in the royalties from his *Little Red Book*. A potent terror
> organisation ensured obedience, a huge gulag swallowed up real
> or imagined opponents and a massive propaganda machine fed the
> myths.

After Mao's death the Communist Party took some time to determine its next direction, but eventually united behind the leadership of Deng Xiaoping (1904–1997). Deng was a great survivor who, like many of his colleagues, experienced the swings of fortune under Mao. Always a competent leader, he learned to confess his 'mistakes' and retreat from power when he was not in favour. He had, however, experienced life outside China. In 1920, at the age of 16, he went to Paris as part of a scheme to teach Chinese students the ways of the West. The scheme collapsed but Deng survived, working in factories, until 1926. Perhaps this experience showed that there was another way and that this was the origin of two of his most famous quotes: 'poverty is not socialism' and 'becoming rich is no sin'. Deng released the Chinese people from the stultifying central planning by party cadres and allowed individuals to profit from their actions. Two initiatives kick-started the process; in the countryside a household responsibility system was introduced whereby land was contracted out to farmers, who were required to pay taxes and hand over a certain amount of output but could sell any surplus. By 1984 a total of 98% of agricultural households adopted the new system and farm output grew as a result by some 8–10% per year. For the first time since the revolution, peasants received some attention and support from the authorities.

The second initiative was to set up special economic zones (SEZs) on a pattern begun in Taiwan, South Korea and Ireland. In these zones, foreign companies were allowed to establish factories, employ cheap Chinese labour, import goods, manufacture and re-export. Special privileges, tax breaks and subsidies were given to attract investment. In May 1980 the fishing village of Shenzhen, across the border from Hong Kong was declared the first SEZ. It was a success from the first; companies in the British colony moved manufacturing over the border, employing millions as a result. More SEZs were set up, particularly on the Yangtze delta around Shanghai, along the Pearl River in Guangdong and in south Fujian in Xiamen. The SEZs output doubled every 3 years and, more importantly, by the end of the decade an estimated 50,000 managers were running factories for investors from abroad. These managers were gaining the experience and skills to understand how to compete in a capitalist world.

International cooperation

World trade grew exponentially in the post-war years. It grew from a very low level in 1948 to 3,500,000 million dollars in 1990, as shown in Figure 11.4. This is a growth rate (which includes inflation) of 10% p.a. The result was a dramatic change in how the world bought and sold goods. Before the war in Britain, most consumer durables were sourced locally; by 1990 it was commoner to source them from abroad, not only from Europe but also from Japan and the USA. Outsourcing manufacturing was also becoming common. Places like Korea, Malaysia and Mexico were becoming manufacturing centres. As the volume of manufacturing increased, formerly very expensive goods, like computers and cars became affordable by most people in the West. The world was becoming interlinked and interdependent, and a great deal richer.

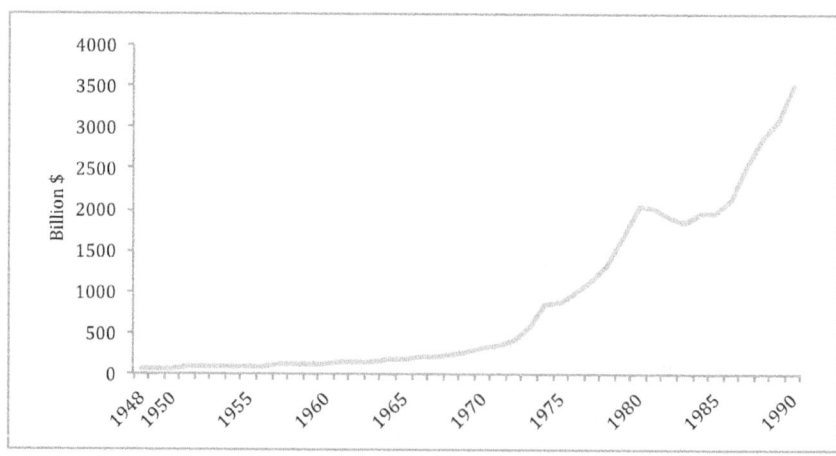

Figure 11.4 Value of world trade at current prices

Source: World Trade Organisation

Lessons were learned from the Great Depression. International cooperation would be necessary if the benefits of international trade were to be maintained. The institutions established at the Bretton Woods conference in 1944 made international financial cooperation feasible. These were the International Monetary Fund (IMF), the International Bank for Reconstruction and Development (IBRD) and the International Trade Organisation.

The initial purpose of the IMF was to stabilise exchange rates and assist in the reconstruction of the world's international payment system after the Second World War. Countries contribute money to a pool in the IMF through a quota system. Contributing countries can obtain loans from the IMF at reasonable interest rates, provided there is an agreed local restructuring of the economy. Originally comprising 45 members, it provided valuable support to member countries in the post-war years and to this day continues to provide useful fall-back support for countries in economic difficulties.

The IBRD was established with the original mission of financing the reconstruction efforts of war-torn European nations following the Second World War. The bank issued its inaugural loan of $250 million (equivalent to $2.6 billion in 2012) to France in 1947. Throughout the remainder of the 1940s and 1950s the bank financed infrastructure projects such as building dams, generating electricity and improving access to water and sanitation. It also invested in France, Belgium, and Luxembourg's steel industry. Following the reconstruction of Europe, the bank's mandate has transitioned to eradicating poverty around the world. In 1960, the International Development Association was established to serve as the bank's concessional lending arm and provide low and no-cost finance and grants to the poorest of the developing countries.

In contrast, the International Trade Organisation never got off the ground. Instead, a patchwork of trade agreements known as GATT (the General Agreement on Tariffs and Trade) was created. Through eight rounds of agreements, tariffs and other restrictions on free trade were gradually reduced across the world. This established a strong and prosperous multilateral trading system between participating countries and ultimately enabled the explosive growth of international trade.

The other major initiative to increase international cooperation was the United Nations. The UN was founded in 1945 to reduce the risk of war by providing a forum for dialogue between the major powers. The main forum in the UN for resolving conflict is the Security Council, which consists of 15 countries. China, France, Britain, USA and Russia are permanent members and have a veto; the remaining 10 are elected to serve a 2-year term. As a centre for resolving conflict, the Security Council has proved to be of limited value, being strong on words but weak in action. However, the UN has given rise to a number of effective international agencies which coordinate activities in areas like nuclear energy, aviation and tourism and provide support for poorer countries in famine relief and refugee management.

After the Second World War, moves towards European integration were seen by many as a chance to escape from the extreme forms of nationalism that had devastated the continent. Economic cooperation began in 1951 with the creation of the European Coal and Steel Community (ECSC) between Belgium, France, Italy, Luxembourg, the Netherlands and West Germany. In 1957 these six countries signed the Treaty of Rome, which extended the earlier cooperation within the ECSC and created the European Economic Community (EEC), establishing a customs union. In 1973 Denmark, Ireland and the UK joined the EEC and Greece, Portugal and Spain joined in the late 1970s. This meant that by 1980, all Western European countries were members of the EEC except for the Nordic countries, Switzerland and Austria, with their committed neutral international political stance.

The community abolished tariff barriers between members and set up common external tariffs with other trading partners. Agriculture was provided protection against the full blast of capitalist trade by the Common Agricultural Policy (CAP). This was aimed at preserving farm livelihoods while providing reasonable prices for agricultural goods to customers. It was one of the least economically successful operations of the EEC. Price interventions in times of a glut soon took over most of the EEC budget and created notorious butter mountains and wine lakes. Nevertheless, it did succeed in maintaining the rural economy, which had been neglected in many countries in the run up to the Second World War.

From the start there was always a tension in the EEC between those who pursued the dream of full European political integration and those who wanted to limit its remit to economic matters. Initially France took a very nationalist approach, seeing the community as a counterweight to English-speaking powers and an opportunity to promote French as the European language of choice. Once Britain joined, however, resistance to integration was strongest from the British. Nevertheless, gradually more powers were devolved to Brussels; a common fishery policy was set up, infrastructure projects were financed by the EEC in the poorer member states and competition law was strengthened. In 1979 elections to the European Parliament were held. However, the European Parliament's powers were limited by jealous national governments and it had little practical impact. For all its faults, however, the EEC was a great economic success, proving a valuable instrument for the mutual economic development of European nations after centuries of national rivalry.

There were few other effective international organisations to encourage, police and control the international economic development of the West. Much relied on international cooperation. Fortunately there was sufficient international trust and goodwill in the secular capitalist world to enable the mutual benefits of trade and economic specialisation to develop, bringing a new world of wealth and economic interdependence among the secular capitalist nations.

Competition between political cultures

The secular capitalist states of West Europe, North America, Japan and Oceania had thrived whereas communism was now a discredited political culture. All secular capitalist states had delivered a massive increase in personal wealth and lifestyle to their citizens. Five new states had adopted the secular capitalist culture. Three European states, Spain, Greece and Portugal, had overthrown dictators and joined the European mainstream. Two former Japanese colonies, South Korea and Taiwan, with American support, copied the successful Japanese economic model. In Korea, particularly, some very powerful multinational companies, such as Samsung, had emerged, which were capable of competing with the best on the world stage.

There had been no wars between secular capitalist states. The desire to compete economically and to gain the economic benefit of increasing international trade had diminished the risk of a competitive war between nations of a similar political philosophy.

Wars had taken place between capitalist and communist countries, notably in Korea and Vietnam. However, both parties had been reluctant to extend the war outside the direct area of conflict. The threat of destruction by the atomic bomb had lessened the risk of all-out war between the major powers. The success of political cultures now depended much more on their economic success than military victory. Countries sought other ways to compete to increase their national prestige. States vied to produce the tallest building, the longest bridge or the largest dam. The USSR and the USA competed in the space race. Russia put the first man in space in 1961 and Americans walked on the moon in 1969. The most important development in international competition, however, was in competitive sports. Without the need to go to war, sport gave a natural outlet to competitive instincts and allowed spectators to share feelings of local and national comradeship.

Competitive sports and games have been a part of man's natural development from ancient times, improving skills, leadership and teamwork, The Greeks had set up formal sport competitions at Olympia and other venues but in modern times organised sport was first developed by the English public schools, which considered sport as an important part of education. The lawnmower was first patented in 1830 and made possible the preparation of suitable pitches, courts and playing fields. Each school began to codify the rules of their sports to ensure fair play. When train connections made competitive matches against other schools practical, schools might play one-half according to the rules of one side and the other half according to the rules of the other. Over time, a common set of rules for identifiable modern sports emerged. Football, rugby, cricket, lawn tennis, badminton and golf all developed in Britain. At the same time a parallel development took place in the USA for baseball, American football and basketball. However, it was the British games which spread internationally as a result of their empire, trade links and historical connections with other European countries. As transport links improved, it became possible to play international matches. Cricket was one of the first sports to organise internationally. The first international cricket match was played as early as 1844 between the USA and Canada, although the sport has not developed since in either nation. The Olympics, focusing on individual rather than team sports, started in 1896 in Athens, where 14 nations and 241 athletes took part in the first Olympic Games. Sport flourished in the twentieth century and became a huge recreational activity for spectators and players alike, satisfying a human need for excitement and bonding.

Eighty-eight new independent states were created between 1945 and 1989 as the old Western colonial empires disbanded. This meant the number of independent states had more than doubled since the end of the Second World War. None of the new countries had a suitably developed cultural consensus to take on the wholesale implementation of secular capitalism. Further, half of these new states were in sub-Saharan Africa and consisted of a patchwork of tribes with no common history of statehood. Some of these states would struggle to exist.

There were several aspiring democracies. These were countries where democracy was practised more or less freely and the exercise of law was mostly independent of government, but there were issues with corruption, religion or racial discrimination. There were several types of political culture in aspiring democracies but it is worth mentioning two in particular: socialist planned economies and ethnically separated states. In socialist planned economies, like India, Turkey and Mexico the government maintained a directive, socialist planning style of government. International companies could not trade freely without government permission.

Corruption and bureaucracy were widespread and capitalism was subject to bureaucratic controls that hindered economic development.

In ethnically separated states, like South Africa (before 1994) and Israel, democracy worked only in support of a ruling ethnic group. These were capitalist countries with a dominant religion (Dutch Reform Christianity and Judaism) that reinforced the rulers' status as a special people. Conquered ethnic groups were unfairly treated and subject to restrictive laws about where they should live.

As a religion, Islam revived and reasserted itself in the political culture of several states. Oil extraction made a group of Islamic states in the Middle and Far East very rich. These states were still ruled by sultans, emirs or kings. An enormously wealthy elite maintained a very traditional Sunni Islamic political culture in which women had a subservient role. Wealth filtered down to its citizens through social programmes and government financed development and industrial projects.

In Shi'ite Iran the last shah was exiled in 1979 and Iran became an Islamic theocracy, enforcing religious practices through state controls, with Ayatollah Khomeini as its supreme leader. Its Constitution is based on the concept of velayat-e faqih; the idea that everyone requires guardianship, in the form of rule or supervision by a leading ayatollah. Khomeini served as the supreme ayatollah until his death in 1989. Iran's economy, which had hitherto been rapidly modernising, was now subject to strict controls by new Islamic inspired policies. Much industry was nationalised, laws and schools islamicised, and Western influences banned. The Islamic revolution in Iran was the first time since the Sikh empire in the nineteenth century that a religious movement had gained power. Its impact reverberated around the world.

Summary: situation in 1990

In the secular capitalist countries, with few wars, better living conditions and doctors who were increasingly effective in combating disease and illness, life expectancy improved dramatically. For the first time in all Western nations and Japan, life expectancy rose above 70 years. Even in India life expectancy was above 50 and in China it rose to over 60 years. Over two centuries life expectation had been transformed and the average life span throughout the world was now approximately double that of 1800.

Medicine had begun to gain a real understanding of the workings of the human body. Huge advances were made in diagnostics, drug treatments and surgery. In diagnostics, the workings of the body could be imaged by X-rays, by a computed tomography scan, magnetic resonance or with ultra-sound. Blood samples could be tested for levels of concentration of compounds down to the pg/ml level (a millionth of a gram per millilitre). A whole battery of new drugs became available to combat infection, relieve pain, reduce inflammation, mitigate depression and even to stimulate sexual activity. Pharmaceutical companies like Glaxo and Pfizer became huge corporations. Surgical procedures now allowed people to have their arthritic joints replaced with synthetic implants. Human organs could be transplanted from one person to another. Even the process of fertilisation could be achieved in a laboratory.

As trade grew and wealth increased, world population soared to 5.3 billion people, more than doubling the level of 1950, with a record growth in population of 1.9% p.a. Figure 11.5 shows that all areas of the world, bar Europe, grew around the 2% p.a. level. In Europe, women were increasingly able to decide when to start a family. As a result, Europe's population only rose by 0.7% p.a., and even this rate was achieved only by net immigration from developing countries.

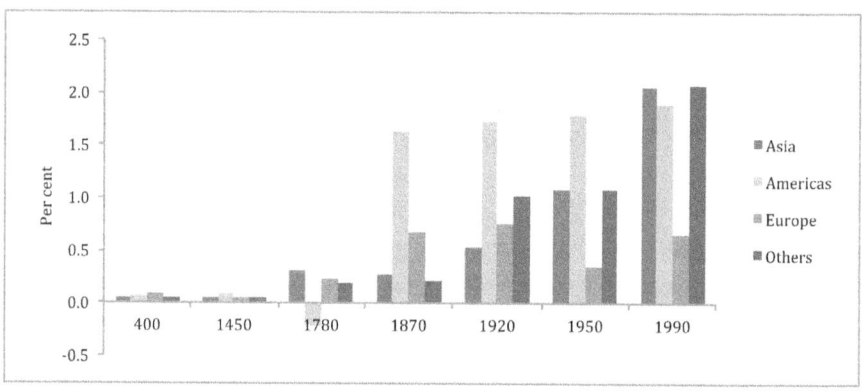

Figure 11.5 Population growth rates (per cent per annum)

The world was becoming increasingly urban as farmers migrated to the towns. Over 40% of the population in 1990 lived in urban areas compared to less than 30% in 1950. Cities with over 10 million people emerged in the less affluent countries; Mexico City, Sao Paolo, Buenos Aires and Shanghai now joined Tokyo, New York and Los Angeles as the largest metropolitan areas in the world.

With the collapse of the Communist approach to economic management, the full benefits of the technological revolution were about to reach all parts of the world. A new era of global trade and prosperity was about to begin.

CHAPTER 12:
GLOBALISATION AND THE LIMITS TO GROWTH
(1990–2010)

The period from 1990 onwards saw companies across the world freed from the straightjacket of central state planning. Individual initiative was encouraged and the level of global trade increased dramatically. As a result, the health and freedom from want of ordinary people improved markedly. Communism was no longer credible as a political philosophy; both Russia and China embraced capitalism but not democracy. The change in China was the most dramatic. China became the manufacturing centre of the world in a state-directed implementation of their own version of the technological revolution. Everywhere the positive effects of increased enterprise were being felt. After decades of stagnation following independence, India finally began to modernise. The economies of many south-east Asian and Latin American countries grew dramatically. Even some governments in Africa began to stabilise and their countries began to grow and develop. Across the world peasant farmers left their lands to take up new opportunities in expanding cities. The largest resistance to change came from traditional Muslim communities, who were reluctant to change their way of life.

The enablers and beneficiaries of this global embrace of capitalism were the **multinational companies**. Multinationals were responsible for making highly technical products available to the world at affordable prices. They were able to source their product from countries with low manufacturing costs and gain economies of scale from marketing their brands globally. As companies became more international they began to escape the control of their parent country. They were subject to no global authority and began to be courted rather than controlled by individual states. International trade transactions were deliberately set up to avoid paying tax. The senior management team were paid enormous salaries and given generous share issues. Democracy started to be subverted by the lobbying power and political campaign contributions of companies. Short-term business results were required to keep the financiers happy, in some cases this led to 'creative' accounting and crooked business practices. All this culminated in the 2008 banking scandal when many key banks and financial institutions were caught holding valueless securities based on irresponsible lending to poor Americans. This scandal came within a whisker of subverting the world's economy in the same way as in the Great Depression of the 1930s.

As more and more people achieved a comfortable lifestyle, the first signs began to appear that there were limits to the wealth that humans could achieve. This was not primarily due to a shortage of food, as predicted by Malthus, but to the inability of the environment to absorb the waste from industrial development. All Chinese cities were enveloped in smog and rarely saw the sun. Acid rain formed from burning sulphurous coal destroyed trees and ruined land. The increasing amounts of carbon dioxide in the atmosphere created global warming which melted the icecaps at the poles and produced more extreme weather conditions. Water shortages began to become apparent and some countries started to rely on aquifers which were not replenished, threatening water supplies for future generations. The sea became more acidic and fish stocks were threatened. Everywhere forests were being cut down and biodiversity was reducing.

To avoid impending disaster governments were encouraged by scientists to promote more **renewable sources of energy**. A new political movement was born, the **Green Party**, which had the aim of encouraging a more ecological approach to the environment. So far both initiatives have had a limited impact outside a few countries in Europe.

The major technological change came from the use of the **internet** and the availability of **mobile phones**. The internet fundamentally changed business communication. Written communication within the office environment could be exchanged in a matter of seconds. The days when secretaries typed out memos that were duly delivered by internal mail were now over. Managing their email inbox became a never-ending challenge for many managers. Search engines such as Google allowed companies to have almost instantaneous access to facts and documents. All companies now had web pages promoting their products. Goods were ordered over the internet using **credit cards**. Companies, such as Amazon and eBay, became huge concerns just on the basis of internet trading.

Mobile phones made the pace of life even faster. Everyone could now interact immediately with everyone else, not just those within hearing distance. Business and public life began to be conducted on a 24-hour basis. For those in authority or those who were the focus of media attention, there was no quiet time available to relax and reflect. Person to person communication, which could formerly be conducted only by direct speech or letter writing now involved messages, pictures and short videos. People set up their own web pages using Facebook or other proprietary software. The most important and famous could now communicate their thoughts and opinions directly with huge numbers of followers.

Internet

The internet is a system of linking computers which started to develop in late 1960s. Crucial to the technology was the communication protocol based on packet switching, developed in the UK by a Welsh scientist Donald Davies (1924–2000). In packet switching data is grouped into discrete units called packets and directed to a specific user by reference to a unique IP address. By this means a telephone line can be shared between many users, each receiving their own packets of information. The most common form of IP address, at present, is IPv4 which can cope with a maximum of 4.3 billion users. The scale of the exponential growth of the internet is shown by the fact that address exhaustion occurred in 2011 and the latest protocol IPv6 allows for the cube of 4.3 billion (80, followed by 27 zeros) users.

For packet switching networks, it is the bandwidth (measured in bits/second) that determines the capacity of the network. Along with improvements in computing power there has been a huge investment to increase bandwidth using optical fibre technology. Line capacity has been transformed; by 2010 it was possible to send huge amounts of data over the internet. It became practical for television companies to allow individual users to choose to view whichever programme they liked from a catalogue. Thus instead of just sending out one television programme at a time, thousands, even millions, could now be sent over the network.

The first use of the internet was for email. Companies established local area networks, enabling staff within their company to exchange memos electronically. It does not sound much of a change but it revolutionised office life. The typing pool disappeared, along with circularised paper memos, long queues at photocopiers and huge mailing rooms for sorting mail. Communication could now be stored electronically, reducing the need for filing cabinets. When software that could invite people to meetings electronically was introduced, the classic role of the secretary also disappeared. Office life could be managed directly from the manager's desk. Answering and sending emails came to dominate the working day. Taking a holiday meant facing an email overload on return.

The next stage in internet development was the establishment of the World Wide Web. Its hero is Tim Berners-Lee (1955–) from south-west London, who had the vision and organising genius to set up this wonderful information exchange system for the benefit of all. The basis of the web is that all information (documents, images or music) stored digitally has its own unique reference, or URL. For example, the URL for text called XXX in Wikipedia is http://en.wikipedia.org/wiki/XXX. Using the common communication protocol it is possible to access any document from anywhere. To connect to the World Wide Web, home users need an internet service provider and software known as a web browser. The result has been an explosion in information exchange, allowing not just knowledge transfer but revolutions in banking, shopping, trading and communications. Amazon has become a huge company selling directly to the public, starting with books and expanding to sell a large range of consumer items. eBay became the world's chosen host for buying and selling second-hand goods. Skype has become a favourite video conversation system, linking families across continents. The biggest company in this sector, however is not an internet service provider, trader or a retailer, but a company that just allows you to search the internet for web pages of interest – Google. Founded in 1998, this company has grown rapidly to become one of the largest in the world by selling targeted advertising space. As of 2013, Google was estimated to process six billion search requests every day.

Mobile phones

Mobile radio telephony has a long history going back to ship to shore radio and the bulky equipment used by signalmen in the Second World War. The modern mobile phone age did not really start until the 1990s.

Mobile phone signals are carried by a network of radio transmitters set in a hexagonal cellular arrangement using shortwave radio frequencies. The signals are not strong enough to be received outside the adjoining hexagonal cells and therefore the same radio frequencies can be used in another part of the network to communicate with different users. A small range of radio frequencies is therefore sufficient to support the whole network. Technology had to be developed to seamlessly support reception during the handover from one cell to another and to provide digital encryption of the signal to ensure security.

The first mobile phones were the size of bricks but as the networks became denser and battery technology improved, they reduced in size down to 100–200 g. Not many predicted the impact this new communication system would make. There are now roughly as many mobile phones as there are people in the world. Remote person to person communication became affordable for almost everyone, even peasant farmers living in mud huts in Africa.

New forms of communication evolved. Networks encouraged short text messages to be sent at very low prices and soon this became the dominant form of mobile phone communication. A whole new language of abbreviations was born as in: 'KNIM' – know what I mean – and new symbols of smiley faces and other emoticons. Mobile phones also doubled up as cameras and the resulting pictures could be sent digitally to others. Young men and women seemed never to tire of taking selfies.

The most profound change came with an extended concept of friendship. Whereas once it was necessary to see and talk directly to friends, new software, from companies like Facebook and Twitter, allowed users to communicate with an extended network. As of October 2012, Facebook claimed it had over a billion people using its software – a seventh of the world's population. This new social media involved sharing intimate details of experiences, thoughts and photographs with hundreds of 'friends'. The young particularly embraced this new technology; many now began to spend more time communicating through their mobile phone than face to face. Celebrities could update millions of fans on recent events by tweeting or posting.

Phone networks have been upgraded from second generation (2G) in the 1990s to up to 4G in the 2010s. In doing so they are integrating with internet technology. As a result it is now possible to view films on your mobile device. We are still at the beginning of its possibilities and I expect many unforeseen twists in the future as remote person-to-person communication grows apace.

Communism takes on capitalism

After the dissolution of the USSR in 1991, the President of the Russian Federation, Boris Yeltsin, took the decision to abandon Russia's socialist command economy and adopt the capitalist model. Acting on the advice of neo-conservatives in Washington, he adopted a radical, high-risk approach. At a stroke, in 1992 price and currency controls were abandoned and foreign trade was liberalised. At the same time he instigated a harsh regime intended to restrict inflation, cutting back on state subsidies to industry and welfare spending. He also raised interest rates to restrict credit. The result was one of the most disastrous economic experiments ever conducted. The old planning structures were abandoned before anything had been established to replace them. Price controls on staple foods like bread were abandoned. Unprofitable state organisations either never paid their employees or paid in kind. Hyperinflation wiped out personal savings. Tens of millions of people were plunged into poverty. Through the 1990s Russia's GDP fell by 50% and industry in many sectors of the economy ceased.

Shares in state enterprises were issued to the public. The most profitable of these were quickly taken up by members of the old nomenklatura (the Communist Party administrators). During the same period, violent criminal groups established capitalist enterprises, ensuring their success by corruption, extortion and assassination. Under a corrupt government outrageous financial manipulations were performed to enrich the narrow group of individuals in key positions in business, government and the Russian mafia. Many took billions in cash and assets outside of the country; in Britain so rich was the oligarch Roman Abramovich that he was able to indulge himself by buying Chelsea football club.

The manner of this implementation of capitalism, without any of the necessary changes in the legal system and any curtailment of corrupt practices, ensured that capitalism soon became a dirty word in Russia. Ordinary Russians yearned for the good old days of communism again, and in 2000 elected a strong man Vladimir Putin (1952–) as President. Putin was a former KGB operative and knew how to ensure that, once elected, he would maintain the reins of power. He was backed by friends from the nomenklatura and he rewarded these friends after his election by giving them powerful positions in government and industry. Putin was lucky. Russia had become reliant on selling commodities like oil and gas on the world market and when world prices increased he was able to stabilise the economy. He formed a party that won elections, fairly at first, then by vote rigging when he became less popular. The KGB came back under a new name, the FSB, to have a sinister influence on events and critics of the government died in mysterious circumstances. The judiciary remained a servant of the government's will and censorship returned. Russia was back where it started: a one-party state, a privileged and corrupt elite, a repressive government and limited individual freedoms. Communism as a philosophy had died but many of the same elite remained in power.

Meanwhile in China, the Communist party under Deng Xiaoping was directing change away from communism rather more successfully. In the countryside, a new, gritty and unglamorous form of capitalism was emerging. In 1984 a new form of business structure was created, known as a town and village enterprise (TVE). This structure had emerged from the old village collectives. But now restrictions on the enterprises were eased and TVEs could manufacture anything, not just farm equipment, they could sell outside their home districts and they were allowed to take loans from state banks. More and more TVEs came to be controlled by individuals or families, as opposed to collectives. They were a huge success. By 1992 there were 20 million TVEs employing 100 million people and increasing at 30% p.a. At the end of the decade these employees were producing a quarter of the goods in China. This new Chinese venture into capitalism was of a raw, cut-throat variety. It harked back to the robber baron era in the USA and the early days of industrialisation in the UK. The state provided little legal support for businesses: property rights were unclear, commercial accounting was weak and commercial law was hazy. Party cadres and local officials often demanded a bribe in return for no interference. There were no local courts of appeal. Tax evasion was prevalent, leaving businesses vulnerable to blackmail by tax authorities. Workers were exploited and health and safety legislation was weak and poorly enforced. The environment was polluted. A new, distinctly unsocialist middle class was emerging. Its members were single-minded, ruthless operators. They had contacts in the right places and were able to grease palms to exploit whatever market opportunity was available, with scant regard for the legalities. The Chinese Communist Party was encouraging a decidedly capitalist road to development.

In 1992 the State Council issued a decree designed to increase the freedom of state enterprises and at the same time to give greater power to the authorities to shut down loss-making concerns. Corruption continued to soar, particularly among the sons and families of state officials. Not to be outdone by the SEZs on the coast, the provinces started providing incentives for new industries to come to their towns. Capitalism took off and all over China young rural men and women flocked to the towns to find new jobs and prosperity. Huge mega-cities were created in the interior. The state invested heavily in infrastructure, in particular in roads, power stations and airports.

If you consider party cadres to be the equivalent of eighteenth century British aristocrats, the situation in China was the same as in Britain at the start of the Industrial Revolution. In the towns the rural men and women were paid low wages, worked long hours and lived in crowded accommodation. The towns grew dramatically and the rich became even wealthier. The only difference is the scale; China is a hundred times larger than Britain at the start of the nineteenth century and as a result the changes are a hundred times larger. Like Britain in the nineteenth century, China has become the manufacturing centre of the world, supplying a huge proportion of global demand for textiles, electronics and consumer goods. But the Chinese leadership can no longer claim any pretensions to socialist government. This is a modern one-party state in which the party cadres maintain their authority and privilege at the expense of the rest of society.

For all China's economic success, it has yet to develop a legal system that supports the development of responsible companies. Its version of capitalism is raw, brutal and exploitative. Honest trading is not an option; palms have to be greased and the desires of party officials have to be accommodated. Successful products are copied freely with no patent or brand protection. Counterfeit goods and medicines are freely available. Fake documents and identity cards are easily purchased. Few Chinese companies have managed to establish a brand that can be trusted globally. As a result the Chinese are not reaping the full benefit of their economic success. This is explained further in *China Shakes the World: A Titan's Rise and Troubled Future – and the Challenge for America*, pp. 160–1 by James Kynge:

> The manufacturer, which is in more and more cases a separate company contracted by the brand owner, gets a relative pittance for his work....This may seem unfair. Certainly, the effort expended by a typical migrant worker who is far from her village, working fifteen-hour days without welfare or union protection under a management regime that may use fines as its motivational strategy is not inconsiderable. But the harsh truth is that this effort is valued less by the consumer than some of the other functions that go into bringing the product to market.... The customer is the ultimate arbiter, and if customers in the USA and Europe think that the greatest proportion of a product's value derives from its logo, then most commercial rewards will continue to accrue to those who build and sustain the brand.

Multinational companies

The organisations that have gained most from the internet and the growth of global trade are multinational companies. They design the product to include all the latest technical innovations, source their raw materials internationally, manufacture their goods in the cheapest location, ensure quality by setting rigorous standards and market the product globally. The consumer across the globe benefits by receiving good quality products at affordable prices and, as owners of the brand, the expertise, power and profit of these multinationals continues to increase.

However the power of multinationals has now grown to such an extent that they have become independent of the states that originally harboured them. They are free to invest in, operate from, or trade with whichever country suits their needs best. They have changed the whole international competitive paradigm that existed over the past two centuries. States now compete to offer the best terms for multinationals to operate in their countries. This is most apparent in taxation. It is a company's mission to create money for their shareholders and this includes finding ways to pay as little tax as possible while trading legally. By arranging to trade their goods through low-tax countries, companies can minimise their tax burden. Often the flow of goods does not reflect the flow of invoices; goods will be invoiced via a low-tax country while making their way directly to the final destination. These tax reducing trades quickly become labyrinthine, making it very difficult for governments to discover what is going on. This is an example quoted from *The Guardian* on 30 May 2013, concerning a US Senate investigation into Apple, the leading electronics company, and its Irish operations. Apple had transferred the development rights of many of the group's products outside the Americas to Irish companies. One company, Apple Operations International (AOI) had accumulated $30 billion of profit over four years despite the fact that it had no physical presence or employees in Ireland or indeed anywhere else. The newspaper reported that:

> What surprised investigators was that at least three of these [Irish] companies, including AOI appeared to have no tax residency anywhere in the world. Their boards have been able to tell the Irish tax authorities that ... important decision-making rests in California. As a result, AOI and others are not deemed tax resident in Ireland. Meanwhile, because these same companies are incorporated at addresses in Ireland, under US law they appeared not to be tax resident in the US either.

Another example is Google, who also uses Irish and Bermudan companies to reduce its corporation tax. During a UK parliamentary investigation into the tax paid by Google in the UK it was established that Google generated £11.5 billion in revenue from the UK between 2006 and 2011, but paid just £10 million in corporation tax. The company's Vice President, Matt Brittin, insisted that Google did comply with UK law and had paid all taxes required. He maintained that anyone purchasing advertising from Google in Europe – including the UK – was buying it from Google Ireland, where all the company's sales outside the USA were billed. Ms Hodge, the committee's chairperson, said they were not singling out Google but believed that its tax avoidance activities were 'illustrative of a much wider problem' among multinationals in the globalised business environment.

Internet companies are particularly difficult to tie down for tax purposes. Amazon sells one in four books in the UK and is Britain's biggest online retailer, with sales of more than £3.3 billion, but currently pays no corporation tax in the UK on any of the profits. It does this by taking advantage of EU law and basing its company in Luxembourg. The point is that as long as states compete freely to attract multinationals, rather than cooperating among themselves to ensure these companies pay tax, the multinationals will always win out. Multinationals can afford the smartest accountants and lawyers and will always outgun states to exploit tax loopholes and pay as little tax as possible.

A more sinister result of multinational power is its ability to influence opinion, in both people and governments. Most media, television, newspapers, magazines and most internet services such as Facebook or Google are paid for by advertisers. This gives advertisers tremendous power to influence the media and, as a result, the public at large.

Large multinational companies are able to operate even more subtly in influencing opinion. They have the financial muscle to generate bogus studies, confuse arguments and generally stand in the way of logical debate to try to tip the balance of argument in a favourable direction. To quote a well-known example around about the turn of the century, ExxonMobil saw the developing scientific argument on climate change theory as a threat to its oil sales. This is an extract from *A Private Empire: ExxonMobil and American Power* by Steve Coll (2012), p. 184:

> Exxon Mobil had persistently funded a public policy campaign in Washington and elsewhere that was transparently designed to raise public scepticism about the science that identified fossil fuels as a cause of global warming. ExxonMobil ran some aspects of its campaign clandestinely; that is, it did not initially disclose the full scope and purpose of contributions it made. What distinguished the corporation's activity during the late 1990s and the first Bush term was the way it crossed into disinformation.

In America companies are legally allowed to directly influence legislators by making financial contributions to their party. In the 2010 decision in the case of Citizens United v Federal Election Commission, the Supreme Court ruled that corporate campaign funding cannot be limited, as this denies companies their rights to freedom of speech, as defined in the First Amendment. This extraordinary decision, essentially conceding corporations the same rights as ordinary citizens, gives large companies all the power they need to subvert democracy. Television advertising is a crucial part of campaigning for election in the USA and it costs a lot of money. Corporations duly support congressmen and senators who are likely to support their interests and expect a quid pro quo after the election. Al Gore writes in *The Future: Six Drivers of Global Change* (2013) pp. 104–5:

> It is now common for lawyers representing corporate lobbies to sit in the actual drafting sessions where legislation is written, and to provide the precise language for new laws intended to remove obstacles to their corporate business plans – usually by weakening provisions of existing laws and regulations intended to protect the public interest against documented excesses and abuses. Many US state legislatures often now routinely rubber-stamp laws that have

been written in their entirety by corporate lobbies.

> ... I have felt a sense of shock and dismay at how quickly the integrity and efficacy of American democracy has nearly collapsed. There have been other periods in American history when wealth and corporate power have dominated the operations of government, but there are reasons for concern that this may be more than a cyclical phenomenon – particularly recent court decisions that institutionalise the dominance and control of wealth and corporate power.

American corporate lobbies have enormous power to influence the government to maintain policies that are not for the overall good of the general public. The pharmaceutical lobby, for example, ensures that American prices for pharmaceuticals are the highest in the world. The CEOs of large multinationals also have enormous power to change and influence events. They meet regularly with state leaders. In 2001 the Indian Prime Minister asked President George W. Bush to speed up a deal between ExxonMobil and India's largest state-owned oil company. The President's response was telling: 'Nobody tells those guys what to do'. The fact that the president of the world's largest economy hesitates to issue instructions to multinationals is an indication of their power.

Multinational shareholders, usually financial institutions such as investment and hedge funds, invest in companies that will deliver increased profits and growth. Company success is highly dependent on the ability of its leadership team, particularly the CEO. Executives with a specific combination of vision, leadership and experience can earn billions of dollars for their companies. All companies are looking for the executives that will allow them to be successful and are prepared to pay extra to attract them. As long as the leadership team deliver the expected overall results the shareholders are largely unconcerned with how they are rewarded. From the 1980s onwards this pressure to attract the best talent combined with the lack of controls from the shareholders has led to a huge growth in executive pay. This developed first in the USA, was then taken up in Britain and spread to Europe. According to the American Economic Policy Institute, in 1978 CEOs pay was 26.5 times more than the average worker and by 2011 this had risen to 206 times. Executive remuneration has been growing many times faster than is justified by economic results and is creating a new class of super-rich in secular capitalist countries.

Banks

The fact that governments need to exercise greater control over multinational companies has been amply demonstrated in the great banking crisis of 2008. This was followed by a worldwide recession, from which secular capitalist countries are still trying to recover. Capitalism generally improves the lot of nations as long as the market place is freely open to competition, sufficient credit is available to support growth and business leaders are focused on improving their product. Capitalism does not work so well when it is focused purely on the short-term acquisition and manipulation of money, irrespective of the product and the market it is supporting. It is also particularly vulnerable to bubbles, where over-investment in new ideas leads to a collapse of the market, as in the dot-com bubble of 1997–2000. The banks were affected by both problems in the early part of the present day.

The end of the previous century had seen controls on banks relaxed and the international movement of money was made much easier. At the same time the development of tax havens made it simpler to hide the movements of money from governments. Tax havens provide anonymous and secure banking at a low tax rate for individuals and companies. As such, they are also interesting to crooks seeking to launder money and to corrupt government officials seeking to hide their illicit gains. These are places where the local political system is dominated by the financial services sector and therefore, there is little risk that local politics will interfere with the business of making money. Hence they have traditionally been small countries like Switzerland or Liechtenstein or the self-governing colonies of a larger power like the Cayman Islands or Jersey. Relinquishing control on the movement of money in secular capitalist countries in the 1980s led to a huge increase in the use of tax havens. This is a quote from *Treasure Islands*, p. 8 by Nicholas Shaxson (2011):

> More than half of world trade passes, at least on paper, through tax havens. Over half of all banking assets and a third of all direct investment by multinational corporations are routed offshore.... The IMF estimates in 2010 that the balance sheets of small island financial centres alone added up to $18 trillion – a sum equivalent to about a third of the world's GDP. And that, it said, was an under estimate. The US Government Accountability Office reported in 2008 that 83 of the USA's biggest 100 corporations had subsidiaries in tax havens. The following year research by the Tax Justice Network, using a broader definition of offshore, discovered that 99 of Europe's hundred largest companies used offshore subsidiaries. In each country the largest user by far was the bank.

Tax havens are not isolated from the world as a whole; they are fully integrated by the banks into an international network of cash flows. There are broadly three important groups of tax havens; British, European and American. The centre of the British system is the City of London, which acts like a spider in the centre of its web, with three layers of tax havens around it with progressively less tight controls. The first layer is the islands around Britain, such as Jersey and the Isle of Man; the next layer consists of the British Overseas Territories such as the Cayman Islands; and the outer layer includes ex-colonies like Hong Kong or the Bahamas. The City of London, right at the heart of the British establishment, has a vested interest in maintaining the status of tax havens.

European tax havens include Switzerland (the most famous), Luxembourg, the Netherlands and Liechtenstein. There are three levels of US tax havens. At the federal level the US government allows foreigners to deposit their money anonymously; US banks may legally accept proceeds of crime as long as the crimes are committed elsewhere. US states also act as tax havens; Florida has a long history of harbouring money from drug cartels. The third level is the offshore islands like the Marshall Islands.

With the relaxing of currency controls, the old image of the high street banks was totally transformed. Banks now controlled a huge network of cash flows from all over the world. Hitherto relatively traditional establishments, banks now began to attract ambitious businessmen who were paid enormous salaries and gained spectacular bonuses whenever their investments paid off. Caution was thrown to the wind; banks that should have controlled and managed the risk of their investments were now incentivised to seek short-term gain and were able to use ordinary citizen's money for speculative investments.

It took only one bubble and the whole banking structure of secular capitalist countries was under threat. The bubble in this case was the US mortgage market. Mortgages were offered in large numbers to low-income families or those with a poor credit history. Banks found a way of selling on these mortgages as bundles of debts, a mixture of secure loans and these risky mortgages. These bundles of debt were accorded an AAA (top) rated credit risk status. To quote from *The Divide: American Injustice in the Age of the Wealth Gap* by Matt Tabbi, p. 39:

> It was a modern take on the Rumpelstiltskin fairy tale. Big banks took great masses of straw (i.e. risky home loans to the poor, undocumented and unemployed) and spun it, factory style into gold (i.e. AAA-rated securities).

Soon, mortgage-backed securities accounted for a third of the whole US

$27 trillion bond market – and of this, at the end of 2007 $1.3 trillion was sub-prime, that is sold to the poor. Rising interest rates burst the bubble. Not surprisingly, the poor could not pay the higher mortgage rates and their houses were repossessed. House prices fell by 25% from the peak in July 2006 to the financial crisis in the autumn of 2008. As a result the value of the mortgage loan fell by a similar amount.

This need not have had a negative effect outside the USA if it had not been for the practice known as leverage. This where debt is used as security to borrow more money and is described in *The Storm: The World Economic Crisis and What it Means* by Vince Cable p. 34:

> In effect, institutions borrowed money in order to buy debt, which was the security for the borrowing, and the money they borrowed was in turn borrowed, sometimes through several institutions. In addition, debt default could be insured against, but the insurers depended in turn on borrowed capital. Derivatives markets also made it possible to hedge (or speculate) against the risk of default. The credit default swap market, for example, which grew on the back of the growth of these debt instruments, achieved a notional value of over $60 trillion. This, in turn, represented about one tenth of the overall size of derivatives markets, which Warren Buffet warned us was the H-bomb to follow the sub prime A-bomb.

By the middle of 2007 it was apparent that serious losses were being made with sub-prime mortgages. The first effects were felt among the hedge funds and other city traders who were highly leveraged. But soon banks themselves began to question the value of their assets. They hoarded their cash and ceased to offer new loans. Banks began to fail and it looked as if the collapse of Western capitalism, as in the Great Depression, was about to happen again.

In the event the banks were rescued by governments, who in turn brought greater debt on themselves. It was touch and go, but sufficient liquidity was maintained in the system to allow capitalism to function with only a modest fall in demand.

This combination of greed and incompetence by banks almost brought capitalism to its knees. It was apparent to all that there must be something fundamentally wrong with the banking system. Many commentators believe it is, in part, due to the reward system. Since the 1980s bankers have been rewarded with massive bonuses when their investments went up and not personally penalised when their investments went down. This changed banking from the traditional risk-adverse approach to that of always looking for a fast buck. Despite government exhortation and public disapproval there is no sign of this lopsided reward system changing. The free movement of capital, exploited by banks and tax havens and supported by wealthy individuals and businesses, continues to undermine and put at risk the economies of secular capitalist countries.

Global industrial society

The newly created global economy changed the competitive nature of human society yet again. This was due to the increasing importance of multinational companies and the growing wealth of elites. I have already discussed the rise in power of the multinationals. Rather than cooperating to monitor and control multinationals, secular capitalist countries appeared to be competing with each other to offer them the most attractive business terms. Multinationals are therefore shown as another group of organisations on the same level as states in the evolutionary structure of the global industrial society. This change is shown in Figure 12.1.

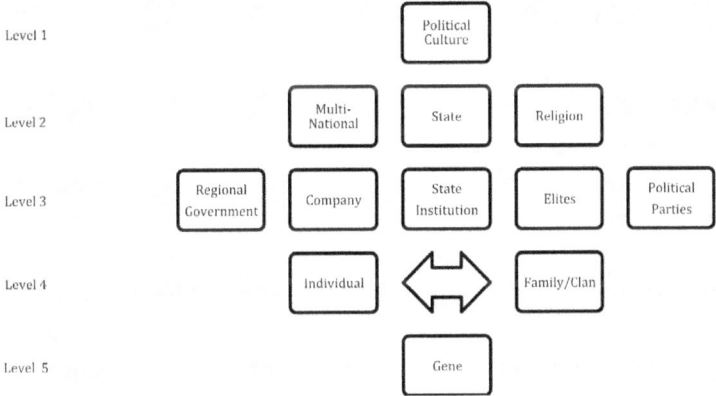

Figure 12.1 Evolutionary structure of global industrial society

On the third level of the hierarchy, class had become increasingly less important since the Second World War. Thomas Picketty, in his book *Capital in the Twenty First Century*, has shown that in the major countries of the world the period from 1950 to 1990 was the most egalitarian period in terms of income and wealth in at least the last two centuries. Indeed, it is probably the first time this level of equality had been achieved since the dawn of civilisation. Education reforms since the war in secular capitalist countries extended state education to 16–18-year olds and, irrespective of their class, most of the brightest children were able to attain a university education. Individuals from any class could now be successful at the highest level. The incomes of the top 1% of earners in secular capitalist countries declined from around 20% of national income before the First World War to 5 to 10% by 1990. The institutions concerned with class issues had declined in power. The aristocracy no longer had privileged rights in society. Trades unions, the traditional representatives of the working class, had become less important. The relationship between political parties and class had become more blurred. Class divisions, which had been present from the time the first states were formed, were gradually losing their competitive importance.

However with the growth of the capitalist culture since 1990 inequality began to grow again. Business leaders' salaries increased at a level far beyond those of the average worker. Sports or entertainment stars amassed huge incomes. The party cadres in China and Russia joined the ranks of the fabulously rich. These new elites were wealthy beyond the conception of ordinary people. They indulged themselves by buying multi-million pound yachts, football clubs and stables of racing horses. Those with a conscience set up huge charitable foundations. This change in incomes was most apparent in the USA; the top 1% of earners increased their share of national income from 9% in 1990 to almost 20% by 2010. All countries, however, showed the same trend.

These new elites, defined by their wealth, often occupied positions of power. Because of their wealth, states competed to attract their presence by lowering taxes and offering generous tax residency agreements. Many were business leaders or politicians. As such they were in a position to influence political developments for their own benefit. This is most apparent in the USA, where the wealthy have managed to occupy the leading positions in the Republican Party. Tax laws have been changed to benefit them and this is one of the causes of the leap in wealth of those of the highest incomes in the USA.

We are only at the beginning of this development. Elites are able to hide their wealth in offshore tax havens and pass it onto their kin, avoiding inheritance tax. The next generations of the elites will have fabulous incomes from their inheritance. The degree of social mobility, which is a measure of equality of opportunity, is on the decline especially in the USA. Money still buys a better education. Young people attending private schools in Britain still have much greater access to the best universities and the cost of the best universities in the USA is beyond the means of many Americans. Inherited wealth will give the offspring of the current elite privileged access to housing, business opportunities and influence on those in power. It is possible the inequality levels seen before the First World War when aristocrats and rich businessmen owned most of a country's wealth will return again.

To reflect this I have changed the title of the class entry on the competitive hierarchy to elites. This is a reflection of the fact that in the decades to come there will be a small proportion of the population determined to maintain their privileged status over the rest.

State competition

Economies right across the globe shed their old restrictions and liberalised manufacturing and trade. The result has been an explosion in trade and wealth spread across all continents. Figure 12.2 is drawn from UN statistics, showing a global volume trade index in which 1980 = 100. It shows global trade volumes grew almost 5% p.a. between 1980 and 2011. In particular, the volume of trade increased by a factor of three between 1990 and 2010.

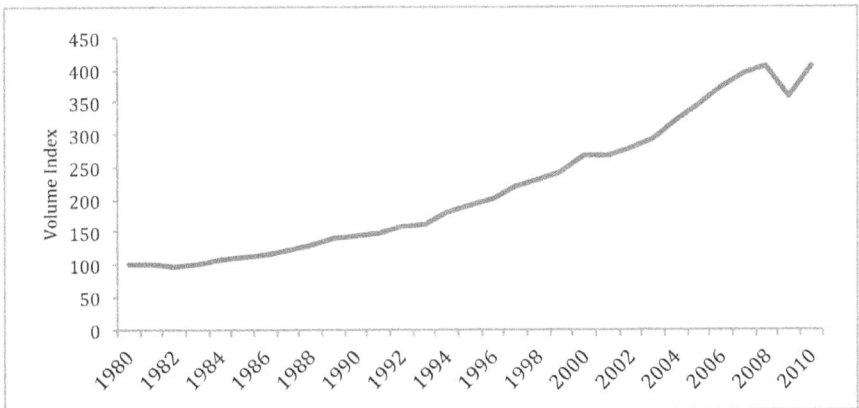

Figure 12.2 World trade volume index 1980–2010

Source: UN

Wealth spread outside secular capitalist countries. Figure 12.3 shows that the percentage shares of GDP of all the older developed countries fell for the first time. Chinese GDP showed a spectacular increase and it was now the second largest economy in the world. India showed a more modest increase. The rest of the world, led by countries in Latin America and south-east Asia, also increased their share of the world's GDP.

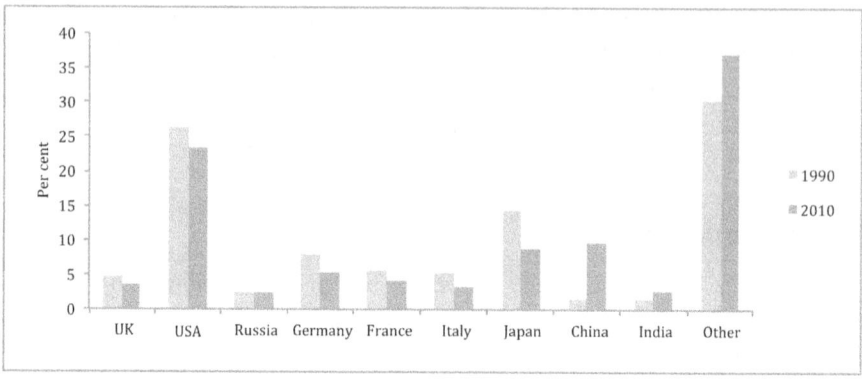

Figure 12.3 Share of world GDP 1990–2010

The era when a few countries dominated the world has clearly come to an end. The benefits of capitalism and technology have begun to spread more widely across the globe. However, the faster growth rates in the developing world are partly due to higher population growth rates. In terms of wealth per capita the average person in secular capitalist countries is still several times richer than the rest of the world. In the analysis below I have compared the average individual wealth of particular countries with that of the USA. Figure 12.4 shows that all the major secular capitalist countries had a GDP per capita at least 70% of that of the USA. Russia was at 24%, China at 9% and India at 3%. The other countries combined were at 14% of the US levels. The significant factor, however, is that many of these other countries have improved their relative wealth in the 20 years since 1990. In the analysis below I examine this change in more detail according to their political culture.

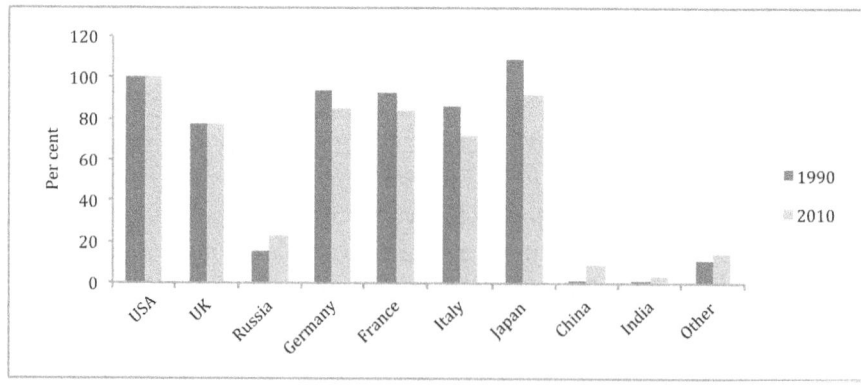

Figure 12.4 GDP per capita as a percentage of USA

The secular capitalist political culture continued to gain adherents. There still had been no wars between secular capitalist states. Competition between these states continued to be expressed practically in economic terms and emotionally on the sports field and in the sports stadia.

Ten East European states had made the transition from communism to secular capitalism in a very short period. They joined the EU and, supported by EU transitional funding, have undertaken the huge task of transforming their economies. There had been many difficulties in making the change and several countries are still struggling to meet the economic expectations of a Western-style standard of living, still being at only 14–49% of US standards (Table12.1). Nevertheless, this period had seen a step change since the communist years in the relative wealth and prospects of the citizens of East Europe.

	GDP/capita (% of USA)		
Country	1990	2010	increase %
Slovenia	38	49	30
Czech Republic	16	40	147
Slovak Republic	10	34	258
Estonia	?	30	?
Hungary	14	28	99
Poland	7	26	258
Lithuania	12	24	92
Latvia	12	23	89
Romania	7	16	125
Bulgaria	10	14	31

Table 12.1 Changes in relative GDP per capita of East European states

Although only North Korea survived as an old-style communist country, an alternative to secular capitalism had emerged in Russia, China and Vietnam that I shall call elite capitalism. These one-party states had embraced capitalism, while ensuring its elites remained privileged. The ruling parties maintained power by decree or, if necessary, by rigging elections. Corruption was rife and the police and courts were controlled by the government. However, individual capitalist initiative was encouraged, state control of industry was being phased out and personal wealth was increasing. It is interesting that, despite the differences in political culture, that the evolutionary structure of the ex-communist countries and the West was beginning to converge again. Both societies now have elites who form a distinct community in the evolutionary structure.

All elite capitalist states had invested heavily in infrastructure. Table 12.2 shows how this change in political philosophy had transformed the economic performance of these countries. China and Vietnam showed spectacular growth during this period, albeit from a low level of wealth. Although the secular capitalist philosophy was still the one most commonly aspired to on the world stage, the economic development of China was a sign that other political cultures might also be successful, as long as they embraced international capitalism.

Country	GDP/capita (% of USA)		
	1990	2010	increase %
China	1	9	596
Russia	15	22	48
Vietnam	0	3	516

Table 12.2 Changes in GDP per capita for elite capitalist countries

Several states were governed in a democratic style but still fell short of attaining the full set of secular capitalist values. I have termed these aspiring democracies. To focus the discussion, I have concentrated only on those countries with a population of over 20 million people. Table 12.3 shows the considerable advance made by these countries as they discarded attempts to develop state-directed economies and allowed capitalism to take its course. Corruption and excessive bureaucracy were, however, still widespread, hindering economic development, and religious issues still had a considerable bearing on the management of these states. Overall, however, their economic growth had been impressive, helped by the fact there had been no significant wars between these states.

Country	GDP/capita (% of USA)		
	1990	2010	increase %
Venezuela	10	29	183
Brazil	13	24	76
Turkey	12	22	78
Mexico	14	20	45
Argentina	19	20	4
Malaysia	10	18	71
South Africa	14	16	13
Colombia	5	13	154
Peru	5	11	115
Thailand	6	10	52
Ukraine	7	6	-7
Indonesia	3	6	134
Morocco	4	6	33
Sri Lanka	2	5	151
Philippines	3	5	47
India	2	3	81

Table 12.3 GDP per capita of aspiring democracies

All these aspiring democracies had managed a dramatic increase in economic performance except Argentina (after a debt default in 2001), South Africa and the Ukraine.

Brazil had become the sixth largest economy in the world in 2010. Its growth potential was finally being fulfilled after years of corruption, incompetent government and rampant inflation. From 1994 the politicians had largely kept inflation under control. Since 2002, under the popular President Luis da Silva, high levels of economic growth were achieved and after a history of being a debtor nation, in 2008, Brazil became a net creditor nation for the first time.

The growth in Malaysia and Indonesia had shown that it was possible to have a thriving economy in countries that adhere to the Islamic religion. Turkey, where the Islamic religion is still strong, had now reached the same level of GDP per capita as Russia.

India remained the poorest of these aspiring democracies, but with its inevitable population increase will be the most populous nation on Earth and will always be a global power. The demise of the communist party in Russia was a profound shock to its major trading partner India. After independence and decades of socialist-inspired bureaucratic rule, India at last initiated an economic liberalisation programme in 1991. This reduced tariffs and interest rates, ended many public monopolies and allowed the automatic approval of direct foreign investment in many sectors. With its educated, English-speaking middle class, India has made a global impact in the service sector, particularly in IT and call centres. Latterly, too, the pace of industrialisation has picked up with major companies such as Tata becoming global players. However, India has not invested in infrastructure in the same way as China and this is limiting its growth potential.

The newly formed Ukrainian state had found the going very tough, with a divided country, split between Russian and Ukrainian speakers, who favour different pro-Western or pro-Russian developments for their countries future.

South Africa became a full democracy in 1994, under its inspirational leader Nelson Mandela. Mandela preached reconciliation between whites and blacks and was largely successful at launching South Africa down a democratic path with black majority rule. Economically, however, the results have been disappointing. Although South Africa remains the most powerful economy in sub-Saharan Africa, it has failed to develop at a sufficient pace to match the expectations of its citizens.

The rest of the world continued to develop a variety of political cultures. There are four major areas: Islamic countries in the Middle East and North Africa, sub-Saharan Africa, countries from the former USSR, and south Asia. By and large these are poor countries with an unstable system of government. It is in these countries that rebellion and war have disrupted economic progress. Table 12.4 shows that, with a few notable exceptions, these countries had disappointing levels of growth during this period.

Country	GDP/capita (% of USA)		
	1990	2010	increase %
Saudi Arabia	31	35	12
Algeria	11	10	-8
Syria	4	6	43
Egypt	3	6	75
Iraq	?	5	?
Sudan	2	3	62
Uzbekistan	3	3	4
Ghana	2	3	63
Nigeria	1	3	110
Cote d'Ivoire	4	2	-34
Pakistan	2	2	40
Kenya	2	2	7
Bangladesh	1	1	16
Nepal	1	1	39
Tanzania	1	1	51
Uganda	1	1	5
Madagascar	1	1	-24
Mozambique	1	1	7
Ethiopia	1	1	-29
Congo, Dem. Rep.	1	0	-62
Iran	9	?	?

Table 12.4 GDP per capita of other countries

The clear exception to this picture of poverty is in the oil-rich Islamic states that were still ruled by sultans, emirs and kings. They continued to thrive; their enormously wealthy elite maintained a very traditional Islamic culture, where women continued their traditional subservient role. Saudi Arabia, the Arab Emirates, Qatar, Brunei and Kuwait remained an important source of oil to the world with an apparently stable political system.

Iran continued to be the only Islamic theocracy. However, as a country, it was failing to keep up with the economic progress of the rest of the world.

By far the most troubled region had been the states of North Africa, Middle East and through to Pakistan, the majority of whose population are Muslims. In this area the local religious traditions still had an enormous influence on people's lives. This had resulted in a culture of resistance to change and restrictions to individual initiative. Only Egypt and Syria had grown significantly during this period and they have since been knocked back by revolution. The so-called secular leadership of Iraq, Syria and Egypt had failed to deliver the growth expected by their peoples due to corruption, bureaucratic controls and the inefficiencies of their state-run industrial organisations. No country had achieved sustained effective democratic government over a long period. Often regime change was only possible by violence. There were two strands to the opposition, one of which wanted to pursue the secular capitalist approach, the other to return to Islamic cultural precepts. By and large the majority are profoundly conservative and the Islamic approach had been more popular. The secular capitalist values of the West has been seen by some as deeply threatening to their culture and religion and led to violence against the West in the form of terrorist attacks, most spectacularly with the demolition of the World Trade Centre towers in New York in 2001. Violence against the West, however, had proved relatively minor, compared to the wars among adherents of Islam in the region itself. Iraq, Syria, Lebanon and Afghanistan had all suffered greatly from invasion and civil wars. The centuries-old antagonism between Shi'a and Sunni had fuelled the flames, with Saudi Arabia and other oil-rich states supporting the Sunni cause and Iran that of the Shi'a. Exacerbating the regions' misery was Israel's success in establishing a nation-state by occupying Palestine. The continued unresolved issue of the displaced Islamic Palestinian community is a continuing threat to the region's stability. Taken together, the whole region had failed to move forward like the rest of the world in the capitalist boom post-1990.

At long last some countries in Africa were showing signs of economic growth, notably Nigeria and Ghana, but overall the region was still struggling to develop a consensual government system that treats all their diverse tribal communities fairly. Across the continent corruption was endemic. Political leadership was associated with promoting a particular kin group. Population growth rates remained high, placing ever-increasing pressure on the environment. The poorest countries in the world were all in Africa. States like Somalia, Yemen and now the Republic of South Sudan had weak central governmental control, with regions ruled by a number of local warlords and their militias.

Environment

Thomas Malthus postulated that population growth will always be checked by war, famine or disease. In this he has so far been proved totally wrong. Wars have not halted human progress, food supply has kept pace with population growth and life expectancy has increased dramatically.

However, there are other checks on the increasing human population of the planet that are not so easily overcome. On land, in the sea and in the air there are signs that the ruthless exploitation of the natural resources of the Earth by humans is beginning to reach its limits. The remaining areas of wild forests and savannah are diminishing rapidly, the waste products of industry are beginning to have an adverse effect on the environment and critical deposits of naturally formed resources are diminishing.

Around 10,000 B.C., when all humans were hunter–gatherers, the world population was approximately a million. Today, the ability of humans to live off the land by hunter–gathering alone is limited to a few surviving tribes in very remote forests and deserts. However, humans still hunt in the oceans; fish provide a significant proportion of human food. We are reaching the limits of what the oceans can provide. Europe has had fishing quotas in its waters since 1983. Recently there have been many instances of over-fishing in other areas, causing massive depletions of fish stocks and thus removing these fish as sources of food for future generations. The Peruvian coastal anchovy fisheries and the Californian sardine industry have collapsed several times during the second part of the twentieth century. In the 1980s the blue walleye fish became extinct in the Great Lakes in North America. In 1992 the Canadian government imposed an indefinite moratorium on fishing cod off the Grand Banks near Newfoundland. With improved methods of electronic detection and fishing technology, many more species of fish are at risk, including the Atlantic blue fin tuna. If humans lived by eating marine wildlife alone we would have probably reached the Malthusian limit.

Before the technological revolution, when most people were farmers the Earth supported a population of around 900 million. Today, the Earth provides food for 7 billion people, less than half of whom are directly involved in farming. This additional food supply has been due to two factors. Firstly, the area of land put to arable use has increased with the destruction of forests and conversion of pasture. Secondly, farming efficiency has increased dramatically. Selective breeding has led to much improved yields; a more scientific approach, including the better use of fertilisers, herbicides and pesticides, has reduced crop losses; and improved powered machinery has greatly increased labour effectiveness.

From the time of the Neolithic revolution mankind has been destroying forests. In the early centres of civilisation (southern Europe, Middle East, parts of India and China) forests have long ago been cleared and replaced by agricultural fields. Even in the Bronze Age there was a shortage of trees in the Middle East, and one of the reasons China failed to develop its maritime fleet, after its successful voyages in the fifteenth century, was a shortage of wood for the construction of ships. As the population has grown further, the dense rain forests of South America, Africa and south-east Asia have been cut down. In these areas, loggers profit by supplying hardwood. After the land is cleared, it is farmed. Rainforests once covered 14% of the world's surface area; today they cover a mere 6%. Some experts say that rainforests could disappear within as little as 40 years.

The destruction of the rainforest has many negative effects. Rainforests are so dense that they contain a large proportion of the world's natural carbon: it has been estimated that one-quarter of the world's carbon stored in plants and animals is held in the Amazon rainforest alone. Thus, burning the rainforest has a large effect on atmospheric carbon levels. The water cycle is also adversely affected. Trees extract groundwater through their roots and release it into the atmosphere. When a forest is cleared the climate becomes drier. As result, soil cohesion is reduced, and this leads to the risk of soil erosion and the loss of arable land.

Declining populations of mammals such as the jaguar, gorilla, lemur, orang-utan and the giant panda have become symbols of the loss of forest habitats in their regions. These are just a few of the vast potential losses of plant and animal diversity. The Amazon rainforest alone contains the largest collection of species of plants and animals in the world; some say it accounts for as much as half. The rainforest is also a natural source of plants that have potential pharmaceutical uses. To give just one example, the native Indians in Peru have long used the plant they call para para, (translated as rise up – rise up) to increase male arousal. Only in modern times was it discovered to contain the same chemical as that used in Viagra.

In addition to converting the forests to arable land, humans have been attempting to farm dry marginal lands. This has resulted in turning the land to dust in a phenomenon known as desertification. About 28% of China is covered by desert and it is expanding rapidly. One of the main causes was Mao's plan to raise grain in areas only previously used for grazing animals, particularly in Inner Mongolia. This deprived the land of the grass that prevented soil from being blown away. The effect has been to destroy the livelihood of both pastoralists and farmers and to create huge dust storms polluting cities and clogging up drainage systems.

Another area at great risk is the Sahel, a huge zone that extends across Africa to the south of the Sahara desert from Mauritania in the west to Eritrea in the east. The Sahel consists of a natural green belt that protects farmers to the south from the Sahara. The area used to be rich in species and hunting was the main source of food. The vegetation protects the environment so that little wind or water erosion occurs. Over the past hundred years, expanding populations and more intensive farming have turned some areas in the Sahel to desert. This has not been caused by the extension of sand dunes from the Sahara but from the slashing and burning of natural forest and bush in order to clear land. A few months after harvest, farmers cut the millet stalks and burn them, leaving their fields exposed to strong winds until the next sowing season. These winds blow away the top soil and uproot seedlings; new plants are suffocated by dust.

As of the year 2000, about 37% of the land on Earth had an agricultural use. About one-third of this area, or 11% of the total, is used for crops. The balance is pastureland, which includes cultivated or wild forage crops for animals and open land used for grazing.

In this 11% of the Earth's surface, farmers are trying to produce a monoculture, where just one crop is produced in a given area. Monocultures do not occur naturally and need constant human intervention to be successful, otherwise the quality of the soil deteriorates and pests and diseases thrive, dramatically reducing crop yields. The most common interventions are the addition of fertilisers to the soil and the use of pesticides and herbicides. The use of fertiliser has increased 500% in the last 50 years. Of the 17 critical nutrients for plant growth, nitrogen and phosphates have been particularly important. However, the use of fertiliser has its environmental risks, as not all of the fertiliser is taken up by the plants. The excess runs off the fields into the rivers, polluting the water sources and the oceans. This surplus fertiliser causes an explosion in the growth of algae and waterborne plants. The algae can use up all the oxygen dissolved in the water as they decompose, suffocating fish and other animals. This creates dead zones where no sizable life-forms can survive. The most notorious dead zone is in the Gulf of Mexico, where the Mississippi River that drains large areas of American farmland meets the sea.

The development of herbicides and pesticides is a never-ending competitive evolutionary battle with natural organisms. Life forms mutate and produce resistant strains which become increasingly successful and can only be overcome by developing more potent chemicals. The difficulty is that it is hard to predict all the effects on nature of new herbicides and pesticides. True, we have come a long way since the effects of the use of DDT was exposed by Rachel Carson in *The Silent Spring*. In her book she argued that DDT was poisoning wildlife and the environment and endangering human health. The use of DDT was subsequently banned across the world. However fully investigating the effects of any new chemical is extremely difficult. A case in point is the recent EU ban on the use of neonicotinoids, used for treatment of seeds and foliage as well as for making soil more fertile. These pose 'high acute risks' for bees, according to a scientific report by the European Food Safety Authority. Bees are pollinators and are vital to plant growth. In 2012 there was a reduction in the number of bee colonies in the USA due to a mysterious disease. No one has conclusively proved that this disease is linked to neonicotinoids; the USA continues to take the risk, while Europe has chosen to be more cautious.

The effects of monocultural farming and the lack of pollinators are vividly demonstrated in the production of almonds in the USA. Almonds are harvested by shaking the tree and picking up the nuts by a suction pump. For this system to work, the area beneath the trees must be barren. The only crop or plant of any sort for miles and miles is almond trees. As a result, bees and any other pollinators cannot survive in this area. However, it is vital that plants are pollinated, otherwise millions of dollars will be lost in almond yields. As a result an industry has grown up to supply bees on demand. In the spring when the trees are in blossom, according to Tony Juniper in *What Has Nature Ever Done for Us: How Money Really Does Grow On Trees*, pp. 113–14:

> Over a million hives are required to pollinate the almond trees in California's Central Valley. They are trucked in from all over the USA to join this annual pollen-fest. To get ready for the great event beekeepers move their colonies into staging areas close to the almond orchards, and then as the blossom breaks, the bees are moved among the trees in an operation mindful of a military manoeuvre.... When the almond flowers have withered and with the seeds set and ready to grow, many of the bee colonies depart for the north, to Montana, North Dakota and to Oregon and Washington. Spring comes a little later there and the bees arrive in time to help pollinate the cherry, apple and pear orchards.

These problems of over-farming and over-fishing have had a deleterious effect on the environment but it is the effects of industrialisation that are of the greatest immediate concern to the future of mankind. From 1800 to 2010 the world's population has increased seven and a half times. People need both food and water and the supply of water is becoming ever scarcer in some areas. Less than 1% of the water on Earth is fresh. This evaporates from the seas and oceans, falls as rain and is available for use in as freshwater in streams, rivers, lakes and underground aquifers. Increases in population, leading to greater demand for water, is depleting stocks of water in dry areas. North Africa and the Middle East are areas where this is a major problem. In Yemen 19 of the country's 21 main aquifers are not being replenished naturally. Jordan has a major problem with water supply; and so little water is now reaching the Dead Sea that its water level is dropping by a metre per year. In addition, six more countries in these areas consume more water than they receive by nature: Libya, Algeria, Tunisia, Morocco, Iraq and Iran.

Only a small amount of the water usage is for human consumption; the largest uses are for irrigation and industry. Egypt is dependent on the Nile to irrigate crops to feed its expanding population. However, further upstream, Uganda, Ethiopia, Sudan and South Sudan also have an expanding population and a resulting need for water. There may well be insufficient water to satisfy the future requirements of all these countries.

Spreading urbanisation is also causing an increased demand for clean water in cities. All major cities in the developing world have problems in getting clean water supplies to their rapidly growing urban populations. Mexico City is one example. It was built on a great lake where the Aztecs founded their island citadel Tenochtitlan in 1325. When the Spanish conquerors took control they drained much of the water, laying the basis for the vast expansion of the metropolis across the entire Valley of Mexico. However, as the growing population continues to suck water out of wells, Mexico City is sinking down into the old lake bed at a rate of about 3 inches a year. This downward plunge puts extra pressure on water distribution pipes, which now leak so much they lose about 40% of the supply. In 2009, as a result of a drought the city authorities were forced to cut off the water supply to areas of the city in rotation, leaving over 20 million people short of water.

In rural areas improved access to water free from harmful microorganisms has been an objective of the World Health Organization since its inception. With the support of charities much good work has been done, particularly in Africa. It is estimated that the percentage of people in the developing world with access to safe drinking water has increased from 30% in 1980 to 84% in 2004. However, access to water does not necessarily mean easy access. In Africa many people (usually women and children) travel several miles every day to collect water.

Providing sufficient food and water for the burgeoning population is one problem of industrial societies; coping with its waste products is just as much a challenge. On land, sea and in the air industrial pollution is now a major problem. The most worrying environmental issue is the warming of the Earth's atmosphere. A proportion of the Earth's atmosphere contains gases that have the property of being able to absorb infra-red radiation and then emit it in all directions. Known as greenhouse gases, they ensure that the energy radiating from the earth's surface does not escape directly into space but heats the atmosphere. Because of this greenhouse effect, the mean temperature of the air around the Earth is 15°C; otherwise it would be 33°C colder. The principal greenhouse gases involved are carbon dioxide and water vapour. Water vapour content varies enormously according to local weather patterns; higher levels of water vapour in the atmosphere magnify the effect of higher temperatures, causing more extreme storm conditions. The levels of carbon dioxide in the atmosphere give most cause for concern. They have risen dramatically since the rate of burning of fossil fuels increased after industrialisation; this rate has accelerated since 1950 and now has reached alarming levels. We have data on atmospheric carbon dioxide going back 400,000 years taken from samples of the polar ice caps. This shows carbon dioxide concentrations varying from 180 to 300 ppm; low levels of carbon dioxide concentrations coincide with the ice ages and higher levels correspond to warmer, interglacial periods. At the time of the last interglacial period, 120,000 years ago, carbon dioxide levels peaked at 280 ppm. Charles Keeling (1928–2005), an American from Pennsylvania, first alerted the world to the issue of the warming of the atmosphere due to greenhouse gases. He started measuring carbon dioxide levels in the Mauna Loa Observatory in Hawaii in 1958. The results are given in Figure 12.5 and show the monthly records and the overall trend.

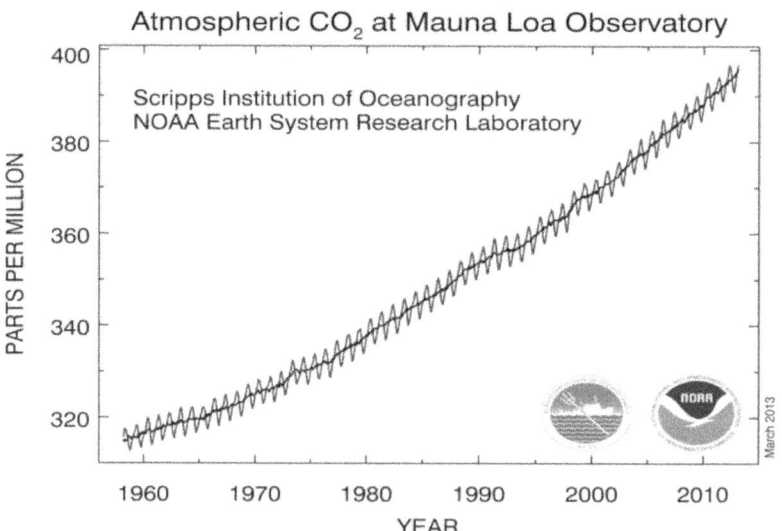

Figure 12.5 The rise in carbon dioxide levels (1958–2010)

Carbon dioxide concentrations are now approaching 400 ppm, having never been above 300 ppm in the last 400,000 years. It is now the accepted view by reputable scientists that global warming caused by this additional carbon dioxide has started; average temperatures have risen 0.6°C since 1950. The resulting ice melt has reduced the extent of glaciers in mountainous areas and the icecaps at the North and South Pole. Higher temperatures have also increased the incidence of more extreme weather events. It is clear that the number of floods, droughts and forest fires have increased dramatically since the war. Examples are given by Stephen Emmott in his book *10 Billion*, which shows that major floods in Asia rose from 50 per decade in the 1950s to nearly 700 in the first decade of the third millennium. In the same period the number of major forest fires in the Americas has increased from two per decade to 90.

Carbon dioxide pollution is cumulative; what is in the air now will be there in the decades to come. Unless there is some change in the method of energy creation, as global industrialisation increases so will carbon dioxide pollution. Global temperatures will rise further.

With more carbon dioxide in the air, more is dissolved in the seas, leading to an increase in carbonic acid in the marine environment. In fact, 30% of the carbon dioxide produced since industrialisation has been absorbed by the oceans in the form of carbonic acid. As acidity increases, the waters become more corrosive to the shells of animals. Corals, mussels and oysters are at a particular immediate risk. Plankton, the bedrock of the oceanic food chain, is also threatened. Photosynthesising plankton are the most important plant food source in the world; all fish ultimately depend on them for food. Forms of plankton with a calcium carbonate structure are vulnerable to acidification. According to Mark Lynas in *Six Degrees: Our Future on a Hotter Planet*, p. 56, in 2006 scientists reported a decline in plankton activity of 190 megatonnes a year as a result of the current warming trend.

One other example of pollution of the seas is the amount of plastic waste in the north Pacific Ocean. This is from *What Nature Has Ever Done for Us; How Money Really Does Grow On Trees* by Tony Juniper (2013), pp. 212–13:

> Sailing into the central part of this vast area of water [the north Pacific Ocean] reveals an expanse of sea that is rather like a plastic soup, now about twice the size of Texas and growing. Held in place by a vortex of ocean currents, the area of plastic now extends from about 900 kilometres west from the coast of California past Hawaii and almost as far as Japan. It is estimated that some 100 million tonnes of debris are now floating there.... Toy bricks, footballs, yoghurt pots, kayaks, bits of cars, bottles and all the other plastic paraphernalia that characterises modern life is floating around in the middle of the ocean. There are several such agglomerations of floating plastic detritus. Another one is awash in the Sargasso Sea, while another huge one has built up in the Bay of Bengal.

Our rivers are also being polluted. This is a particular problem in developing countries like China. About one-third of the industrial wastewater and more than 90% of household sewage in China is released into rivers and lakes without being treated. Nearly 80% of China's cities (278 of them) have no sewage treatment facilities and few have plans to build any. Underground water supplies in 90% of the cities are contaminated. Half of China's population lacks safe drinking water. Nearly two-thirds of China's rural population – more than 500 million people – use water contaminated by human and industrial waste. In the summer of 2011, the Chinese government reported that 43% of state-monitored rivers are so polluted they are unsuitable even for human contact.

Over-fishing, over-farming, lack of water and industrial pollution are all a result of the human success story. The planet is suffering as a result. The delicate balance of nature that provides the environment in which man has thrived is beginning to be threatened. It is starting to look as if our species, *Homo sapiens*, has been too successful for its own good.

The threat that has occupied most attention is global warming. Scientists and engineers have been charged with finding ways of avoiding the use of fossil fuels by producing energy using natural sources of power. Known under the collective term of green energies, these involve the use of water, wind, solar, biomass or geothermal sources of energy.

Water power harnessed by the water mill had been used for centuries. During the twentieth century turbine technology and the use of reinforced concrete allowed the development of hydroelectric power from water stored in huge dams. Now the power of tides and waves are also considered as a possible source of energy.

Windmills were once common throughout Europe. Improvements in technology have made wind farms producing electricity through wind turbines a possible source of power.

The heat of the sun could be harnessed in two ways: firstly by directly heating water to become steam in order to power turbines to produce electricity; secondly by using semi-conductors to convert solar radiation directly into electric energy.

Burning crops (biomass) as fuel does not add to the carbon in the atmosphere. It merely returns the carbon that has been extracted from the atmosphere during plant growth. Either the biomass can be burnt directly to produce electricity or it can be converted into replacement fuels. For example, sugar cane can be directly converted into ethanol, which can be used as a fuel for cars.

Geothermal energy comes from pumping water deep into the ground until it meets hot rocks, where it is heated, and then converted to steam which drives turbines to produce electricity.

The difficulty is that all these forms of energy creation except hydroelectric power are generally more expensive than using fossil fuels. Many are also environmentally damaging in their own right and implementation leads to public protests. Dams for hydroelectric power create water bodies that flood farmland. Tidal dams damage the delicate ecology of the shoreline; wind farms are rejected as unsightly. Growing plants for energy takes away valuable agricultural land for growing food and so on. As a result, while some governments have subsidised green energy initiatives, as yet the use of green energy has not caused a major reduction in fossil fuel usage.

The problem has been compounded by the development in the USA of fracking, a new way of accessing fossil fuels. This allows fluids to be accessed by injecting water mixed with sand and chemicals at high pressure into a wellbore to create small fractures (typically less than 1 mm). Along these fractures petroleum, water and particularly natural gas, can migrate to the well. Fracking has allowed the extraction of gas from porous rocks and shales which were previously inaccessible. This, in turn, has reduced the cost of energy creation and made green energies even less price competitive.

Health

Relative freedom from major wars and famine and the continued effectiveness of health care has improved life expectancy across the planet. Figure 12.6 shows life expectancy in 2010 was now over 60 for all countries outside Africa and around 80 in Europe and Japan.

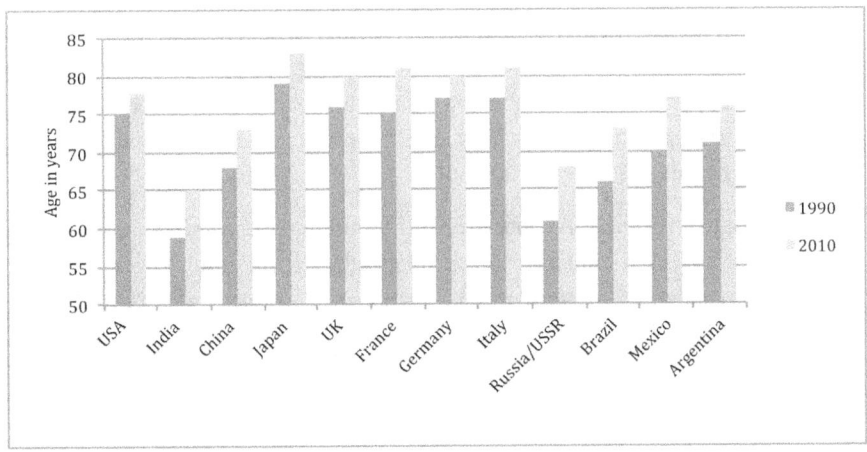

Figure 12.6 Life expectancy (1990–2010)

There are, however, a few clouds on the horizon. The first is the acquired immunodeficiency syndrome (AIDS) epidemic. AIDS is a sexually transmitted disease cause by HIV (the human immunodeficiency virus). The disease mutated from primates in West Africa in the twentieth century. During the initial infection, a person may experience a brief period of influenza-like illness. This is, typically, followed by a prolonged period without symptoms. As the illness progresses, it interferes more and more with the immune system, making the person much more vulnerable to cancer, TB, viral, bacterial and protozoal infections. As of 2010 AIDS had led to around 30 million deaths and approximately 34 million people are living with HIV globally. Sub-Saharan Africa had been the region most affected by AIDS, and it has had a huge impact on life expectancy. There is currently no cure for AIDS, but life expectancy can be prolonged with drugs.

AIDS is a reminder that mankind remains vulnerable to infectious diseases that mutate and cross over from animals. AIDS is just the most recent lethal example. In the last decade, there have also been scares about the SARS virus, which may have mutated from bats, and an influenza virus, bird flu, both of which have been confined to a local outbreak. At the time of writing the Ebola virus has developed into a new threat.

The second cloud is obesity. Historically, only the rich had access to sufficient food to get fat; plumpness was seen as a sign of success. In the late twentieth century, as food became more available and convenient, not only could people eat at set meals, but also snack at any time. As a result ordinary people gained weight. There is no natural, built-in feedback to prevent humans overeating. The extent of the current obesity epidemic is truly alarming. The US Center for Disease Control and Prevention has reported on how obesity has developed in the last 25 years in the USA. In 1990, among states participating in the behavioural risk factor surveillance system, no states had an obesity prevalence greater than 15%. By 2000 obesity levels were over 20% in 23 states and by 2010, no state had a prevalence of obesity of less than 20%. At that time obesity levels had reached higher than 25% in 36 states and in 12 of these states, mostly in the mid-West, obesity rates were greater than 30%.

This is not confined to America. Both Britain and Mexico have obesity rates of over 20% and obesity is on the rise, albeit from much lower levels, right across the world,. In addition to low levels of fitness and general health, obese people suffer increased medical problems. The most significant illness is type II diabetes, which has become the seventh leading cause of death in the USA.

A third cloud is the increase in the prevalence of super-bugs that are resistant to a wide range of antibiotics. Pathogens naturally mutate and over time will develop resistance to an antibiotic. Unfortunately, new antibiotics are difficult to find. It is inevitable, therefore, that pathogens will eventually develop resistance to a wide range of antibiotics. These wonder drugs, that have done so much to increase life expectancy, will eventually be rendered useless. One example is MRSA *(Methicillin-resistant Staphylococcus aureus)*, which is any strain of the bacterium *Staphylococcus aureus* that has developed resistance to a class of antibiotics which includes penicillin. MRSA causes problems, particularly in hospitals and nursing homes, where patients with open wounds, invasive devices and weakened immune systems are at a greater risk of infection than the general public. In addition there is now also a form of TB that is multi-drug resistant. In 2010 it was estimated that 5% of all cases of TB were of this drug resistant form.

If patients do not use the full course of antibiotic treatment, the chances of successful mutation increase considerably, as resistant bacteria are more likely to survive and thrive. There are continued instances of indiscriminate antibiotic usage, particularly in developing countries such as India. With its poor levels of sanitation, Indian doctors have been relying on antibiotics to keep infections on new born babies at bay. But now according to a recent study some 70% of babies in Delhi hospitals have been found to be immune to multiple antibiotics. The situation has been made worse by a recent government programme that pays women to have babies in hospital. Bed capacity is limited and in maternity wards two or three women often share a bed, allowing infections to spread rapidly. The super-bugs causing these infections in India have already been found in other countries, including Japan and the USA.

The worst example of the unnecessary continued use of antibiotics in the developed world is in the USA where it is used as a growth stimulant in livestock. This is despite the proven risk that antibiotic-resistant bacteria will develop in animals, mutate and cross over to affect humans.

Summary: situation in 2010

The spread of the global capitalist economy had transformed the world. For the first time many ordinary people in Latin America and Asia were experiencing the benefits of increased health and freedom from want that had started to be seen in the West over a century ago. The world's population had reached almost 7 billion. It was now more than seven times higher than in 1780 at the start of the Industrial Revolution. The overall growth rate from 1990 to 2010 was down to 1.3% compared to 1.9% from

1950 to 1990. Birth rates were falling but the increased human life span has meant that it will take at least until the end of the twenty-first century to level off the post-war surge in population. Growth rates in Europe slowed substantially, down to just 0.3% p.a. and Europe now only accounts for 8% of the world's population, compared to its peak of 19% in 1870.

Figure 12.7 shows that there are four major bands of population growth:

< 0% in East Europe, Russia and Japan

0–1% in West Europe, North America, China (held back by the one-child policy) and South Africa (limited by AIDS)

1–2% in Latin America, North Africa, most of the Middle East, Indian subcontinent and south-east Asia

>2% in central Africa and parts of the Middle East

The poorest countries are growing fastest in the less developed southern part of the world. Here there is a tradition of large families and there is worse health care and higher rates of infant mortality.

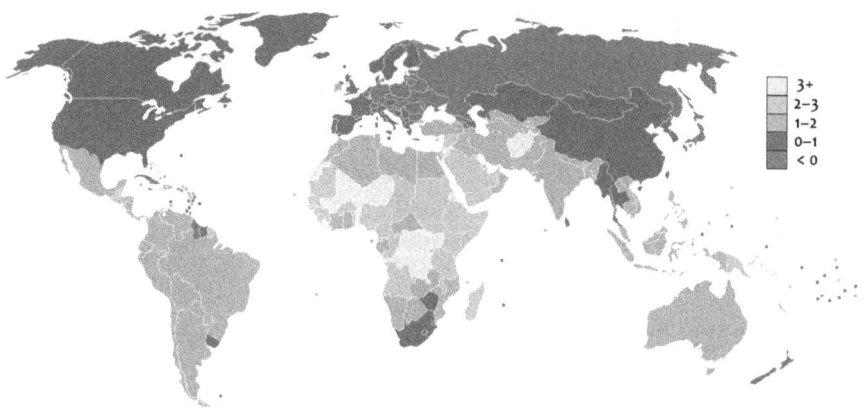

Figure 12.7 Growth in population 2005–2010 p.a.

Source UN

The world was continuing to become more urban and in 2010, a tipping point was reached. For the first time in human history more people lived in towns than the countryside. Cities became even larger. Outside Europe and North America several mega-cities had developed with urban populations of over 20 million, including Jakarta (Indonesia), Shanghai (China), Mumbai, Delhi (India), Seoul (South Korea), Mexico City, Sao Paolo (Brazil), Karachi (Pakistan) and Manila (Philippines). Tokyo remained the largest of all, containing over 30 million people.

However there were downsides to this picture of increasing wealth and health. The excesses of capitalism were not being controlled. Multinationals and banks were avoiding paying tax by using tax havens. Society was becoming less equal; business managers, sports and entertainment stars were paid huge salaries. Corruption was endemic outside secular capitalist counties. The rich hid their money in tax havens and would set up new dynasties of wealthy dependents living off their acquired capital.

It had become apparent that the rate of consumption of the Earth's resources was unsustainable. However the competitive of forces of evolution continued to operate, driving the global industrial society forward to even larger levels of population and wealth. The consequences of this will be discussed in the next chapter.

CHAPTER 13:
WHERE ARE WE GOING?

This chapter takes a look forward at the likely condition of humanity if we continue our present course. It is not to be doubted that technologies will continue to develop, particularly in electronics, materials and health care. However the emphasis in this chapter will be about how we can come to terms with the Earth's limited capacity to support the inexorable growth in human population and wealth.

Humans have become the most successful animal species, because they can communicate and implement a multitude of ideas. These ideas or memes have grown and become increasingly sophisticated over time so that now we can understand many of the workings of the world around us and can manipulate nature to suit our purposes. The technological revolution transformed the material comfort and longevity of human existence. This started in secular capitalist countries, but now has extended to most of the countries of the world. Everywhere, population and standards of living are increasing. The whole Earth is being converted to directly serve our purposes. To feed the expanding population ever-increasing areas of fertile land are being converted to agricultural use and forests are disappearing. As an illustration of this, Borneo was up until recently an island covered by tropical rain forest. Now it has become a land of palm-tree plantations. There are more orang-utans than there are virgin forests to support them. The species survives only by being regularly fed by humans and providing entertainment for tourists.

I hope the past chapter has made it clear that the Earth cannot endlessly support the growth in wealth and population of the human species. Unless human behaviour changes we are now approaching a tipping point in which environmental factors will adversely affect our lifestyle. Our behaviour is determined by our shared memes. Just as other animals, we are bound to an ongoing struggle for survival in evolutionary competition. The process of evolution has no inbuilt mechanism for taking precautions against dangers that are decades away. To solve the issue requires coordinated global action. There is no world organisation in place that can force change and our current organisations have limited scope for cooperating to resolve long-term issues. If human memes don't change we will inevitably pursue our own short-term interests until we reach the limits in population and wealth that the Earth can support. Then we shall suffer a setback. The traditional Malthusian causes of setbacks, the grim reapers, are war, famine, natural disaster and plague. In our current situation, where as a global economy we are testing the environmental limits of the planet, we have more potential causes of disaster: economic malfunction, scarcity of raw materials and environmental disaster. I have also considered internal revolution separately from war as it has different causes and a higher possibility of occurrence. In the sections that follow this one I shall review each of this extended list of grim reapers, consider the risk of them happening and examine their potential impact. But first we must look at the likely population growth in the next few decades and see how this will test the planet.

In 2012 the UN published its population projections for 2050. Forecasts this far ahead are always tenuous and there are already signs that they may be underestimates. However, as the average human life span is now over 60 years, population forecasts 40 years ahead should be at least in the right ballpark. Figure 13.1 shows the UN prediction that world population will grow a further 2.5 billion, reaching 9.5 billion people. The global impact of industrialisation will have resulted in an increase of over 8 billion people in the last two centuries.

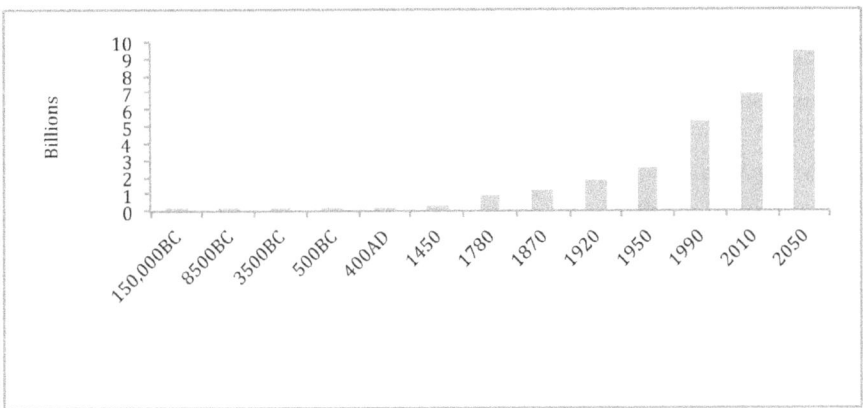

Figure 13.1 Human population to 2050

However the rate of increase in population is declining. Figure 13.2 shows that the rate of increase will be down to 0.8% p.a. in 2050, a level last achieved in the early twentieth century. Nevertheless, this is still a high rate of growth and the UN predicts that human population levels will reach 11 billion by 2100, before growth begins to totally flatten out.

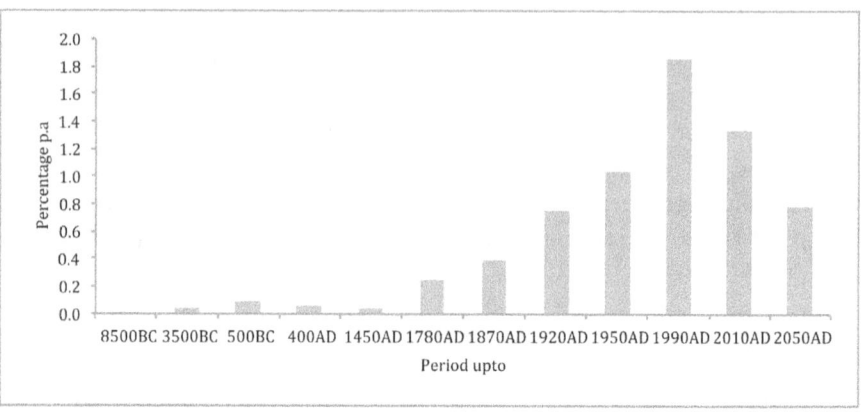

Figure 13.2 Population growth rates to 2050

It is generally thought to be true that declining population growth rates are associated with the education of women, the availability of contraception and the involvement of women in the workplace. Certainly, the population growth rates are lowest in the most developed countries, which have the highest percentage of educated women at work. Table 13.1 shows population growth rates for the major countries. I have shown only figures for countries with a current population of over 20 million and grouped them according to the four subdivisions defined in chapter 12: secular capitalist, elite capitalist, aspiring democracies, with the rest combined as 'others'.

Country	Pop.(M) 2010	Pop.(M) 2050	Growth Rate %	Country	Pop.(M) 2010	Pop.M) 2050	Growth Rate %
Romania	22	18.1	-0.5	Russian Federation	139.4	109.2	-0.6
Poland	38.5	32.1	-0.5	China	1337.8	1310.5	-0.1
Japan	127.6	107.2	-0.4	Korea, Dem. Rep.	24.3	27	0.3
Germany	81.6	71.5	-0.3	Vietnam	89.6	111.2	0.5
Taiwan	23	20.2	-0.3	Total Elite Capitalist	1591.1	1557.9	-0.1
Korea, Rep.	48.6	43.4	-0.3	Uzbekistan	27.9	35.1	0.6
Spain	46.5	42.5	-0.2	Algeria	34.6	44.2	0.6
Italy	60.7	61.4	0	Iran	76.9	100	0.7
France	64.8	69.8	0.2	Myanmar	53.4	70.7	0.7
UK	62.3	71.2	0.3	Syria	22.2	33.7	1
Canada	33.7	41.1	0.5	Saudi Arabia	25.7	40.3	1.1
Australia	21.5	29	0.8	Pakistan	184.4	290.8	1.1
United States	308	422	0.8	Nepal	29	46	1.2
Total Secular C	938.8	1029.5	0.2	Tanzania	41.9	66.8	1.2
Ukraine	45.4	33.6	-0.7	Bangladesh	156.1	250.2	1.2
South Africa	49.1	49.4	0	Ghana	24.3	40.2	1.3
Thailand	66.3	69.6	0.1	Egypt	80.5	137.9	1.4
Sri Lanka	21	25.2	0.5	Kenya	40.8	70.7	1.4
Colombia	44.2	56.2	0.6	Cote d'Ivoire	21	37.1	1.4
Peru	28.9	37	0.6	Iraq	29.7	56.3	1.6
Indonesia	243	313	0.6	Yemen	23.5	45.8	1.7
Argentina	41.3	53.5	0.6	Congo, Dem. Rep.	69.9	144.8	1.8
Brazil	201.1	260.7	0.7	Afghanistan	29.1	63.8	2
Turkey	77.8	101	0.7	Sudan	43.9	97.2	2
Mexico	112.5	147.9	0.7	Nigeria	161.6	402.4	2.3
Morocco	31.6	42.1	0.7	Mozambique	22.4	59	2.5
India	1173.1	1656.6	0.9	Madagascar	21.3	56.5	2.5
Venezuela	27.2	40.3	1	Ethiopia	88	278.3	2.9
Malaysia	28.2	42.9	1.1	Uganda	33.4	128	3.4
Philippines	99.9	172	1.4	Total Other	1341.5	2595.8	1.7
Aspiring Democratic	2290.6	3101	0.8	Grand Total	6162	8284.2	0.7

Table 13.1 Population growth to 2050 for countries with population of over 20 million

The total forecast population growth rate of secular capitalist countries is only 0.2% p.a. Many European countries such as Germany and most far eastern countries such as Japan have negative population growth rates. Only the democracies in the New World (USA, Australia and Canada) which have the space and a history of receiving large numbers of immigrants have a growth rate of over 0.5% p.a.. Russia also has a negative growth rate, as does China, with its one-child policy. All these countries with a declining population will find it difficult to support an increasingly aged society as the proportion of those of working age is decreases. These aging issues are discussed further below.

At least for these low-growth countries there will be relief from the pressures of population increase and they provide hope that eventually the growth in human population will cease. However secular capitalist countries, together with Russia and China, will only account for a quarter of the world's population by 2050. Aspiring capitalist countries are forecast to grow 0.7% p.a. and the 'other' countries, which are on average the poorest, will grow double that at 1.5% p.a. India will have the largest population in the world, with 0.3 billion more people than China. And there are some truly frightening population growth rates in Africa. The predicted top 10 most populous countries in the world in 2050 are shown below. Nigeria will have almost the same population as the USA. Of the remainder of the top 10, only Brazil is moderately wealthy and Ethiopia. Pakistan and Bangladesh are among the poorest countries in the world (Table 13.2).

Rank	Country	Pop.(M) 2010	Pop.(M) 2050	Growth Rate % p.a.
1	India	1173	1657	0.9
2	China	1338	1311	-0.1
3	United States	308	422	0.8
4	Nigeria	162	402	2.3
5	Indonesia	243	313	0.6
6	Pakistan	184	291	1.1
7	Ethiopia	88	278	2.9
8	Brazil	201	261	0.7
9	Bangladesh	156	250	1.2
10	Philippines	100	172	1.4

Table 13.2 Top 10 countries by population 2050

It is especially instructive to look at the list of countries with growth rates of over 2% p.a.: Afghanistan, Sudan, Nigeria, Mozambique, Madagascar, Ethiopia and Uganda. These are all countries where the predominant livelihood of their citizens is subsistence farming. I have visited Uganda and Ethiopia and seen the small sizes of the plots and heard of their large families. Ugandan farmers depend on intensive banana cultivation and Ethiopian farmers produce a small-grained cereal crop called teff, which is their staple diet. It is traditional for a family's land to be divided between the offspring, but the plots are already so small that they scarcely support one family. Ethiopian farmers are also dependent on fickle summer monsoon rains. The failure of these rains in 1983–1984 led to widespread famine. As industrial development is very limited in these countries, it is hard to conceive how these extra people may earn a livelihood. It is inevitable that this will create further pressure on the natural environment. The remaining forests will be cleared; plant and animal diversity will further diminish. Animals like the mountain gorilla and the gelada baboon will lose more of their natural habitat and face extinction.

In summary, the UN forecast of population growth to 2050 paints a picture where

- the world population reaches 9.5 billion

- population growth is slowest in the developed world, where women have been well-educated and have jobs

- many wealthier countries will experience declining populations and face the problems resulting from having an aging population with fewer people of normal working age

- population growth will be highest in the poorest countries, putting increased pressure on the remaining natural grasslands and forests as well as water resources and fisheries

- sub-Saharan Africa and many Islamic countries will experience substantial population growth, As these are already some of the poorest countries in the world, this will place enormous pressures on their governments and inevitably lead to further political instability.

We have seen that with a global population of 7 billion we are already beginning to test the limits of the Earth's resources. How will we fare with 9.5 billion? While population is growing, it is the increase in wealth that is particularly worrying. The OECD is predicting a growth of around 3% p.a. in GDP over the period to 2050. That is crudely, an increase of 2% p.a. in individual wealth, in addition to the population growth. A growth of 3% p.a. over 40 years implies that the world's GDP will be 3.3 times larger by 2050. This means that by 2050 humans could be consuming over three times the Earth's resources that we are consuming now!

The sections below explore how, as we reach the Malthusian limits, the eight grim reapers defined previously are likely to impact on human development.

War

Gradually, over the 200,000 years of *Homo sapiens'* existence, the level of death from violence involved in human competition has diminished. We have progressed from bands fighting bands with high individual death rates to a situation where, within a state, violence is largely controlled and limited. When most of the world became governed by states not tribes, the threat of violent death became largely restricted to either wars between states or internal revolutions. In his book *War What Is It Good For* Ian Morris calculates that violent death rate in the twentieth century was 1–2% of the population. This is taking into account the First and Second World Wars and the mass murders of Hitler, Stalin and Mao. This is a massive change from the 10–20% rate of hunter–gatherer and tribal communities. In the second part of the twentieth century the level of violent death from war diminished further. One reason for this is the fear of mutual destruction by the nuclear bomb. Another reason is that the growth of the global economy has meant nations are highly dependent on each other; global war would also lead to huge losses in wealth. The last major war between the great powers was between America and its allies against China in the Korean war of 1952–1953. Since then we have had 60 years in which conflicts have been limited in regional scope. The level of violent death in the early part of the twenty-first century is now down below 1%. I think the chances of another global war like the Second World War remain small. There is simply too much to lose by the major powers. Apart from the death and destruction caused by the war itself, the disruption to the global economy would be too catastrophic to make any global war pay dividends. This will not prevent big powers from exercising their military might in their own spheres of interest. We have had several examples of that in recent years: Russia in Georgia and Ukraine, USA in Nicaragua and Grenada, China in Vietnam, as well as USA with allies in Iraq and Afghanistan.

There also could be local wars not involving direct conflict between the great powers. Fortunately, traditional wars fought for military glory or territorial gain are becoming less frequent. The last of these was Sadam Hussein's disastrous wars against Iran in the 1980s and Kuwait in 1991. However, groups bound by religious and ethnic ties each contesting the ownership of lands have an undiminished capacity to generate conflict. Places such as Palestine, which has seen three wars between Arabs and the Israelis, or Kashmir, the location of four wars between India and Pakistan, are likely to continue to be centres of conflict.

My bet would be that while there may be a severe setbacks due to war in particular areas, global war is not likely to threaten our species as a whole in the next decades.

Famine

The broad conclusion of the FAO publication, *Food Prospects towards 2030/2050*, revised in 2012 by Nikos Alexandratos and Jelle Bruinsma, is that there is enough scope for the continual improvement in crop yields and enough spare land to provide sufficient food for the world's future population to 2050. They do, however, have three caveats: the demand for biomass to generate energy is unknown, the effect of global warming on agricultural yields is uncertain and the ever-increasing population growth rates in sub-Saharan Africa give cause for concern.

These are large caveats. Global warming, itself, will mean more extreme weather events, floods, forest fires and droughts; biomass is an important potential part of the picture for solving global warming; and sub-Saharan population growth rates are increasing, not declining.

Subsistence farmers are still an important part of the world's population and they are still dependent on the weather and nature to bring in the harvests. Population increase and competition between farmers for land will lead to increasing encroachment on forests, marginal lands and nature reserves. Deforestation, desertification and depletion of the remaining natural reserves will undoubtedly continue. The countries with the highest population growth rates are among the poorest and least able to afford or adapt to new techniques and hence gain advantage from increasing crop yields.

Stephen Emmott in his book *10 Billion*, published in 2013 takes a much more alarming view and believes that food production is going to decline over the coming decades. He predicts that this will be due to three main factors: climate change, soil degradation and water shortages. Climate change will increase the frequency and severity of extreme weather, which will destroy more crops. Soil degradation will come from pollution, salination from irrigation and over-grazing. Finally water shortages will arise from more climate–change-induced droughts, the growing population and the growing demand from agriculture and industry. He gives recent examples of these challenges on pp. 126–7:

> If we want to get just a glimpse of what we can expect this year or next, certainly over the decades to come, we need only look again at the impact of heat waves in Australia (2008), Russia (2010) and the United States (2012), which destroyed up to 40 percent of grain and corn harvests, and in which livestock died in their tens of thousands.
>
> In the heat wave of 2010, the Russian government placed an embargo on grain exports, which caused chaos in the commodities markets, an

unprecedented food price spike and consequently food riots across Asia and Africa – unrest that led to the violence of what we now refer to as the Arab Spring.

It is reasonable to be worried about the supply of food in the next 40 years, particularly in the poorest countries. Even if the world has enough food there will always be local famines due to local climate fluctuations. There is a significant risk, particularly in sub-Saharan Africa with its high level of population increase, that these famines may become more frequent and more severe.

Natural disasters

We have already seen that the limited amount of global warming experienced so far has increased the number of forest fires, droughts and floods. Rising sea temperatures will also increase the frequency, power and range of hurricanes. We all remember Hurricane Katrina that devastated New Orleans. The rate of generation of the largest hurricanes has been increasing and there are signs that the range of hurricanes will not remain limited to tropical areas around the Caribbean and the South China Sea. There was a hurricane in 2004 just off Brazil and in 2005 one off Madeira. This raises the prospect that Europe and Latin America could also be hit by hurricane winds in the future.

Melting glaciers also cause problems. Huge lakes build up behind natural dams of boulder rubble which eventually break, flooding the valleys below. Instances of this occurrence have already wiped out villages south of the Himalayas. Melting glaciers also provide a regular source of water flow in the summer seasons. Once glaciers disappear, the summer flow of fresh water will cease. This will be a huge problem in the Andes. Large cities like Lima have only 23 mm of precipitation a year, most of it falling as a thin mist that immediately evaporates. Pre-Columbian civilisations, such as the Moche (200–800 A.D.) perished, it is believed, due to their mountain water supplies drying up. Modern civilisation in Peru is also at risk from the same danger.

Rising sea levels threaten the very existence of many Pacific atolls. For example, it is virtually certain that the Pacific island of Tuvalu, with a population of 10,000, will disappear.

There will be the same chance as now of earthquakes, tsunamis and volcanic eruptions and they will all cause death and distress. It is very likely that the number of other natural disasters will increase. What is not clear is whether their impact will be confined to the affected local area or whether they will have major implications for everyone.

Plague

There is always a risk of new viruses sweeping the world. We have seen the effect of AIDS, particularly in southern Africa, where it has severely limited population growth and longevity and increased infant mortality. Recently there have been scares about the SARS and bird flu viruses. However, because we are now such an interconnected world, there is less opportunity for viruses to develop in isolation before they are released into the world as a whole and therefore less risk that a virulent virus will wipe out large numbers of humans. If a virus is too potent it will kill its host before it has the chance to spread. Initially, therefore, a virus must allow a fair chance of human survival, in order to allow its own propagation. A virus which develops over a long time in an isolated community will gradually increase its virulence as the community develops its own immunity. If this virus is then released to the world as a whole it can be devastating. This is exactly what happened when the Americas were colonised by the Europeans in the sixteenth century and they were exposed to old world diseases for which they had no immunity.

As I write, this theory is being tested by the spread of the Ebola virus in West Africa. This has had severe consequences in several states and there has been a huge international scare. So far, however, the outbreak has been limited to the local region, it appears be coming under control and casualities are measured in thousands, not millions. The danger to the world as a whole will become greater if the virus remains endemic in West Africa without being eliminated, as it will then increase its potency in the manner described above.

The greatest health risk, I think, comes from infections from bacteria that have developed immunity to antibiotics. We have already seen a growth in forms of TB resistant to most antibiotics. It is depressing that the practice of using antibiotics as a growth stimulant in cattle continues in many countries, including the USA. Yet again, the commercial pressures of short-term profitability have succeeded against a common-sense precautionary approach. *The Future: Six Drivers of Global Change* by Al Gore (2013) p. 227 illustrates how industry can subvert common-sense leadership:

> Again, the U.S. government's frequently obsequious approach to regulatory decision making when a powerful industry exerts its influence stands in stark contrast to the approach it takes when commercial interests are not yet actively engaged. In the latter case, it seems to be easier for government to sensitively apply the precautionary principle. But this controversy illustrates the former case: those who benefit from the massive and reckless use of

antibiotics in the livestock industry have fought a rear-guard action for decades and have thus far been successful in preventing a ban or even, until recently, a regulation limiting this insane practice.... In a democracy that actually functioned as it is supposed to, this would not be a close question.

Since the Second World War antibiotics have been a magic elixir, protecting humanity from potentially fatal infectious diseases. Should more antibiotic resistant strains appear then the effect could be considerable. To quote from the Gore book again: 'Things as common as strep throat or a child's scratched knee could once again kill'.

Economic malfunction

The collapse of the monetary and banking system that occurred in the Great Depression in the 1930s resulted in massive unemployment leading to hunger and famine. In 2008 a similar disaster was narrowly averted. In today's global economy, countries are much more dependent on each other than ever before. A collapse of world trade on the scale of the Great Depression would now have an even greater impact.

The world's economy depends on confidence in the value of traded currencies and the ready availability of credit. As regards currency, the world has proved it can survive collapses in currencies and defaults on national debts of even major countries. One of the worst examples was the collapse of the German mark in the early 1920s. In more modern times both the Argentinean and Russian currencies have collapsed; they have defaulted on their debts and the economies have subsequently recovered.

There is a greater potential problem with currencies that are widely used for international trade. The eurozone crisis showed how the economic mismanagement of one country in the Eurozone could affect all other member countries. For the time being the crisis has been mitigated but the underlying causes have still not been addressed and massive loans have still to be repaid by Greece, Portugal and others, with a continued threat to future credit in Europe. The currency that the world's economy depends on is the US dollar. So far there has been no threat to its stability. However, the US national debt is large; it is importing more than it is exporting and it is dependent on countries with a trade surplus such as Saudi Arabia and China to purchase its bonds. This does present a risk, but the level of US debt is still far from that seen during the Second World War. It therefore seems that a failure in the US currency is unlikely.

The ready availability of credit is much more of an issue. The banking system provides the credit on which the capitalist economy runs. Withdrawal of this credit, as happened in the Great Depression, threatens the economy of the whole world. All the factors that precipitated the 2008 banking crisis are still present. Bankers are still paying themselves huge bonuses, which encourage risky short-term financial investments. They continue to trade non-transparent financial products and they are still being fined for mis-selling and manipulating interest rates.

Despite all the problems caused, countries are still reluctant to fully control the banks and instead compete with each other to attract financial institutions. The UK, for example, is caught on the horns of a dilemma. On the one hand it wants to rein in the excesses of the banks; on the other hand it wants to maintain the competitiveness of London as a financial centre. If more restrictions are placed on banks in the UK, then other financial centres that have more stringent rules would inevitably gain.

Global banking is now such a big international business that the collapse of one bank can threaten the economy of a state. We have seen recently, that the Irish, Cypriot and Icelandic states were bought to their knees by the collapse of their national banks. While controls on the international movement of capital between secular capitalist countries remain weak, states are powerless to act on their own to tackle this problem. The issue can only be solved by all secular capitalist countries working in unison. As yet there is little sign of any major change. Secular capitalist countries have allowed themselves to be cowed by the banking industry, multinational companies and wealthy individuals. Instead of controlling the financial markets, they are competing among themselves to attract large banks, global companies and the rich by offering the best tax breaks and incentives. Companies and individuals still avoid tax in tax havens and they continue to use trusts and shell companies to hide their ultimate financial ownership. It is in the joint interest of all secular capitalist countries to stop these practices and, sharing the same political culture, it is possible that they could cooperate to achieve this goal. There is currently an initiative by the OECD to develop new tax rules that would force multinationals to pay their fair share of tax. It remains to be seen whether states will agree to these proposals or whether the initiative will founder as individual countries try to maintain a competitive advantage spurred on by multinational lobbying. As long as individual states continue to compete, even though cooperation is more rational, the risk of another global economic crisis will remain high.

Shortage of raw materials

The supply of three raw materials is expected to be limited in the next decades. These are fossil fuels (particularly oil), water and phosphates. There may well be many others that will be significant as we switch to other technologies, for example, lithium for batteries could become a scarce resource as green sources of energy develop. However, I shall concentrate on these key three materials.

Analysts have long been speculating about the year when peak oil will occur, that is, when the low remaining reserves result in a fall in oil production. Oil is vital, indeed, currently irreplaceable in the transport industry. The year when peak oil will occur was seen as the point where we would have to definitely cut back our energy-wasteful lifestyle and we would no longer be able to use that emblem of human affluence, the car. This perspective has now changed, due to the ready availability of natural gas through fracking. Natural gas can replace oil for heating and power generation and leave the remaining oil for transport and chemical uses. Hence, the date when low supplies of fossil fuels (including oil) will limit human activity has been postponed further into the future. Indeed, we have now reached the point where it is global warming caused by burning fossil fuels, not their supply, that will limit economic growth. Hence, the risk to humanity will now come from using too much of the existing fossil fuel reserves for a short-term gain in affluence.

Global water usage is growing at around 1.5–2% p.a., about double the rate of population growth. Of this, 70% of water is used for irrigating agricultural land, which is driven by the increased demand for food. Industrial growth is creating further demand. As an illustration of the industrial demand for water, consider the computer chip. It takes 72,000 litres of water to produce one computer chip. As we use over 2 billion chips per year, that is a massive 145 trillion litres of water used on computer chips alone. This growth in water usage is bound to create problems for dry countries. We have already noted that aquifers in many Middle Eastern countries are not being replenished. In addition, sharing the waters of rivers such as the Nile, Jordan, Tigris and Euphrates, is a potential cause of international conflict.

Producing fresh water by desalination has limited potential because of the difficulties of disposing of the resulting brine. If it is returned to the ocean, it is immensely destructive to the coastal ecology.

The shortage of water will inevitably affect agriculture in countries with a low rainfall, in particular the Middle East, the Sahel and northern China.

That said, water supply in 2050 will be a significant local problem, but not one that will affect the whole planet.

Phosphates and nitrogen are the two major fertilisers that are essential for the green revolution and hence for providing food for the burgeoning population. Phosphates are found in a limited number of countries; the USA and China are the largest suppliers but Morocco holds 40% of all known reserves in its western Sahara territory. With limited sources of production, the current problem is security of supply. The western Sahara was a war zone up to 1990 and there is still a stong local preference for independence from Morocco; any long-term disruption to phosphate supply from that area would cause a major problem. Some forecasters are saying that in the period after 2050 phosphate supply will start to diminish. However, this should encourage the more efficient use of fertilisers rather than immediately effect food production.

Revolution

I have considered revolution separately from war. I am taking revolution as an internal military conflict within a state, whereas war is a military conflict between states. Historically, revolutions are less likely to happen in democracies. They occur more often in states where there is no fair and open process for the transfer of power. In such states revolutions provide the only means of ousting the ruling families and elites.

Currently we can see on our television screens the horrific results of the revolution in Syria, with a huge destruction of buildings, loss of life and livelihood and massive problems for neighbouring countries dealing with over a million refugees. Such revolutions do not just have implications for the destabilised country. Regional economies are adversely affected; refugees flee the conflict creating humanitarian challenges on the other side of the country's borders. Wealthier nations are increasingly inundated with desperate illegal immigrants and global economies can be affected by shortages of raw materials. Syria is a relatively small country. Revolutions in major manufacturing centres such as China or oil suppliers such as Russia and Saudi Arabia would have major implications for the global economy. Currently these countries are stable; their processes for the transfer of power to new leaders are well-established. The situation may be different if living standards should decline in future. As has been mentioned, the Arab Spring came about partly due to a temporary food shortage after poor weather in Russia. Even democracies may be at risk. In the autumn of 2000 a wave of protests against high fuel prices started in France and moved across Europe. The protests started in late August with a blockade of English Channel ports by French fishermen, and were later taken up by disgruntled truck drivers and farmers as they blocked oil refineries and distribution depots to complain of high fuel costs. Within a week, concessions by the French government ignited similar protests in Belgium, Germany, Italy, the Netherlands and the UK, creating disruption in cities and cutting off essential services by blocking roads and highways – reducing the flow of fuel to a trickle. The consequent fuel shortages affected millions of Europeans. All this was the result of a change in the price of fuel that now looks relatively minor. More recently, as a result of the Eurozone crisis and the subsequent austerity measures and high unemployment rates there have been huge protests in Greece and Spain. I think we can conclude that if living standards decline in future or food shortages become significant, then revolutions would become more frequent. This is more likely in authoritarian states that democracies, but the latter are also at risk. My guess is that revolutions could become one of the most active grim reapers.

Environmental disaster

There are many possible causes of environmental disaster. In recent years industrial accidents such as the Deepwater Horizon oil spill and the nuclear emissions from Chernobyl have badly affected local economies. However there is one overwhelming environmental problem coming our way that will affect us all – global warming. All the more so, because there is no sign yet that mankind has the capability to address the issue. Scientists have raised the red flag and most educated people agree that there is a problem. A lot of pious words have been spoken and some vague promises have been made, but carbon is relentlessly accumulating in the atmosphere at the same rate as before. In *The Burning Question: We Can't Burn Half the World's Oil, Coal and Gas. So How Do We Quit?* by Mike Berners-Lee and Duncan Clark (2013) the authors discuss the reduction in greenhouse gas emission required to limit the rise in global temperatures to 2°C. They compare the reductions currently pledged by countries, the results so far and what is ultimately required. So far, no reduction in emission rates has been achieved, a 25% reduction has been pledged but has still not been delivered and a 75% reduction is needed by 2050.

The average level of carbon dioxide emissions per person in the UK in 2004 was around 10 tonnes per year and in the USA 20 tonnes. The world average was 4 tonnes a year. Since then, China and other less developed countries have been industrialising fast, increasing their carbon dioxide output. A sustainable level of emissions, one that will not cause global warming, is about 2.5 tonnes per person. This level of reduction will not be achieved by minor tinkering, such as changing to long-life light bulbs or implementing the odd wind farm. It means a major change to the way we live. Scientists blame the politicians, but politicians want to to be re-elected and must in the end react to public opinion. We all appear to cling to the blind hope that somehow the scientists, having raised the issue, will find a technological solution that will allow us to continue our current lifestyle with few changes. It will not happen. If we are to realistically address the issue we have to make hard choices that will change the way we live. At present the public are not driving any change. The principle alternative to burning fossil fuels is to use electric power created by nuclear or renewable means to drive our processes and machines. Now, it is possible to produce electric cars, but to my knowledge electric aeroplanes are not practical as a method of mass transport. If we are serious about global warming, one thing that should stop immediately is mass travel by air for holidays. How many of us are ready to forego our foreign travel for the sake of the planet?

As human animals, our overriding motivation is to survive from day to day and to look after our families. Sacrificial behaviour for the good of the planet in 50 years' time is not memetically programmed into human behaviour. Politicians are as confused as the rest of us. In 2014 George Osborne, the British Chancellor, offered favourable terms to companies who wish to exploit the country's natural gas reserves by fracking. On the one hand the government is saying they want to reduce the use of fossil fuels and on the other they are making every effort to encourage their exploitation.

My guide as to what is practical in the use of renewable energy sources is the splendid book, *Sustainable Energy – Without the Hot Air*, by David J.C. MacKay. He looks at all the possible sources of renewable energy: geothermal, tidal, wave, offshore wind, biomass, photoelectric, solar heating and land-based wind. His conclusion is that it is theoretically possible to provide almost all the UK's power by renewable means. In practice, however, wave and photoelectric power generation is too expensive and we are simply not prepared to accept the massive changes to the countryside and our coastal areas required to achieve the full potential of the other sources of power. The consensus given by a number of experts is that the most we can hope for is that 10–20% of our power needs would be met by local renewable sources.

Another problem with renewable energy is its variability. Tidal power has a daily cycle. Power generation from wind turbines varies according to wind speed and stops altogether when the winds are too low or too high. Solar energy can be generated only in daylight hours and in temperate countries like Britain is very irregular. Renewable energy therefore needs to be backed up by an efficient energy storage system which can be fed into the grid at times of low renewable energy supply. So far, the energy storage technology allows small-scale solutions but no-one has a solution that would allow energy storage on the scale required if all our energy requirements were to be met by renewable sources.

If we exclude fossil fuels there are only two sources of power that can provide enough electricity to support UK power consumption at somewhere near its present rate without massive changes to our countryside and coastal areas: nuclear and imported solar energy.

Nuclear power has many detractors, it is a very emotive subject and so far, despite making favourable noises, the UK government has been slow to commit to a substantial nuclear power station building programme. The difficulty in this approach is illustrated by the fact that when in October 2013 the government finally committed to its first new nuclear power station for decades, the price of the electricity to be generated was fixed at twice the current levels.

As regards solar energy, it is possible, in hot dry conditions, to concentrate solar power sufficiently strongly to drive a steam turbine to act as a source of electricity. To supply the UK's requirements it would be necessary to cover an area of the Sahara desert about the size of Wales (150 km2) with reflecting and concentrating mirrors. So far, this idea is only at the conceptual stage. There are many issues, not least, the issue of security of supply, which would have to be addressed before it was a practical consideration.

It is also theoretically possible to prevent carbon dioxide escaping into the atmosphere by capturing it after combustion and storing it underground. The technology is unproven and so far no one is prepared to invest in a capital project that would actually add to the cost of electricity production by a predicted minimum of 25%. In summary, there are only faint signs that we have a plan yet in the UK that will allow us to produce electricity in a way that does not increase the amount of carbon dioxide in the atmosphere.

What about other countries? Germany is leading the way in the move towards the use of green energy; in October 2014, 23% of Germany's power came from renewable sources, principally wind power, biomass (from wood), photoelectric and hydroelectric power. This comes at a cost. Traditional power stations have to be run inefficiently to cover for the variability in supply of wind and photoelectric energy. Industry is subsidised but German domestic users pay almost 50% more for their energy than the average European. The German problem with the cost of power has been exacerbated by their actions following the nuclear accident at Fukushima caused by the Japanese tsunami in 2011. This accident made countries across the world more nervous than ever about the use of nuclear power. Japan has reduced its reliance on nuclear energy and Germany has announced it will withdraw from nuclear power completely. As a result, Germany is opening up nine new coal-fired power stations between 2010 and 2015. To quote the *Financial Times* in October 2014:

> The paradoxes of Germany's energy policy are impossible to ignore. This is a country committed to reducing CO_2 emissions but which is building more coal power stations. It does not see much of the sun but has pinned much of its future on solar power. It is closing down its well-run nuclear power plants yet relies on nuclear energy imports from neighbouring France.

An agreed international solution is further away than ever. The problem is compounded by the development of fracking which has provided American industry with a cheap source of power. More readily available natural gas has alleviated pressure on oil reserves and hence reduced oil prices. As a result, the immediate financial penalty of the adoption of green sources of power has increased.

It is inevitable that the accumulation of greenhouse gas in the atmosphere will continue. Although humans are amazing animals and are capable of planning a way out of the current crisis, there is nothing in our make-up or organisation that allows us to make sacrifices now, in order to avoid adverse effects more than 5 years in the future. Eliminating global warming will become a priority only when its effects become clear. The difficulty is that greenhouse gas accumulation in the atmosphere is cumulative and once it is in the atmosphere it cannot be retrieved except at high cost and the expenditure of even more energy. We are on a one-way street to global warming.

So what will happen? According to experts we are heading for a 2°C temperature rise by 2050 and a 3°C to 4°C rise by 2100. However, no one can be totally sure of the details of how climate change will affect each region. Many scientists are involved in modelling the climate but the complications are immense. One very critical factor is what will happen to the ice cap of Greenland. The ice cap covers roughly 85% of the land surface of the island and rises to an average height of 2.3 km. If all Greenland's ice were to melt, global sea levels would rise 7m. Cities such as London, New York and Shanghai would flood, lowland areas like Bangladesh and Florida would be mostly underwater and atolls like the Maldives and the

Andaman Islands would disappear under the sea. Higher sea levels would cause increased erosion, salt water intrusion and storm surge damage in coastal areas, directly threatening over one-third of the global population. All the economic and agricultural activity currently taking place in coastal zones would be at risk. The global effect of this rise in sea levels would be catastrophic enough; however, the ocean currents and the climate would change as well. Greenland's melting glaciers add fresh water into the ocean, affecting the balance of fresh to salt water and the circulation of water in the oceans. As Greenland and the Artic become warmer the temperature difference between the equator and the poles will shrink, making equatorial heat move much more slowly to the poles. The flow of heat through the atmosphere from the equator to the poles powers the circulation of air in the atmosphere. The path of air currents such as the jet stream would be altered, which would result in new storm tracks and precipitation patterns.

Between 1993 and 2012, the sea level rose by about 3 mm per year. At this rate it would be over 2000 years before the Greenland ice totally melted and a full 7m rise in the oceans would be seen. The 2007 Intergovernmental Panel on Climate Change (IPCC) expected the melting of the Greenland ice sheet to occur over about a 1000 year period. However, there is recent evidence that the rate of flow of the glaciers to the sea is speeding up. Melt water forms rivers between the ice and the bedrock and this lubricates the flow of glacial ice. It is especially worrying that during the warm period before the most recent ice age, 120,000 years ago, roughly half of the Greenland ice sheet melted. At this time the global temperature rise was only 1°C. At the end of the last ice age 14,000 years ago, when the huge continental ice sheets finally disappeared, global seas levels shot up by a metre every 20 years for four centuries. This shows that dramatic accelerations in ice loss are possible. There are many opinions. Edalin Michael and Jeff Masters from the website wundergound.com state:

> [The] risk later this century [of the melting of Greenland's Icecap] needs to be taken seriously. Because of their complexity, many of the processes governing ice melt and formation (especially sea ice) are not incorporated fully into the models that we currently have. This means that agreed-upon estimates of sea-level rise could be too low. In the latest IPCC document released in November 2007, the group acknowledges that their estimated range of sea level rise by 2100 of 0.18–0.59 metres (0.6–1.9 feet) does not 'provide an upper bound for sea level rise,' and that uncertainties in changes in ice sheet flow could lead to higher sea level rises.

Based on this analysis there is a risk of a great rise in sea levels by the end of the century. This is just one of the possible consequences of global warming; the same degree of uncertainty exists for sea temperatures, currents, air flow and seawater acidity and the consequent effects on biodiversity and agriculture. The risks are huge for everyone on the planet. Nations will act on climate change only when catastrophes start to happen. By then it will be too late.

The aging population

Before summarising this analysis of the prospects for the human race, there is one other factor that needs to be taken into account. Life expectancy at birth has been increasing dramatically since the Industrial Revolution and as yet shows no sign of slowing down. The UK is typical; you can see from Figure13.3 that 2–3 years per decade is being added to British life expectancy. If this trend were to continue, male life expectancy, now at 78, could be 88 in 2050.

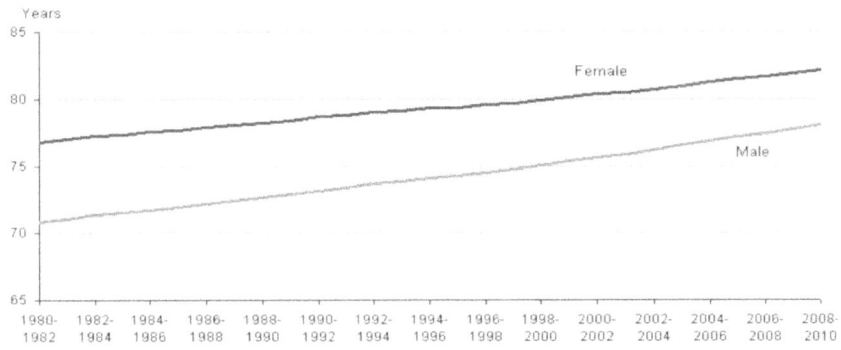

Figure 13.3 Life expectancy at birth, UK, 1980–1982 to 2008–2010

Source: Office of National Statistics

There will be many more centenarians in the years to come. There are already 12,000 of them in the UK but, according to UN forecasts, there will be 82,000 centenarians in the UK by 2050. There are three problems with an aging population. Firstly, as we become healthier and live longer, we have a capacity to continue to work longer, so the age of retirement should increase as well. However, countries have been understandably reluctant to make this unpopular change.

Secondly, older people are less able to do physical jobs; while they have more experience, they have less energy and dynamism. Thus, if the normal age of retirement rises significantly, as people age they are going to have to take on different types of work. It is not clear that suitable jobs can be created.

Thirdly, as medical knowledge increases, the opportunities and demand for medical intervention will increase as well. Medical costs have been inexorably rising as a proportion of GDP. An aging population will further strain medical resources. In the past there was no such thing as retirement age. People worked until they were incapable of working further. This is still the case today in many countries which do not have social security systems. In the Western world the retirement age was fixed at 60–65 merely because people could not be expected to live much longer and that was their age of incapacity. In 1950 in the UK, after the welfare state had been set up, life expectancy was only 69 years. In total, 29% of the population was under 20 years of age, 60% were of working age, and only 11% were over 65 years of age. The working population primarily worked to support the nurture and education of the young and supporting the elderly cost relatively less.

Although by 2010 the proportion of the population of those over 65 had increased dramatically to 19%, the age of retirement for men remained static, This was affordable, because as the post-war baby boomers matured, the proportion of those aged below 20 fell to compensate. In fact, Figure 13.4shows that the proportion of those of working age (20–65) in the UK remained fairly constant at around 60% throughout the post war period. This figure of 60% of working age is the same as in Germany and Japan after the 1970s. Even Russia and China, which were heavily affected by the war and social disruption, have now reached the stage that over 60% of the population are of working age.

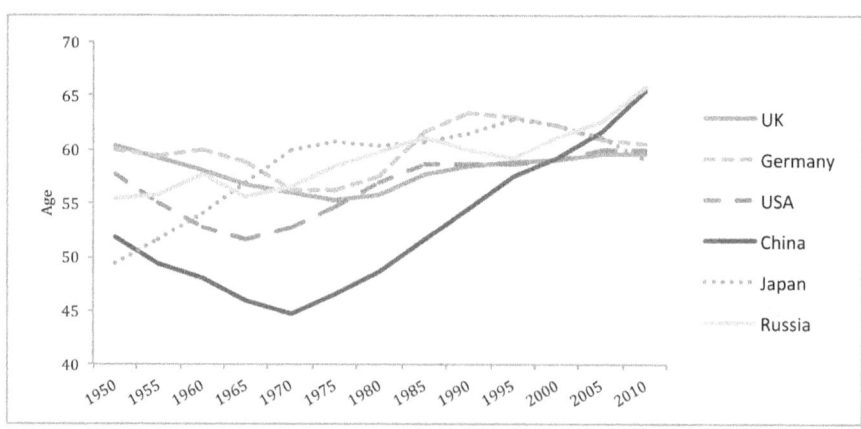

Figure 13.4 Proportion of population aged 20–65 years

The median UN population projection predicts that the proportion of those over 65 in advanced countries is set to soar from 5–10% in the 1950s to 20–35% by 2100. Figure 13.5 shows the retirement age necessary to maintain 60% of the population as working. It shows the retirement age has to approach 75 in Germany, USA and UK and will even have to be 78 in Japan by the year 2100.

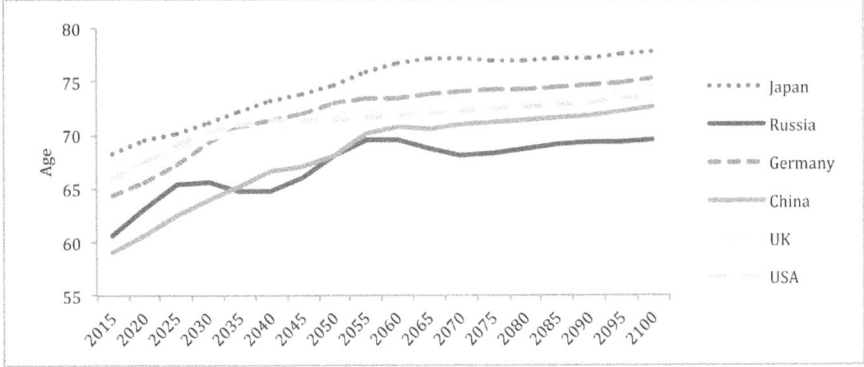

Figure 13.5 Age of retirement to maintain 60% of working population

Source: UN population forecasts based on medium fertility estimates

Countries will have the difficult choice of whether to increase taxes on the working population to pay for the elderly, or to increase the age of retirement dramatically. I suspect that economics will dictate that retirement age will have to increase. The difficulty is that, despite all the recent medical advances, the aging process is one of inexorable decline. Strength, fitness, energy, quickness of thought all decline as we get older. There is no magic pill to stop aging. Although the elderly have the benefit of experience, many of them will simply not be dynamic or flexible enough to maintain their current job in the face of younger competition. Most people are going to have to find second careers with less demanding work, just as the military and the police force do today. At present there are no mechanisms for encouraging people who are too old for their job to retire and find other work. Neither is there a career path for suitable jobs for the elderly. Both issues will have to be addressed but neither has an obvious solution.

Moreover, as technology improves, as we get to know more of how the human body functions, the opportunity to have successful medical treatment improves and health care costs inevitably rise. Demand for health support is rising generally across the Western world. Figure 13.6 shows the rise in the proportion of GDP spent on health care in Western countries. In 16 years growth in spending as a proportion of GDP has grown 10–25% in all the wealthiest countries in the world.

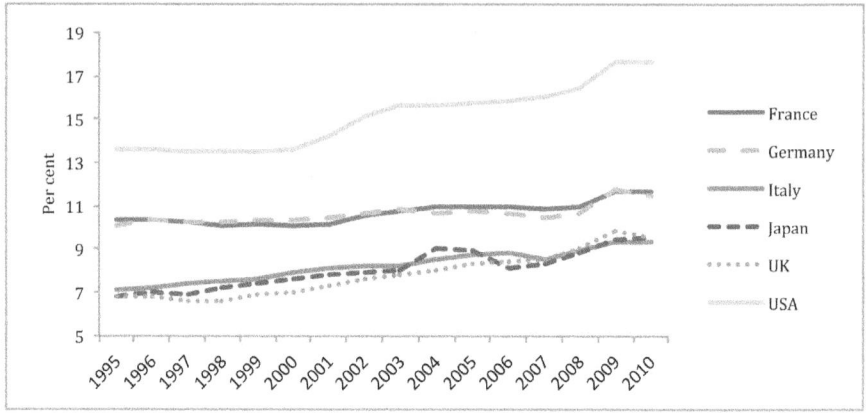

Figure 13.6 Health care expenditure as a percentage of GDP

Source: World Bank

As medical knowledge grows, people have opportunities to ameliorate the physical effects of aging. For example, there has been a huge increase in joint replacement surgery in recent times and the ability to manage heart disease has improved massively. The potential to spend even more on keeping the aged active is growing all the time. Table 13.9 shows how medical costs varied by age in Canada in 2007. It shows that per capita expenditure on those between 70 and 74 was double that of those of working age and the expenditure on those over 80 was about five times as great. Overall health care costs are likely to mushroom as we all get older. Currently over 70% of NHS funding is spent on long-term care, largely for the elderly. This proportion will rise and does not include the cost of keeping the elderly in care homes. It is not at all clear yet how states are going to be able to fund this degree of expenditure.

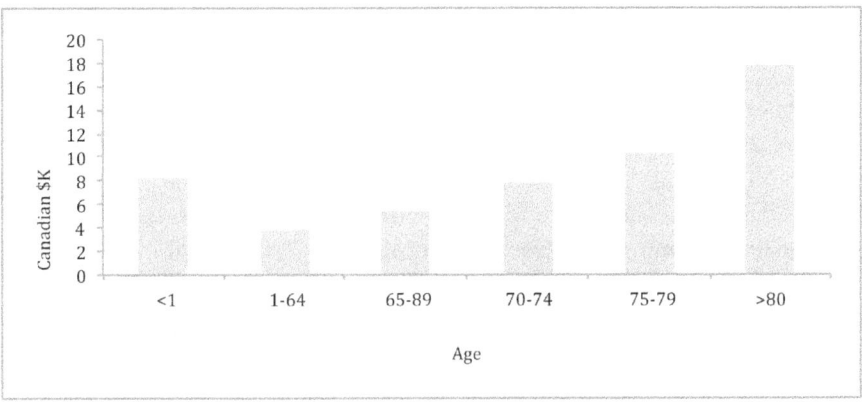

Figure 13.7 2007 Canadian health care expenditure per head by age group

Source: Canadian Institute for Healthcare Information

Difficult questions are already being asked whether some of the new cancer therapies are affordable; and whether the high cost of treatment justifies in keeping a person alive for a limited period of time. This problem will grow. According to UN estimates 8% of the UK population will be over 90 years of age in 2100. While medical science may keep them alive, their mental faculties are going to be diminished and the cost for the state will be considerable.

In my view, even more difficult questions about the value of living longer will have to be asked. In this book I have considered that increased life expectancy is a measure of the improvement in the quality of human life. It is undoubtedly true that living longer gives more opportunities for humans to enjoy themselves and live a more fulfilling life. However when a longer life merely prolongs incapacity during the dying process, this measure may not be so appropriate. Do people really want to extend their life beyond the time they can enjoy their existence? Is it right that the whole of humanity should spend so much resource caring for people who experience more pain than pleasure from existing? The question of euthanasia will have to be seriously addressed. I am not the only one who does not want to hang on to the last breath of life whatever my mental and physical incapability.

The current direction of human evolution

Humans have now come to a crucial point in the development of their species. The dramatic increase in human population and health since the Second World War is now threatening the balance of the Earth's ecology.

Unless we change our behaviour, some of the eight grim reapers mentioned above will cause a catastrophic setback.

We have seen the behaviour of communities of humans is determined by their genetic make-up and their shared memes. Furthermore, these communities can be characterised in a five level evolutionary structure. These evolutionary structures have changed and developed as new technologies have enabled humans to interact in different ways. The current evolutionary structure of the global industrial society is shown in Figure 12.1.

If humans are to avoid these ecological setbacks they need to become the first animal on Earth that plans its future far in advance. They will need to implement a long-term strategy that looks more than 20 years ahead. Such a plan may well involve some decrease in the standard of living now in order to have a more secure future in the decades to come. Currently, scientists are telling politicians what needs to be done but little is happening. It is my contention that this should not be a surprise; this is a problem that mankind has not solved before and no community in our current evolutionary structure has the meme set to enable this to happen.

Consider each of the five levels in turn. There are two principal political cultures that compete at the highest level; secular capitalism and elite capitalism. Classical capitalist economics do not take account of any long-term resource or pollution issues. Addressing these issues would involve governments interfering with the operations of the market by, for example, charging companies for polluting the environment. These changes and the implementation of green policies would almost certainly involve short-term reductions in standards of living for significant sectors of the population.

Neither of the two principle cultures is structured to be able to focus on a very long-term objective. Secular capitalist countries elect their leaders every 4 to 7 years. Democratic politicians know that in order to be re-elected they have to demonstrate success during their term of office. This inevitably sets a time limit on their planning. The expectations of the citizens of secular capitalist cultures are high. They want a thriving economy, excellent state support and an improving lifestyle. In the current situation it would be next to impossible for a democratic government to be elected with a programme that proposed a decreased standard of living in return for some benefit decades into the future. Governments who dare to give a negative message are immediately unpopular. This was illustrated during the recent European financial crisis. European governments caught up in the global financial storm had to implement strategies which involved cutting back government expenditure. The governments were immediately unpopular and most of them were deposed at the next election.

Leaders of elite capitalist countries are under less pressure to achieve immediate success and can take unpopular long-term decisions. A good example is birth control. The Chinese were able to restrict their population growth rate by implementing the one-child policy, whereas in India the government programme in the 1970s to limit population growth failed. The Indian programme encouraged (or coerced, according to its opponents) men to have vasectomies. The result was an electoral disaster and led to a change of government, whereas the deeply unpopular Chinese policy has continued until the present time.

While they do not need to court popularity, authoritarian regimes have to look after their backing elite, whether they are from the army, the priesthood, the administration or the governing party. Authoritarian governments would find it very difficult to propose a long-term strategy that decreased the wealth of its elite. Indeed, authoritarian regimes currently ignore ecological issues in pursuit of financial benefit for their elites.

Apart from the one remaining communist state and a few Islamic countries, there is no sign that any of the other state cultures is anything other than materialistic. Almost all participate in the global capitalist economy. It is therefore unlikely that any of the existing political cultures will help humans cope with this long-term challenge.

At the second level there are multinational companies, states and religions. We cannot expect multinational companies, by their very nature, to act in the long-term interests of the planet. Companies always have to demonstrate growth in profit and are looking for a return on their investments in as short a time as possible. Indeed, their effect may be negative. Multinational companies have considerable influence over the economic policy of states; they will lobby against anything that will affect their immediate earnings potential. We have already seen an example of how the oil supplier ExxonMobil opposed green policies in the USA. Indeed, the way capitalism works will make adoption of green policies more difficult. Any strategy to eliminate greenhouse gases would lead to a surplus of fossil fuels. This would, in turn, result in lower fuel prices and make the relative cost of the green approach even higher.

No states have any urgent incentive to look beyond the next few years. The immediacy of current communications in this over-connected world means that politicians constantly have to deal with the issues of the day. In the 1960s the British Prime Minister, Harold Wilson is reputed to have said that 'a week is a long time in politics'. This is even truer today. Ecologists struggle to get their voice heard in the maelstrom of current political issues.

States do try to coordinate activities and cooperate with each other: the UN, the IMF and the EU are prime examples, but self-interest always limits the scope for difficult decisions. In practice, states cooperate only by mutual agreement. The UN Security Council passes fine-sounding resolutions but rarely intervenes to punish or coerce individual states. The IMF can impose conditions for providing a financial loan, but only if a country asks for the loan. The EU can exercise only the authority that has been given to it by all its 27 European members. If a course of action is unpopular in a democratic country or is against the interest of a ruling elite in an authoritarian country it is extremely unlikely to be adopted.

Examples of the ineffectiveness of international cooperation to address environmental issues are shown by the lack of results from these three 20-year old UN initiatives:

- The UN Framework Convention on Climate Change, whose aim was to ensure the stabilisation of greenhouse gases in the Earth's atmosphere

- The UN Convention to Combat Desertification, whose purpose was to stop land degrading and becoming desert

- The Convention on Biological Diversity, whose target was to reduce the rate of biodiversity loss.

None of the difficult decisions required to overcome ecological issues are likely to result from cooperation between states.

Neither is change likely to come about from the established religions. The world's major religions are stuck in the past, bound by ancient texts and traditions, and trying to deny and prevent social change. The ecological dangers now facing us were not apparent at their founding and therefore absent from their theological canon.

Catholicism with its hierarchical, dogmatic and aged male leadership has failed to adjust to the changed role of women in the world. The church refuses to accept equality of opportunity for women and has ruled out the possibility of women priests and bishops. Despite the proof that many of its priests are homosexuals, it has opposed rights for gay men. Worse, its attitude to contraception and abortion has specifically encouraged the population explosion that has overwhelmed the planet.

Fundamentalist Protestant religions, based largely in the USA, are the fastest growing religions in the world. Their faith is founded on literal interpretations of the Bible. This makes it difficult for them to accept evolution, which contradicts the story in Genesis that God created the world in seven days. They are opposed to gay rights and encourage population growth by mounting vehement campaigns against abortion.

Despite the political demise of Confucianism, Chinese traditional culture remains strongly ingrained in the East. Traditional Chinese medicine and the apparent pitiless attitude by the Chinese to the animal kingdom are a significant threat to biodiversity. Among the endangered species threatened by demand for Chinese medicine are Asiatic black bears, pangolins, tigers, seahorses, rhinoceroses, saiga antelopes and musk deer. The apparently insatiable demand for body parts from these animals is a disgrace to the nation.

Hindu religious beliefs and the resulting class divisions have held back development in India for centuries. Hinduism has created an ossified society which has restricted teamwork and enterprise. Castes and varnas are still immensely important in India today. Read any Indian paper and you will see advertisements for marriage partners. Almost all the ads specify the varna of the partner sought.

Islam is the religion that is finding it hardest to adapt to the modern world. Throughout the Middle East the tension between Islamic traditions and secular capitalism is a cause of discord and violence. The Koran was supposedly directly dictated to Mohammed and hence is literally God's word. Muslims therefore emphasise the literal truth of their principal religious text; as a result many Muslims find it hard to accept the concept of evolution. Shar'ia law has laid out specific roles for men and women in society and has thus institutionalised the inequality of the sexes. Alone among the main political cultures, modern Islamic states have not experienced religious pluralism. Chinese neo-Confucianism was always a mixture of religious ideas. Indian Hinduism covers a broad variety of beliefs and it has co-existed with Sikhism, Buddhism and Jainism across the centuries. Europe had to undergo violent religiously inspired wars in the sixteenth and seventeenth centuries before the secular principle was established and it took further centuries before Catholics and Protestants could live comfortably side by side. As of now the principles of separation of authority between church and state and freedom of individuals to pursue the religion of their choice have not been established in many Islamic countries. Many Muslims feel that their core values are threatened by the growth of secular ideas. Based on the European experience, it may be some time before Islamic states can comfortably embrace the concept of a secular society, yet alone address the issues of the environment.

The third level of the evolutionary hierarchy contains companies, state organisations, regional government, political parties and elites. Companies have the same issues as multinationals, regional governments have the same issues as states, and state organisations are dependent on state support. Hence none are in a position to address the long-term problem.

Political parties are by their nature trying to make their political message as popular as possible. That environmental issues are not of immediate general concern is emphasised by the lack of impact of the Greens outside a few countries in Europe. This party's political platform, placing particular importance on green issues, social justice and non-violence, has had limited appeal in today's tough materialistic environment.

Elites have the money and opportunity to encourage long-term action. While some rich people may act as philanthropists for good causes, as a whole they are unlikely to look favourably on any proposal that would limit their wealth. Unfortunately, it has been shown time and again that just because you are rich does not mean that you cease wanting to acquire more wealth. Indeed, their overall influence may be negative because a large proportion of the elite are in an excellent position to distort the political process and influence policy development for their own benefit.

Finally at the individual level nothing in human evolutionary development prepares them to exhibit sacrificial behaviour for the long-term good of the species. There may be a small proportion of the population who can rationalise their behaviour and alter their lifestyle for the good of all. However most of the world's population are focused on immediate issues; many are poor. At present for most people the long term good of the planet comes a distant second to their own and their families' immediate concerns.

The evolutionary process whereby each gene, individual and community of individuals competes and cooperates for its own benefit has created a species of animal that has dominated planet Earth. This same evolutionary process is now driving *Homo sapiens* towards an ecological catastrophe. If the human race is to avoid this coming disaster the competitive instinct will have to be overcome. Humans will have to be able to make material sacrifices for the good of others. In the final chapter I explore how such a change could occur. But before that I want to retell the story of Easter Island, the ultimate metaphor for the future of man's life on Earth. Easter Island is the most remote habitable land on Earth. The nearest land is the Pitcairn Islands 1300 miles to the west; Chile is 2300 miles to the east. Its area is around 50 square miles and it is triangular shaped, with the craters of extinct volcanoes on each corner. When it was occupied by Polynesians around 900 A.D., it was covered with subtropical forest and woody bushes. There were six native species of birds and it was one of the most important breeding sites for seabirds in the entire Pacific. Initially, the settlement thrived. The Easter Islanders built ocean-going canoes and caught porpoises and tuna; they ate palm nuts, Malay apples and other fruit from the forest and caught wild seabirds and shellfish from the coast. Like all Polynesians they were a tribal society with about 12 chiefdoms. They built stone platforms (ahu) and cremated their dead. On these stone platforms they erected huge stone statues (moai) each with a weight of about 12 tonnes.

The population grew to around 20,000. The chiefdoms competed and over-exploited their environment. Around 1600 A.D. the last of the forests was cleared. There was no longer wood to make ocean-going canoes, no bark for clothing, no rope to haul the statues into position, and the dead could no longer be cremated. Most sources of wild food disappeared. No porpoises or tuna could be caught because large canoes could no longer be built, land birds had been hunted to extinction, wild seabirds only nested on offshore islets, the shellfish they ate became smaller. The only wild food available was the rats they inadvertently brought with them from Polynesia. Deforestation led to land erosion.

People starved. The political organisation collapsed; chiefs and leaders were overthrown. The giant statues were toppled. People turned to cannibalism. New military leaders emerged and people turned to living in caves which could be better defended. The population declined, maybe by as much as 70%.

And yet the Easter Islanders survived. When the Dutchman Jacob Roggeveen landed on Easter Island in 1722 he saw a land with not a single tree or bush over 10 feet tall. The islander's only watercraft were no more than 10 feet long and had to be bailed out while the Islanders paddled. Chickens provided a source of meat and specialist agriculture had been developed to preserve water and stop plants being dried out by the strong winds. As a great testament to the ingenuity of man, in 1864 2000 Easter Islanders still survived.

Just as Easter Island is isolated in the oceans, so the Earth is isolated in space. There is no other planet suitable for human habitation. We will have to survive on what the Earth has to offer. Yet if we continue on the same path we will suffer the same setbacks, deforestation, loss of biodiversity, hunger and soil erosion as the Easter Islanders. There will be revolutions and population decline. Humans will survive but they will have a poorer, more miserable existence. What happened on Easter Island is a warning to us all.

CHAPTER 14:
IS THERE ANY HOPE?

In the previous chapters I have told the story of how the human species thrived by increasingly effectively exploiting the resources of planet Earth. This has been achieved by the serendipitous process of evolution in which new memes were developed by different communities. These communities competed and spread their memes. Successful memes were copied or absorbed by other communities until eventually humanity achieved the globally interconnected society we have today. However, there is no aim to evolution. As a species we are not predestined to succeed. Indeed, at present it appears that the species is heading for ecological disaster. So how do we escape the seemingly inevitable consequences of the competitive battle of the survival of the fittest? We have the technical knowledge to determine our own destiny but it seems short-term self-interest makes it impossible for us to unite behind a programme for the good of the species. What can stop the human race from over-exploiting the Earth and condemning our descendants to a more miserable, poorer existence?

One difficulty is that effective action requires global agreement. Individual action by people or governments will not be effective enough. We can stop it only if everyone wants to stop. That is not just scientists, New-Age travellers and the liberal middle classes but American red-necks, Indian peasants, Chinese communist party members, African subsistence farmers, the urban poor of Mexico City and all the peoples of the world. In short it will only happen if human behaviour as a whole changes. It sounds impossible, but history shows there is just one chance.

Humans are able to demonstrate altruistic behaviour to members of their own communities. The most obvious example of this is soldiers fighting for their country, but it is also apparent when workers exhibit solidarity during a strike and when individuals are motivated to excel for the good of their sports team. Unfortunately, few are able to regard the population of the whole world as part of their community. There are too many examples of poor regimes, corrupt officials, violent and self–destructive behaviour for many of us to behave altruistically for the benefit of the whole of mankind. There is one type of community, however, that has a proven ability to appeal to a wide variety of national groups and that is a religion or philosophy of life. Buddhism, for instance, spread from India to China and on to Japan; Islam from Spain to Indonesia, communism from Russia to China and Christianity from Europe to Latin America, Asia and Africa. Moreover religions and philosophies of life have a proven record of persuading their followers to fast, give money to good causes and to make other personal sacrifices according to the customs of the movement. In addition, their congregations have maintained common rules of behaviour that are defined by their doctrines. Is it possible that an ecological philosophy of life could create such a fervent support that it could unite the whole of the planet in its cause? It sounds a very long shot but it may be our only hope.

It would have to be a completely new philosophy of life. Religions have enough trouble adjusting their ancient religious canons to the modern age, let alone trying to resolve future issues. In the section below I try to outline a philosophy of life called Eco-humanity that is not only ecologically sound but builds on all the advances that mankind has made over the last 200,000 years. If this philosophy is to change the way people behave, it has to not only take on the arguments against fundamental religious tenets but inspire the same kind of fervour that religion can attract. This means not just explaining the science but enthusing people to act differently.

Most religions have a long history. Building on natural superstitious instincts, they developed meme sets that provided a structure and purpose for human existence. Initially they were a crucial factor in establishing a common community-based culture, but as states began to become secular in approach, their importance to society as a whole began to diminish. Nevertheless, the meme sets of religions have survived into the industrial age with only a modest amount of adjustment. In modern times, apart from communism, no philosophy of life has seriously challenged the hold religion has on peoples' lives. To attract and hold their congregations over such a long period of time religions have learnt to appeal primarily to human emotions and instincts. Even though it is based on scientific principles Eco-humanity will have to match the emotional elements of a religion if it is to be successful. To do this I will outline the concept of Eco-humanity based on the classic seven elements that characterise a religion.

Doctrine and philosophy

Eco-humanity is a philosophy of life which is rational, humane, egalitarian and forward-looking. Its primary focus is to ensure that future generations of humans are able to live a fulfilling existence. Eco-humanists do not believe that there is a heaven in which blissful rewards are given to the virtuous or a hell punishing the wicked in the afterlife. They believe that on death our bodies are reclaimed by the earth but our genes and memes live on. The theory of evolution provides the creed for Eco-humanity.

Mankind is an animal. All animals have evolved from other life-forms during the 4.5 billion years of Earth's existence. The structure of our beings is defined by our DNA. Segments of DNA form genes which are each responsible for different facets of our makeup. Our purpose in the scheme of life is to preserve our genes through the creation and nurture of our children and our cooperation with other humans. When we die, our bodies decay but our genes live on through our offspring and our fellow human beings. The ideas, words and pictures we have created also live on in the minds of our family and friends and those we have touched.

We believe that the world is governed by natural forces, not gods, demons or spirits, and our behaviour should be governed by rational argument, not religious dogma, superstition or untoward physical force.

We believe that mankind is one species that evolved out from the same African origins and all the human family is entitled to our respect regardless of race, age, gender, sexual orientation or physical ability.

Community norms and customs

Community norms should be built around four commandments for ecological living.

- Look after yourself so that you can reach your full potential.

Giving people a sense of purpose in life is one of the most important roles of a religion. Eco-humanity should provide a support network and a belief system that enables individuals to take responsibility for their own physical and mental well-being. Everyone should be able to make a positive contribution to society.

- Respect other humans, cooperate, and support them in advancing the lot of mankind as a whole.

Here the important concept is community. This is something that has been lost in our modern materialistic world with its individually centred aspirations. If Eco-humanity is to work, appropriate ecological behaviour has to be developed and honed through peer pressure in a community. We will need to learn from and help each other. I have in mind something similar to the parish organisation of the church where the community comes together to support each other. This may be one area in which the retired but still fit could play a major role by using their experience and knowledge.

Eco-humanity would have to emphasise that we come from a common gene pool and all people, whatever their background, should participate together in their local community. This sense of community must be strong enough to override racial and other forms of kinship. In Britain with its multicultural background, this would mean former Christians, Jews, Muslims, Sikhs and Hindus coming together in one congregation.

- Recognise the integrated world of nature, respect how it supports our lives and preserve its full diversity for the benefit of our children

The important principal is to preserve the natural habitat and biodiversity. It would be a central tenet of Eco-humanity that the fundamental cause of the pressure on nature is the number of humans on the planet. Voluntary control of family size would be a core principal. The aim would be to reduce human population to a more sustainable level in the centuries to come.

Adherents would be expected to campaign against the major threats to the Earth's ecology: encroachment on the world's forests and other natural reserves, animal poaching, over-fishing, desertification by over-grazing and so on. It would also expect adherents on a more mundane level to be against factory farming and restrict cruelty to animals. Those Chinese medicines which are based on the use of endangered species would be anathema.

- Preserve the Earth's resources for the benefit of our offspring

Here the principal approach would be to encourage sustainable development to ensure sufficient resources remain on the Earth for our successors to thrive. It means looking after the soil, the atmosphere, the rivers and the seas to ensure that they are not polluted and overused. It means estimating the reserves of minerals and making sure that these reserves are husbanded. On a personal level adherents would be consciously conserving energy and water and recycling all possible materials. Adherents would be expected to encourage their governments to think and act in an ecological way; minimising global warming would be uppermost on their agenda.

The detail of all this needs considerable development, but I think forms the basis of a new set of community norms and customs which is different from the current, predominantly materialistic attitudes of many in secular capitalist countries. Population control is vital. We have already seen that education for women and freeing them from an exclusively homebound way of life has been a major step forward. There are other significant issues. For example in many societies without a social security net, children, particularly boys, are a form of insurance to provide support for parents in old age. If people are going to be persuaded to have fewer children, the community will have to be motivated to provide appropriate and acceptable support to the aged.

Experiences and emotions

One of the most important areas for the religious devotee is the feelings that it generates. To quote from my introduction to religion course at the Open University, to understand a particular religion you have 'to feel the sacred awe, the calm peace, the rousing inner dynamism, the perception of a brilliant emptiness within, the outpouring of love, the sensation of hope, the gratitude of favours received'.

I predict that Eco-humanity would not be short of emotional feeling, primarily due to its human and animal friendly context and the bonds that develop between humans and animals. Wildlife experiences, whether they be at a game park, a coral reef, or in the forest can be very uplifting. For me, seeing the gorillas in Uganda was a deeply emotional experience. The picture of a polar bear and its cubs starving as a result of the melting Arctic icecap is an immensely powerful image of the effects of global warming. Each religion develops different sorts of emotion: Buddhism emphasises calmness and inner contentment; charismatic Christianity, joy and orthodox Christianity, mystical serenity. I hope Eco-humanism would develop feelings of wonderment for the workings of nature, of love for all creatures and of hope for the future.

Rituals

Rituals are vital to any religion or philosophy of life; they bind the community together by engendering a common emotional experience through music, singing, dancing, and inspirational stories. All religions have rituals for

- the passage of life, births, marriages and deaths

- initiation into the religion, such as baptism or confirmation

- anniversaries of important religious events such as Easter, Eid and Diwali

- regular meetings affirming the faith such as Friday prayers for Muslims or a mass for Christians.

Appropriate rituals would have to be developed for Eco-humanity. They could build on existing traditions adapting them for their own use. Humanists already offer excellent support for funerals. Maybe a form of existing Chinese ancestor veneration could be developed, giving thanks for the genes and memes passed on from our forebears. Christians adapted the natural celebrations of the winter solstice and the vernal equinox to become Christmas and Easter. I see no reason why Christmas and Easter could not be re-adopted as primary celebrations of life and the natural order of the world.

However, in developing rituals it is important to make them relevant and effective to the local community. If the rituals are artificial, just an ineffective clone of current religious practice, they will fail. The ritual that is most difficult to see continuing is the regular weekly meeting with its advice to its adherents through the weekly sermon, as in the Sunday service for Christians or Friday prayers for Muslims. There have been attempts in the past to develop a humanistic weekly service which celebrates life by singing folk songs to guitar music. This always seems to me rather forced and inept. Maybe in this sophisticated electronic age, the weekly sermon is a thing of the past.

Myths of its foundations and its heroes

Eco-humanity is not short of inspiring stories. For its foundations we have the story of Darwin: the voyage of the Beagle, the Galapagos Islands, his experiments, the rush to publication with Wallace, the threat from the religious establishment. There is the discovery of the double helix by Crick and Watson and the important and under-rewarded role of Rosalind Franklin. We also have the tracking of carbon dioxide levels in the atmosphere by Robert Keeling, which was the first measurement of the greenhouse effect, and the subsequent struggle for the world to accept his data.

Its scientific heroes are legion: Faraday, who came from lowly beginnings to become the greatest experimental scientist of his time, Tesla, demonstrating the safety of high voltage transmission at great personal risk, Marie Curie, discovering the radioactive properties of radium at the cost of her health. There are also many heroes from the world of nature preservation and ecology. New heroes would be born in the initial struggle to develop the popularity of the Eco-humanity movement.

Organisation

It would require great leadership to successfully launch such an organisation. Existing religions have very deep roots; their tenets are ingrained into children from an early age and they engender strong emotional ties of kinship and tradition. Richard Dawkins was attacked for his aggressive atheism when he dared to question the basis of some religious beliefs. Both the establishment and the British press are scared of saying anything negative about Islam. Anyone attacking Jewish beliefs is accused of anti-Semitism. The idea of launching a new philosophy of life in competition with existing religions is truly daunting.

Since communism was discredited, no major philosophy has sought to change the hearts and minds of people in the West. Promotion will have to be very active; people's beliefs will have to be challenged; the establishment will have to get upset; minority groups will have to be encouraged to join the mainstream. It would require enormous persistence and a truly charismatic approach to be successful.

As the organisation grows it would need to develop a supportive infrastructure. The organisation would need the equivalent of the parish priest, looking after his community by providing support and advice and leading the community rituals.

Places of worship

At start up the organisation would have no access to churches, chapels or mosques to inspire its congregation. In this electronic age this need not be a disadvantage. Eco-humanity could exploit digital media to reach its congregation. This would have the added benefit of engaging the young. It would be fantastic if the young could be inspired to move on from the depressing modern cult of celebrity with its heroes of overpaid football stars, self-obsessed models and actresses and pop-idols basking in their brief moment of fame. In the past 50 years there has been very little in the form of political or religious ideals to inspire the young. Maybe Eco-humanity could provide such inspiration. After all, it is the young and their children who are going to be most affected when the limits of the Earth's resources are reached.

This then is the outline of a philosophy of life called Eco-humanity. Is it credible? I leave that to you. Can it convert the planet? It looks immensely difficult.

Political culture

Unfortunately this is just stage one of the process. Reaching a sustainable level of development for the planet cannot be achieved just by individuals converting to a new philosophy of life. Governments need to change policies and set new laws. Religions or philosophies of life no longer determine the way states behave. States would need to adopt a new eco-friendly political culture. Eco-humanity would have to have a political wing capable of gaining power in democracies and converting the state to a new political culture. For convenience I shall call this new political culture **ecological government**. This would be based on the tried and tested secular capitalist culture but adapted to take account of the ecological limits to growth. The **belief** systems of ecological government would be those of Eco-humanity.

Ecological government would adopt the same **human rights** as secular capitalism. However, there should be additional rights to euthanasia. People should have the right to determine their own time of death. With life expectancy moving forward to 90, it is unreasonable to insist that everybody should endure a slow and probably painful deterioration in their body and mental powers. Everyone should have the right to experience a dignified death surrounded by their loved ones.

As regards **commercial policy**, I think it would be important to state that Eco-humanity would not be against capitalism per se. Capitalism has delivered so much benefit to the world, that it would be important to continue to use its underlying philosophy. However, the use of scarce minerals and natural occurring resources would have to be restricted by pricing or allocation. The cost of disposing of waste would either have to be factored into the price of the product or, if it has deleterious consequences for the planet the product would have to be prohibited. The precise mechanism for such actions needs to be developed but the overall operation of capitalism would continue with mass production and global trade giving us the benefits of accessible goods.

Eco-humanity would also support the democratic system of **governance**. As Churchill said 'Democracy is the worst form of government, except for all those other forms that have been tried from time to time'. The principal of election of government is still the best system to ensure rule by consent. States all round the world have developed their own forms and traditions of democratic government. Like all institutions, however, these are slow to change and often bound by tradition. The British Parliament has many eccentric survivals of ancient rituals such as the office of Black Rod, the speaker in his silks and the use of the term 'honourable' to address other MPs. Having no superior authority, the world's legislative bodies are often unable to critically examine their own performance and improve the way they operate. Further, the vested interests of the leadership frequently prevent change. As a result, the world's democratic governmental systems are not as effective as they should be. To implement a challenging, ecological friendly political agenda our democratic systems will need to provide stronger leadership and better public engagement and restrict the influence of outside lobbying.

Most democratic politicians are good communicators but usually have no training in managing large organisations and often lack experience relevant to the government departments they lead. Both David Cameron and Ed Milliband, the UK leaders, for example, have been professional politicians most of their lives but they have never held a high position in anything other than the peculiar area of party politics. In business, candidates for job vacancies without a relevant track record would have little chance of success. Frequently the electorate are asked to elect new political leaders, knowing nothing of their governmental competence. In the UK ministers are often catapulted into managing departments of government for which they have no practical knowledge and frequently act in an amateur knee-jerk fashion. Education and the national health service, in particular, have suffered greatly with ministers micro-managing organisations in a totally inappropriate fashion in response to the latest crisis or their own particular foibles. Political ability is no guarantee of any management ability. We need to select better leaders based on their proven aptitude in order to make the changes required for a an ecologically friendly society.

Across the democratic world there is a dangerous, falling level of interest in the democratic process. Politicians have a bad press and are not respected. Membership of political parties is falling. People are failing to exercise their voting rights. There must be many reasons why this is happening. My own view is that the presentation of politics in the media turns people away. All public democratic debate appears to be confrontational. New ideas cannot be discussed in a calm and rational manner. The press is usually accusatory, looking always to pick up on mistakes. Media interviews consist of trying to find fault on one side and avoiding the question on the other. This makes for tedious viewing and listening. The act of creating a new political movement could rekindle political interest. Politics is more engaging when there are two clear sides to a debate, as in the conservative versus liberal debate in the nineteenth century and the capitalist versus socialist debate in the twentieth. Either way, if Eco-humanity is to succeed as a political movement it has to find better ways of engaging with the public than those presently used.

Rich elites and global businesses are a major source of funding for political parties. They are in a strong position to ensure that governments enact policies that support their interests. This is a particularly important problem for the USA. Even in the UK, the recent phone-tapping scandal has indicated how a wealthy press proprietor can influence the day to day actions of government. Ways need to be found to ensure that government, independent of businesses and rich elites, is free from corruption and able to work for the common good.

Could this political culture come to be accepted across the world? It is theoretically possible for ecologically friendly political parties to gain power in secular capitalist countries and aspiring democracies. It looks a formidable task, as the existing political parties are entrenched with strong traditional support. Should, however, Eco-humanity become a popular philosophy of life, existing political parties will embrace its ideas. Ecological government would then be implemented by the existing political organisations.

However this leaves the issue of how elite capitalist countries could be converted. If Eco-humanity is allowed to operate as a popular movement in Russia and China, then perhaps internal pressure can force a change. However, if the elites remain in power and continue to implement non-environmentally friendly policies, the future of the species will remain in danger.

Final thoughts

I have presented a sketch of a philosophy of life and an associated political culture, born out of the advances humans have made over the last two hundred thousand years and dedicated to allowing humans to survive in comfort much longer than a few generations. It needs more thought and development; however I trust I have given sufficient detail for the concept to be grasped. I see these initiatives as the only hope for mankind to avoid an eventual, catastrophic fall in living standards and life expectancy. Whether it can or will happen will depend on others. If it does not, our children's children will be worse off. I have been fortunate. I have lived in a golden age of prosperity and opportunity in a rich secular capitalist nation. I have not seen a major war; I have travelled all over the world and had many experiences. My life is full and comfortable and, despite my age, I am in good health. Just two hundred years ago such a lifestyle would be unimaginable. I have lived a more rewarding and enjoyable life than many princes or kings in ages past. I would love my successors to live a life which is just as happy as my own; but I fear it will not happen. The laws of economics and evolution have their inevitable logic which is propelling mankind towards a setback. We have just a short window of time to change course. I would love someone to take up the challenge and spread this new philosophy and political culture. *Homo sapiens* is the most successful animal species ever, but its very success is threatening its survival. Over aeons of time all species eventually become extinct. Our challenge is to delay the demise of *Homo sapiens* for the benefit of all our offspring.

FOLLOW UP

For those who are attracted to the idea of developing a new ecologically friendly way of life as outlined in this book I have set up a web site called eco-humanity.co.uk. If you are interested please visit the site, make your comments and, if you want to be involved in creating a new movement, leave your email address. Please also tell your friends about the book. It has been self-published and has thus not had the benefit of promotion by a publishing house in the national media; its success will depend on personal recommendation.

NOTES ON SOURCES

Population statistics

Angus Maddison: *The World Economy, A Millennial Perspective* (2001)

Massimo Livi-Bacci: *A Concise History of World Population* (2007)

Introduction

Alex Needham: Paxman's A-Z is hot ticker as the Edinburgh festival fringe, *The Guardian* 19 August 2014

Edward O. Wilson: *The Meaning of Human Existence* (2014)

Chapter 1

Richard Dawkins: *The Selfish Gene* (1976)

Desmond Morris: *The Naked Ape: A Zoologist's Study of the Human Animal* (1967)

Jane Goodall: *My Life with the Chimpanzees* (1988)

Wikipedia entry on chimpanzees http://en.wikipedia.org/wiki/Chimpanzee

Robin Dunbar: *Human Evolution – A Pelican Introduction* (2014)

Elaine Morgan: *The Aquatic Ape Hypothesis* (1997)

Wikipedia entry on man's origins http://en.wikipedia.org/wiki/Human_evolution

BBC2 Horizon 8 August 2001: Do you see what I see? http://www.bbc.co.uk/programmes/b013c8tb

BB2 Horizon 11 October 2011: The origins of us; bones

Chris Stringer: *The Origin of our Species* (2011)

Raymond Tallis: *Aping Mankind; Neromania, Darwinitis and the Misrepresentation of Humanity* (2011)

Steven Pinkerton: The false allure of group selection: http://edge.org/conversation/the-false-allure-of-group-selection (2012)

Susan Blackmore: *The Meme Machine* (1999)

Chapter 2

Azar Ghat: *War in Human Civilisation* (2006)

Jared Diamond: *Guns,Germs and Steel* (1998)

Luigi Luca Cavalli-Sforza: *Genes, Peoples and Languages* (2000)

Richard Dawkins: *The God Delusion* (2006)

Hugh Brody: *The Other Side of Eden* (2001)

Jared Diamond: *The Rise and Fall of the Third Chimpanzee* (1991)

Wikipedia on the Ice Age Extinctions http://en.wikipedia.org/wiki/Pleistocene_extinctions

Wikipedia on tanning http://en.wikipedia.org/wiki/Tanning

Ian Morris: *War: What Is It Good For?* (2014)

Chapter 3

Azar Ghat: *War in Human Civilisation* (2006)

Jared Diamond: *Guns, Germs and Steel* (1998)

Luigi Luca Cavalli-Sforza: *Genes, Peoples and Languages* (2000)

S. M. Channa: *Religion and Tribal Society* (2002)

Elman R. Service: *Origins of the State and Civilisation* (1975)

Thomas and Dorothy Hoobler: *Confucianism* (2004)

Wikipedia entry on lactose intolerance http://en.wikipedia.org/wiki/Lactase_persistence

Wikipedia entry on the history of textiles http://en.wikipedia.org/wiki/History_of_textiles

Chapter 4

Azar Ghat: *War in Human Civilisation* (2006)

Jared Diamond: Guns, Germs and Steel (1998)

Luigi Luca Cavalli-Sforza: *Genes, Peoples and Languages* (2000)

Richard Cohen's teaching notes http://mygeologypage.ucdavis.edu/cowen/~GEL115/index.html

Mark Collier and Bill Manley: *How to Read Egyptian Hieroglyphs* (1998)

Paul Kriwaczek: *Babylon* (2010)

Wikipedia entry on Babylonian numerals http://en.wikipedia.org/wiki/Sexagesimal

Wikipedia entry on education in ancient Greece http://en.wikipedia.org/wiki/Education_in_ancient_Greece

Wikipedia entry on the six arts of Chinese nobles http://en.wikipedia.org/wiki/Six_Arts

Wikipedia entry on Athenian democracy http://en.wikipedia.org/wiki/Athenian_democracy

Bill Manley: *The Penguin Historical Atlas of Ancient Egypt* (1996)

Ian Morris: *War, What is it Good for* (2014)

Yuval Noah Harari: *Sapiens – A Brief History of Humankind* (2011)

Chapter 5

Wikipedia entry on Roman Roads: http://en.wikipedia.org/wiki/Roman_roads

Damien Keown: *Buddhism, a Very Short Introduction* (1996)

Norman Solomon: *Judaism, a Very Short Introduction* (1996)

Kim Knott: *Hinduism, a Very Short Introduction* (1998)

Linda Woodhead: *Christianity, a Very Short Introduction*(2004)

Karen Armstrong: *The Case for God* (2009)

Tom Holland: *Persian Fire* (2005)

Paul Kriwaczek: *In Search of Zarathustra* (2002)

Justin Wintle: *China, Rough Guide Chronicle* (2002)

Dilip Heard: *India, Rough Guide Chronicle* (2002)

Wikipedia entry on the Han Dynasty: http://en.wikipedia.org/wiki/Han_Dynasty

Angus Madison: *Contours of the World Economy 1–2030 AD* (2007)

Chapter 6

Top cities in history http://geography.about.com/library/weekly/aa011201a.htm gives a list of the world's largest cities over the ages, based on Tertius Chandler's compilation of the population of cities throughout history, *Four Thousand Years of Urban Growth: An Historical Census*

Wikipedia entry in Genghis Khan: http://en.wikipedia.org/wiki/Genghis_Khan

Malise Ruthven: *Islam a Very Short Introduction* (1997)

Hugh Kennedy: *The Great Arab Conquests* (2007)

Hugh Kennedy: *The Court of the Caliphs* (2004)

John Man: *Atlas of the Year 1000* (1999)

Ian Morris: *Why the West Rules for Now* (2010)

Max Weber: *The Religion of China* (1951)

Chapter 7

Wikipedia on urban community sizes over the ages: http://en.wikipedia.org/wiki/Historical_urban_community_sizes

Wikipedia on scholasticism: http://en.wikipedia.org/wiki/Scholasticism

Wikipedia on renaissance humanism: http://en.wikipedia.org/wiki/Renaissance_humanism

Wikipedia on purgatory: http://en.wikipedia.org/wiki/Purgatory

Niall Ferguson: *The Ascent of Money* (2008)

A.R. Disney: *A History of Portugal and the Portuguese Empire, Vol. 2* (2009)

Hugh Thomas: *Rivers of Gold* (2003)

Jonathon Israel: *The Dutch Republic; Its Rise, Greatness and Fall 1477–1806*(1995)

Angus Maddison: *The World Economy, A Millennial Perspective* (2001)

David Sobel: *Longitude* (1995)

Frederic William Maitland and Francis C. Montague: *A Sketch of English Legal History* (1915)

Keith Dawson: *The Industrial Revolution* (1972)

Paul Johnson: *A History of the American People* (1997)

William Bynum: *A Little Short History of Science* (2012)

Wikipedia on janissaries: http://en.wikipedia.org/wiki/Janissary

Robert K. Massie: *Peter the Great* (1980)

Chapter 8

P. J O'Rourke: *On the Wealth of Nations* (2007)

William Doyle: *The Oxford History of the French Revolution* (2002)

Wikipedia on Brunel: http://en.wikipedia.org/wiki/Brunel

Wikipedia on Watt: http://en.wikipedia.org/wiki/James_Watt

Wikipedia on Joseph Aspdin: http://en.wikipedia.org/wiki/Joseph_Aspdin

Paul Kennedy: *The Rise and Fall of Great Powers* (1988)

Michael Freeman and Derek Aldcroft: *The Atlas of British Railway History* (1985)

Eugen Weber: *Peasants into Frenchmen* (1976)

Simon Winchester: *The Map That Changed the World* (2001)

Wikipedia on James Hutton: http://en.wikipedia.org/wiki/James_Hutton

Peter J. Bowler: *Evolution, the History of an Idea* (2009)

Wikipedia on Georges Cuvier: http://en.wikipedia.org/wiki/Georges_Cuvier

Wikipedia on Carl Linnaeus: http://en.wikipedia.org/wiki/Carl_Linnaeus

William Bynum: *The History of Medicine, a Very Short Introduction* (2008)

Wikipedia on Roland Hill: http://en.wikipedia.org/wiki/Roland_Hill

Wikipedia on the history of telegraphy: http://en.wikipedia.org/wiki/Electric_telegraph

Wikipedia on William Russell and the Crimean War: http://en.wikipedia.org/wiki/William_Howard_Russell

Chapter 9

E.P Thompson: *The Making of the English Working Class* (1963)

Alistair J. Reid: *United We Stand* (2004)

Leslie Holmes: *Communism a Very Short Introduction* (2009)

Michael Newman: *Socialism a Very Short Introduction* (2005)

Roger Magrav: *France 1815–1914* (1983)

Edgar Feuchwanger: *Imperial Germany 1850–1918* (2001)

Andrew Gordon: *A History of Modern Japan* (2009)

Wikipedia on Faraday: http://en.wikipedia.org/wiki/Michael_Farady

Wikipedia on electric power transmission: http://en.wikipedia.org/wiki/History_of_electric_power_transmission

Wikipedia in internal combustion engine: http://en.wikipedia.org/wiki/History_of_the_internal_combustion_engine

Wikipedia on Niklaus Otto: http://en.wikipedia.org/wiki/Nikolaus_Otto

Wikipedia on Gottlieb Daimler: http://en.wikipedia.org/wiki/Gottlieb_Daimler

Wikipedia on the diesel engine: http://en.wikipedia.org/wiki/Diesel_engine

Chapter 10

Orlando Figues: *A People's Tragedy* (1996)

Orlando Figues: *The Whispers: Private Life in Stalin's Russia* (2007)

Emma Goldman: *My Disillusionment in Russia* (1923)

Ruth Heinig: *The Weimar Republic* (1998)

Leslie Iversen: *Drugs a Very Short Introduction* (2001)

Jonathon Fenby: *The Penguin History of Modern China, the Fall and Rise of a Great Power 1850-2009* (2008)

Kevin Passmore: *Fascism, a Very Short Introduction* (2002)

Wikipedia on Marconi: http://en.wikipedia.org/wiki/Marconi

Wikipedia on the history of the telephone: http://en.wikipedia.org/wiki/History_of_the_telephone

Wikipedia on the history of photography: http://en.wikipedia.org/wiki/History_of_photography

Wikipedia on the Douglas DC-3: http://en.wikipedia.org/wiki/Douglas_DC-3

Wikipedia on Imperial Airways: http://en.wikipedia.org/wiki/Imperial_Airways

Wikipedia on the atom bombs dropped on Japan: http://en.wikipedia.org/wiki/Atomic_bombings_of_Hiroshima_and_Nagasaki

Jim Ali-Khali: *Quantum: a Guide for the Perplexed* (2003)

M. Kumar: *Quantum: Einstein, Bohr, and the Great Debate about the Nature of Reality* (2008)

Chapter 11

Neil Shubin: *Your Inner Fish* (2008)

Sean B. Carroll: *The Making of the Fittest* (2008)

Mark Mazower: *Governing the World, the History of an Idea* (2012)

Stephen White: *Communism and its Collapse* (2001)

Gosta Esping-Andersen: *The Three Worlds of Welfare Capitalism* (1990)

Piers Brendon: *The Decline and Fall of the British Empire 1781–1997* (2007)

Maria Misra: *Vishnu's Crowded Temple* (2008)

Richard Trillo: *The Rough Guide to Kenya* (2010)

Wikipedia on the history of television: http://en.wikipedia.org/wiki/History_of_television#Overview

Wikipedia on computers: http://en.wikipedia.org/wiki/History_of_computer_hardware

Wikipedia on integrated circuits: http://en.wikipedia.org/wiki/Integrated_circuits

Georgia Institute of Technology Paper on car reliability: http://smartech.gatech.edu/bitstream/handle/1853/22665/yang_ling_200805_mast.pdf.txt?sequence=2

Wikipedia on vehicle registration history: http://en.wikipedia.org/wiki/Motor_vehicle

Wikipedia on containers: http://en.wikipedia.org/wiki/Containerization

Wikipedia on Bretton Woods: http://en.wikipedia.org/wiki/Bretton_Woods_system

Wikipedia on the history of nuclear power: http://en.wikipedia.org/wiki/History_of_nuclear_power#History

Wikipedia on Huguenots: http://en.wikipedia.org/wiki/Huguenot#England

Samuel P. Huntington: *The Clash of Civilizations and the Remaking of World Order* (1996)

Chapter 12

Wikipedia on packet-switching: http://en.wikipedia.org/wiki/Packet_switching

Wikipedia on the world wide web: http://en.wikipedia.org/wiki/World_Wide_Web

Wikipedia on cellular networks: http://en.wikipedia.org/wiki/Cellular_network

James Kynge: *China Shakes the World: the Rise of a Hungry Nation* (2009)

Steve Coll: *Private Empire* (2012)

Stephen White: *Communism and its Collapse* (2001)

Al Gore: *The Future* (2013)

Nicholas Shaxson: *Treasure Islands: Tax Havens and the Men Who Stole the World* (2012)

Vince Cable: *The Storm* (2009)

George Soros: *The Crash of 2008 and What It Means* (2008)

Wikipedia on AIDS: http://en.wikipedia.org/wiki/AIDS

Obesity data in the USA: http://www.cdc.gov/obesity/data/adult.html

Roger Blackmore, Maggie King, Robin Roy and Joe Smith: *Open University Course U116, Environment: Journeys through a Changing World, Block 1, Setting Out from Home*

Wikipedia on atmospheric carbon dioxide levels: http://en.wikipedia.org/wiki/Carbon_dioxide_in_Earth's_atmosphere

Tony Jupiter: *What Has Nature Ever Done for Us?* (2013)

What happens when the water runs dry?: *Observer*, New Review 20 February 2011

Richard Dowden: *Africa, Altered States, Ordinary Miracles.* (2009

Matt Taibbi: *Divide – American Injustice in the Age of the Wealth Gap* (2014)

Gardiner Harris: 'Superbugs ' spread globally – Babies die by the thousand as India battles drug resistant bacteria. New York Times article 14 December 2014

Chapter 13

UN 2010 revision of population estimates: http://esa.un.org/wpp/Excel-Data/population.htm

Food Prospects towards 2030/2050 –the 2012 FAO revision

Stephen Emmott: *10 Billion* (2013)

Mark Lynas: *Six Degrees* (2007)

Tim Flannery: *Here on Earth* (2010)

James Lovelock: *The Vanishing Face of Gaia* (2009)

Al Gore: *The Future* (2013)

Chinese water problems: http://factsanddetails.com/china/cat10/sub66/item391.html

Mike Berners-Lee and Duncan Clark: *The Burning Question* (2013)

David J.C. MacKay: *Sustainable Energy- Without the Hot Air* (2009)

Roger Blackmore, Maggie King, Robin Roy and Joe Smith: *Open University Course U116, Environment: Journeys through a Changing World, Block 1, Setting Out from Home*

Edalin Michael and Jeff Masters: http://www.wunderground.com/climate/greenland.asp

Canadian Health Care costs by age group: http://upload.wikimedia.org/wikipedia/commons/d/d7/Spending_on_health_care_per_capita_by_age_group.png

Health care expenditure: http://data.worldbank.org/indicator/SH.XPD.TOTL.ZS?page=6

Jared Diamond: *Collapse- How Societies Choose to Fail or Survive* (2005)

Thomas Homer-Dixon: *The Upside of Down – Catastrophe, Creativity and the Renewal of Civilisation* (2007)

Thomas Piketty: *Capital in the Twenty-First Century* (2013)

Simon Bowers: International tax rule updates to be agreed by G20 countries, *The Guardian* 16 September 2014:http://gu.com/p/4xtzd

The costly muddle of German energy policy: *Financial Times* Editorial 6 October 2014: http://www.ft.com/cms/s/0/ffa462f2-4d4b-11e4-bf60-00144feab7de.html

Karen Anderson: The myth of religious violence, *The Guardian* 25 September 2014 http://gu.com/p/4xzq6

Chapter 14

Marion Bowman: *Open University Course A217 Introducing Religions: Introduction* (2005)

Naomi Klein: *This Changes Everything – Capitalism vs the Climate* (2014)

INDEX

www.ingramcontent.com/pod-product-compliance
Lightning Source LLC
Chambersburg PA
CBHW070849180526
45168CB00005B/1750